Structure and Evolution of Galaxies

NATO ADVANCED STUDY INSTITUTES SERIES

*Proceedings of the Advanced Study Institute Programme, which aims
at the dissemination of advanced knowledge and
the formation of contacts among scientists from different countries*

The series is published by an international board of publishers in conjunction
with NATO Scientific Affairs Division

A Life Sciences   Plenum Publishing Corporation
B Physics   London and New York

C Mathematical and   D. Reidel Publishing Company
   Physical Sciences   Dordrecht and Boston

D Behavioral and   Sijthoff International Publishing Company
   Social Sciences   Leiden

E Applied Sciences   Noordhoff International Publishing
                                  Leiden

Series C – Mathematical and Physical Sciences

Volume 21 – Structure and Evolution of Galaxies

Structure and Evolution of Galaxies

*Lectures Presented at the NATO Advanced Study Institute
held at the International School of Astrophysics at the
'Ettore Majorana' Centre for Scientific Culture in Erice (Sicily)
Italy, June 22–July 9, 1974*

edited by

GIANCARLO SETTI

*University of Bologna, Laboratorio di Radioastronomia CNR,
Via Irnerio 46, 40126 Bologna, Italy*

D. Reidel Publishing Company

Dordrecht-Holland / Boston-U.S.A.

Published in cooperation with NATO Scientific Affairs Division

Library of Congress Cataloging in Publication Data

Main entry under title:

Structure and evolution of galaxies.

      (NATO advanced study institutes series : Series C, Mathematical and physical sciences ; v. 21).
      "Lectures delivered at the 2nd course of the International School of Astrophysics at the 'Ettore Majorana' Centre for Scientific Culture in Erice, Sicily, June 22–July 9, 1974."
      Includes bibliographical references and index.
      1. Galaxies – Congresses. 2. Stars – Congresses. I. Setti, Giancarlo.
II. Series.
QB857.S86   523.1'12   75–22261
ISBN-13: 978-94-010-1720-6   e-ISBN-13: 978-94-010-1718-3
DOI: 10.1007/978-94-010-1718-3

Published by D. Reidel Publishing Company
P.O. Box 17, Dordrecht, Holland

Sold and distributed in the U.S.A., Canada, and Mexico
by D. Reidel Publishing Company, Inc.
306 Dartmouth Street, Boston, Mass. 02116, U.S.A.

CONTENTS

FOREWORD

This volume contains a series of lectures delivered at the 2nd
course of the International School of Astrophysics at the "Ettore
Majorana" Centre for Scientific Culture in Erice (Sicily) from
June 22 to July 9, 1974.

The course was jointly planned by L. Woltjer and myself and
was fully supported by a grant from the NATO Advanced Study Institute
Programme. It was organized with the aim of providing students and
young researchers with an up to date account of the structure and
evolution of galaxies and was attended by 94 participants from
20 countries.

The study of galaxies is one of the most important areas of
contemporary astrophysics both for its intrinsic interest and
because it is a prerequisite to a quantitative understanding of the
structure of the universe. Recently, a qualitatively new understanding has become available through both observational and theoretical progress. On the observational side, new techniques applied
at the large optical telescopes and, in particular, the impressive
results obtained with the high resolution radio telescopes have made
it possible to obtain a detailed mapping of the structure of galaxies
other than our own. At the same time, new theoretical insights and
the availability of powerful computers to construct models holds
out the hope that a full understanding of the structures of galaxies
may be within our reach. The various aspects of this exciting
subject, from the study of our own galaxy to the problem of the
formation of galaxies, have been covered in a series of lectures
totalling 58 hours and 15 topical seminars given by the participants.

I wish to express my gratitude to the Scientific Affairs Division
of the North Atlantic Treaty Organization for the generous support
given to the course and to Prof. A. Zichichi for providing the
facilities of the "E. Majorana" Centre. I am also deeply indebted
to Dr. G. Zamorani for his invaluable assistance in the editing of
this volume; to Mr. L. Baldeschi for helping with the organization
of the meeting and for preparing a number of the diagrams contained
in this volume; to Mrs. B. Mandel for the patient typing of much of
this material; to Mrs. P. Zanlungo-Greiger and Miss P. Savalli for
secretarial assistance in Erice; and to Mr. R. Primavera for the
photographic reproduction of many of the figures.

A special thanks is, of course, due to all of the lecturers and participants who have contributed so much to the success of the course and have provided the manuscripts for this volume.

Finally, I want to express my sincere gratitude to Prof. L. Woltjer, with whom it has been a pleasure to work.

Giancarlo Setti
University of Bologna
Laboratorio di Radioastronomia CNR
Via Irnerio 46
40126 Bologna, Italy

GALAXIES INCLUDING GLOBULAR CLUSTERS AND LOCAL GROUP

S. van den Bergh

David Dunlap Observatory, University of Toronto
Richmond Hill, Ontario, Canada

1. STELLAR POPULATIONS IN THE GALAXY

According to Baade (1944) the stellar population in the central
bulge of M31 resolves at the same magnitude level as do the globular clusters associated with that galaxy. This observation led
Baade to conclude that the nuclear bulge of the Andromeda Nebula
consists of globular cluster-like stars. He immediately realized
that this idea could be tested by studying the stellar content
of the nuclear bulge of our own galactic system. In particular it
should be possible to observe the RR Lyrae variables in the nuclear
bulge of the Galaxy which would be associated with such a globular
cluster-like population. Baade (1951) therefore proceeded to search
for such cluster-like variables in a low-absorption window centered
on the globular cluster NGC6522 at $\ell=1°$, $b=-4°$. The expectation
that the nuclear bulge should contain large numbers of RR Lyrae
stars was fully confirmed by Baade (1951) who found more than 100
such variables in the field surrounding NGC6522. At first sight this
observation appeared to give strong support to the view that the
nuclear bulge of the galaxy consists of a globular cluster-like
population.
    The first doubts about this population picture, in which ellip-
tical galaxies and the central bulges of spiral nebulae consist of
globular cluster-like stars, arose in 1957. Morgan and Mayall (1957)
were able to show that the integrated spectrum of the nuclear region
of M31 is dominated by cyanogen giants. This observation shows that
the dominant contribution to the blue-green region of the spectrum
is provided by strong-lined (i.e. not metal-poor) stars. Subsequently
Nassau and Blanco(1958) were able to show that the RR Lyrae stars
in the field surrounding NGC6522 were outnumbered two to one by

G. Setti (ed.), Structure and Evolution of Galaxies, 1–11. All Rights Reserved.
Copyright © 1975 by D. Reidel Publishing Company, Dordrecht-Holland.

M-type giants. Since M giants do not occur in halo-type globular
clusters it follows that the dominant stellar population in the
nuclear bulge of the galaxy cannot be pure Population II. This
conclusion was strongly confirmed by Morgan (1959) who was able to
show that the integrated spectrum of the nuclear bulge is dominated
by strong-lined cyanogen giants. This conclusion is confirmed by
observations of the colour-magnitude diagram of the nuclear bulge
at b=-4° by van den Bergh (1971). These observations show that the
dominant stellar population at |Z|=0.6 kpc resembles that in the
old metal-rich cluster NGC188. Metal-poor halo stars account for
at most a few percent of all giant stars. UBV photometry of sky-
free star patches in the nuclear bulge also shows that most of the
brightest main sequence stars in the nuclear bulge must be metal-rich.
Within the nuclear bulge the fraction of metal-poor stars is observed
to increase with increasing distance from the galactic plane. At
|Z|=1.2 kpc van den Bergh (1974) finds a significantly larger frac-
tion of Population II giants than are observed at |Z|=0.6 kpc.

    The model which emerges from these studies is that the dominant
stellar population in the nuclear bulge of the Galaxy consists of
old metal-rich stars similar to those in the old open cluster
NGC188 and that it contains only a sprinkling of true globular
cluster stars.

    The colour-magnitude diagram of stars near the Sun shows that
these objects constitute a mixed population that contains young
spiral arm stars, large numbers of old disk stars and a sprinkling
of metal-poor stars. The radial variation in the relative importance
of old and young stellar populations in the Galaxy is dramatically
shown in spectra published by Morgan and Osterbrock (1969). These
show that the spectrum of the integrated light of the nuclear bulge
of the Galaxy observed through the transparent window near NGC6522
has a spectral type near K0 with the metallic line indicator near
λ4390 being strong. This indicates that most of the light is contributed by metal-rich stars. A spectrum obtained from a partially
obscured region about 2° from NGC6522 shows a strenghtening of Hγ
and Hδ yielding an overall integrated spectral type near F8 for the
galactic disk. This observation shows that the integrated spectral
type of the galaxy exhibits a radial variation in the same sense as
that which is observed in the Andromeda Nebula.

    The first evidence for a possible metallicity gradient within
the galactic disk was provided by van den Bergh (1958) who showed
that Cepheids with periods in the range two to four days are predominantly located in the anti-centre direction whereas Cepheids with
periods of seven to nine days mainly occur in the direction towards
the galactic centre. This conclusion has recently been strenghtened
by Hartwick (1970) who finds that the ratio of the number of red to
blue supergiants increases with increasing distance from the galactic
centre as it does in M33 (Walker 1964). Finally Janes and McClure
(1972) find that K giants with unusually strong cyanogen absorption
probably originated in regions interior to the Sun. These observations of a possible abundance gradient among the stars in the

galactic disk are entirely consistent with the results of Peimbert (1968) and Searle (1971) who find strong evidence for radial composition gradients in the interstellar gas in spiral galaxies.

2. STELLAR POPULATIONS IN THE MAGELLANIC CLOUDS

2.1 Introduction

A number of important differences between the Galaxy and the Magellanic Clouds have emerged from the wealth of new observational data that have been accumulated during the last two decades. Most of these differences can be understood in terms of differing evolutionary histories of the Galaxy and the Clouds: (1) In the Magellanic Clouds the present rate of star formation is similar to its average rate during the last $\sim 1 \times 10^{10}$ years whereas the present rate of star formation in the Galaxy is much lower than it was when the Galaxy was young. (2) For all age groups, stars in the Magellanic Clouds have lower heavy element abundances than do their galactic counterparts. (3) The history of cluster formation in the Galaxy and in the Magellanic Clouds was radically different.

2.2 Star clusters in the Magellanic Clouds

The formation of star clusters in the Galaxy is characterized by two quite distinct eras. During the collapse phase of the proto-Galaxy massive globular clusters were formed. Subsequently only low mass star clusters were formed in the galactic disk. The masses of these open clusters are $\sim 10^2$ times smaller than are those of typical globular clusters. In the Magellanic Clouds no such dichotemy exists. Not only do the Clouds contain massive old globular clusters, but they also contain massive young clusters such as NGC330 in the Small Cloud and NGC1866 in the Large Cloud. Following a suggestion by Hodge we shall adopt the nomenclature "populous clusters" for massive young and intermediate-age objects.
    The fact that both stars and massive clusters have been forming more or less continuously in the SMC suggests (Freeman and Munsuk 1972) that there may not have been a great burst of star formation when the Small Cloud collapsed. If this is so then, because the number of massive clusters in the SMC is small, none of the Small Cloud clusters may in fact be as old as the SMC itself. This might explain why (Rood 1973) the oldest Small Cloud clusters have stubby red horizontal branches despite the fact that they appear to be relatively metal-poor. The evolution of the LMC appears to have been somewhat different from that of the SMC. The oldest Large Cloud clusters that have so far been studied all have blue horizontal branches. This suggests that these clusters are older than the oldest SMC clusters and that they were formed during the collapse of the LMC. The fact that such old clusters are observed in the Large

Cloud suggests that the LMC (which is more massive than the SMC)
started its evolution with a more violent burst of star (and
cluster) formation than did the Small Cloud.

The reddest giant stars in the SMC clusters are observed to be
redder than are the reddest stars in the LMC clusters. The reddest LMC
giants are, however, still very much redder than are the stars at
the tips of giant (asymptotic) branches of galactic globular clusters.
This observation suggests that the giant stars in the Cloud clusters
are younger, and hence more massive, than their galactic counterparts
and therefore were able to mix carbon to their surfaces.

The clusters NGC419 and K3 are almost certainly massive intermediate-age clusters. This conclusion is supported by the observation
(Tifft 1963) that field stars similar to those on the giant branch of
the true globular cluster NGC121 occur throughout the halo of the
SMC whereas bright red giants like those in NGC419 are only observed
in the vicinity of the SMC Bar.

Hagen and van den Bergh (1974) have shown that the young SMC
cluster NGC330 has brighter and bluer giants than do clusters of
similar age in the Galaxy. A similar, but slightly less pronounced,
difference is observed between young galactic clusters and young
populous clusters such as NGC1866 in the LMC. Intercomparison of
these results with theoretical evolutionary tracks is consistent with
the view that the heavy element abundance in young SMC clusters is
significantly lower than in clusters which are currently being
formed in the Galaxy. The heavy element abundance in the LMC is,
presumably, intermediate between that in the Galaxy and in the
Small Cloud.

2.3 Distribution of stellar populations

Intercomparison of blue and infrared photographs of the Small Magellanic Cloud shows that the young blue stellar population in this object is much more concentrated to the centre of the SMC than is the
old red population on which it is superimposed. The conclusion
that a burst of star formation has recently taken place in the SMC
Bar is confirmed by two lines of evidence: (1) The B-V colour of
the core of the SMC is significantly bluer than is the integrated
colour of the outer region of this galaxy and (2) The long period
(young) Cepheids are more strongly concentrated in the core of the
SMC than are the older short period Cepheids. In this respect the
Small Cloud differs radically from the Large Cloud. In the LMC
regions of active star formation, as outlined by HII regions,
long period Cepheids and Wolf-Rayet stars are currently distributed more or less uniformly over the face of the Large Cloud.
As judged by the distribution of short period Cepheids and of the
less luminous B-type stars a totally different pattern of star formation prevailed in the Large Cloud only $\sim 3 \times 10^7$ years ago. At that
time star formation was, apparently, strongly concentrated in the
Bar of the Large Cloud.

2.4 Variable stars

The variable stars in the Magellanic Clouds exhibit a number of striking differences from their galactic counterparts. These differences are of particular importance because they provide information on the effects of differing age and chemical composition on the evolutionary tracks of highly evolved stars.

a) Cepheids. Intercomparison of the period frequency distribution of Cepheids in the Large and Small Magellanic Clouds shows that the mean periods of Cepheids in the LMC are significantly longer than are those of Cepheids in the SMC. Since the search for variable stars in the two Clouds has mainly been carried out on similar Harvard plate material, and in many cases by the same investigators, this difference is certainly real. The fact that the mean periods of Cepheids in the Galaxy are even longer than those of the LMC Cepheids suggests that the mean period of Cepheid variables is longest in those objects having the highest metal abundance. This conclusion is considerably strengthened by the observation (Gascoigne 1969, Madore 1974) that the SMC Cepheids are ∿0.1 mag bluer in B-V than are those in the Galaxy. According to Bell and Parsons (1972) the observed colour difference between galactic and SMC Cepheids might be accounted for in terms of line blanketing if the metal abundance of the Small Cloud Oepheids is ∿4 times lower than it is in their galactic counterparts.

b) W Virginis stars. The most remarkable fact about the W Virginis stars in the Magellanic Clouds is that they are so rare. In this respect the Population II component in the Clouds resembles that which occurs in dwarf spheroidal galaxies in which no W Virginis (P>10 days) variables have so far been discovered. This similarity between Population II in dwarf spheroidals and in the Magellanic Clouds is strengthened by the following two observations: (1) A single BL Herculis variable in the SMC (Tifft 1963) lies on the period luminosity relation for these objects in dwarf spheroidals (van Agt 1967) which is significantly brighter than is the period luminosity relation for BL Herculis stars in galactic globular clusters. (2) The stars at the tips of the giant (asymptotic) branches of globular clusters are much redder (B-V≳2.0) than are the giants in any galactic globular cluster. In this respect the cluster giants in the Clouds are similar to those in the Sculptor dwarf spheroidal system (Hodge 1965).

c) Variable M-type supergiants. Lloyd-Evans (1971) finds that ∿15% of the M-type supergiants in the SMC are large amplitude variables. Large amplitude M supergiant variables, such as S Persei, are quite rare in the Galaxy and in the LMC. The fact that an object of similar type has been found in the faint dwarf galaxy IC1613 suggests that low metal abundance might favour the occurrence of large amplitude red supergiant variables.

d) Novae. Observations of novae in the Galaxy and in M31 (van den Bergh 1975) show a relatively tight correlation between the absolute magnitude of a nova at maximum light and its rate of decline.

Application of this relation to the novae in the LMC yields an
apparent distance modulus $(m-M)_V = 19.3$. This value is significantly
greater than is the distance modulus obtained from all other distance indicators. The discrepancy between the observed and the
expected magnitudes of novae is particularly glaring in the case of
Nova Doradus (1971a) for which excellent photoelectric observations
(Ardeberg and de Groot 1973) are available. From the period versus
rate of decline relationship derived from the novae in the Galaxy
and in M31 a value $(m-M)_B \simeq 20.0$ is obtained. The blue colour of
Nova Doradus (1971a) near maximum light shows that this discrepancy
can not be accounted for by interstellar absorption.

Somewhat fragmentary observations of SMC novae also give a
larger distance modulus than do other distance indicators. It is
therefore concluded that the novae in the Magellanic Clouds are similar to each other but differ systematically from those in the Galaxy
and in M31.

2.5 Abundance determinations in stars.

Data on the abundance determinations of individual stars in the
Magellanic Clouds are summarized in Table I. The data in the table
suggest that the stars in the LMC are probably slightly deficient
in heavy elements whereas those in the SMC show a substantial heavy
element deficiency. These conclusions are entirely consistent
with the observations of Cepheids and star clusters which also suggest
that LMC objects are intermediate between those in the Galaxy and
in the SMC. Additional support for the hypothesis that the Small
Cloud is metal-poor compared to the Large Cloud is provided by
observations of absorbing clouds in the LMC and in the SMC (van den
Bergh 1974, Hodge 1972, 1974). These investigations show that the
LMC and the SMC both contain dark nebulae. The dust clouds in the
Large Cloud are, however, much more prominent than are those in the
Small Cloud. These observations suggest that the interstellar gas
in the LMC has been able to produce dust much more efficiently
than has the gas in the SMC. Since interstellar grains are largely
composed of heavy elements it seems quite probable that the observed
difference between the dustiness of the LMC and the SMC is due to
a difference in the heavy element abundance of the gas in these two
galaxies. Strong support for this speculation is provided by the
observations discussed by Peimbert (1973) who finds that the oxygen
abundance in the SMC emission nebulae is ~ 5 times lower than it is
in LMC emission nebulae.

In summary it is concluded that the observed differences between stellar populations in the Small Magellanic Cloud, the Large
Magellanic Cloud and the Galaxy may be understood in terms of (1)
differing chemical composition and (2) differences in the rate
of star formation as a function of time.

Table I

Abundance Determinations in the Magellanic Clouds

	Star	Sp	$[\text{Fe/H}]$ [†]	Reference
LMC	HD 32034	B9Ie	-0.8	Przybylski (1971)
	HD 33579	A3Ia-0	-0.2 ± 0.2	Przybylski (1968)
			~0.0	Wares et al. (1968)
			-0.3 [††]	Wolf (1972)
SMC	HD 5045	B3Ia	$\lesssim-0.6$	Osmer (1973)
	HD 7099	B2.5I	$\lesssim-0.6$	Osmer (1973)
	HD 7583	A0Ia-0	-1.0	Przybylski (1972)
			-0.3 [††]	Wolf (1973)

[†] $[\text{Fe/H}] \equiv \log(\text{Fe/H})_{\text{star}} - \log(\text{Fe/H})_{\text{Galaxy}}$

[††] From best determined elements (Mg,Si,Ca,Ti,Cr,and Fe) relative to α Cyg.

3. STELLAR CONTENT AND EVOLUTION OF OLD STELLAR SYSTEMS

3.1 The central bulge of the Andromeda Nebula

Low dispersion spectra obtained by Morgan and Mayall show that the blue region of the spectrum of the central part of M31 is dominated by cyanogen giants. This conclusion is confirmed by the photoelectric spectrum scans of van den Bergh and Henry (1962). From multi-colour photometry Faber (1972) finds that an acceptable fit to the broad and intermediate band colour data is provided by astrophysically reasonable stellar population models with 15<M/L<60. This result is consistent with the value M/L≈13 which Morton and Thuan (1973) obtain from observations of the velocity dispersion in the nucleus of M31. A relatively low mass-to-light ratio for the central region of M31 is also favoured by Whitford's (1972) observations of the Wing-Ford molecular feature at λ9910 which gives no positive indication of a dwarf contribution to the total light. This conclusion is strongly confirmed by observations of the CO band at 2.3μ (see Table II) that have been published by Baldwin et al. (1973). It should be emphasized (Faber 1972) that a good fit to the observed photometric parameters of the nuclear region of M31 can only be achieved if the giants in this region are very strong-lined and hence presumably "super metal-rich".

Table II

Infrared CO Band Index

Observed	0.098±0.013
Model with M/L=15	0.079
Model with M/L=44	0.028

3.2 Luminosity dependence of colours and spectra of E galaxies

Baum (1959) first noticed that intrinsically faint elliptical
galaxies tend to be bluer than are more luminous ones. Furthermore Spinrad (1961) has shown that luminous early-type galaxies
have stronger lines than do objects of lower luminosity. According
to Faber (1973) the difference in colour and line strength between
M32 and the most luminous elliptical galaxies is consistent with
an increase in [Fe/H] from the solar value in M32 to twice the solar
value in the most luminous elliptical galaxies. McClure and van
den Bergh (1968) have shown that galaxies of the same absolute
magnitude have similar colours, and hence presumably similar
compositions in clusters and in the general field. This result
may be interpreted in either one of two ways: (1) The evolutionary
history of an elliptical galaxy is independent of the environment
in which it is formed, i.e. a galaxy is essentially insulated from
its neighbours after star formation and heavy element enrichment
begins or (2) field E galaxies have escaped from clusters.

3.3 The evolution of elliptical galaxies

a) <u>Star formation in dwarf ellipticals</u>. Ultraviolet plates
by Baade (1951) show that NGC205, which is a dwarf elliptical
companion to M31, contains a few dozen bright blue stars. The
clumpy distribution of these stars and the presence of a number of
dust patches in NGC205 suggest that these objects are stars which
have recently formed from interstellar material. A sprinkling of
young blue stars is also found in NGC185. No evidence for star formation is, however, observed in NGC147 and in M32, which are also dwarf
elliptical companions to the Andromeda Nebula. The fact that the rate
of star formation in the dwarf elliptical companions to M31 differs
from system to system might be accounted for if star formation in
these objects occurs in bursts. Searle and Sargent (1972) have
previously suggested that the star formation in dwarf <u>irregular</u>
galaxies might also take place in bursts. Another low-luminosity
nearby elliptical that appears to have undergone a recent burst of
star formation is NGC3077. Perhaps the most striking example
of a low-luminosity elliptical in which a huge burst of star formation

is presently taking place is NGC5253. Inspection of 48-inch Schmidt
plates shows that the underlying luminosity distribution of NGC5253
resembles that of an E or SO galaxy. Superimposed on this smooth
distribution of stars is a conspicuous concentration of knots that
are grouped around the centre of this galaxy. About a dozen such
knots are recognizable on the best plates of this galaxy. A widened
image tube spectrum of the brightest of these knots has recently
been published by van den Bergh (1972). This spectrogram shows
an early-type absorption line spectrum resembling that of the bright
knots (super starcluster) in the nucleus of M82.

b) The evolution of giant elliptical galaxies. In giant
elliptical galaxies the total amount of matter ejected by evolving
stars probably amounts to $\sim$ 1M$_\odot$ per year. It is therefore somewhat
surprising that, with the exception of a few unusual objects such
as NGC1275 and NGC5128, such objects give no evidence for the
existence of large quantities of interstellar material. This apparent lack of gas and dust may be accounted for in a number of ways:
(1) Mathews and Baker (1971) suggest that a high supernova rate
might set up a galactic wind that sweeps gas out of the galaxy and
into intergalactic space. (2) Gunn and Gott (1972) have shown
that elliptical galaxies might be swept clean of gas as they plough
through intracluster gas at high velocity. (3) A significant fraction of the gas ejected by the stars might eventually collapse
into the nuclei of giant elliptical galaxies giving rise to violent
explosions. Such explosive events might sweep galaxies clean of
any remaining gas (van den Bergh 1963, 1972). The view that bursts
of star formation may accompany violent explosive events in the
nuclei of luminous elliptical galaxies receives strong support
from recent new observations of NGC5128 (=Centaurus A). These
observations show that the integrated colours of the regions near
the equatorial dark band in this galaxy are somewhat bluer than
are the areas near its poles. This suggests that the observed
integrated colours of NGC5128 are affected by a young stellar
population which is superimposed on the old metal-rich population
which constitutes the dominant component in giant elliptical galaxies. Strong support for this view is provided by the observation that a number of bright blue knots (see photograph published
by J.J. Rickard in the ESO Annual Report for 1971) are situated
just outside the equatorial dark band that crosses NGC5128. UBV
observations of the brightest of these knots through a ten second
diaphragm yield V=14.17 (M$_V \approx -13$), B-V=0.41 and U-B=-0.58. These
observations show that this object is either a reddened OB cluster
or a knot of emission nebulosity.

Additional evidence for the occurrence of bursts of star formation in giant elliptical galaxies is provided by observations of
NGC1275 (=Perseus A). Spectroscopic observations show that the
nucleus of this object emits a strong continuous spectrum on which
broad emission lines are superimposed. An image tube spectrum of a
region located 5" to the south of the nucleus shows a well-developed

A-type absorption line spectrum. The most straightforward interpretation of this observation is that the active nucleus of NGC1275 is surrounded by a region in which a burst of star formation has recently taken place. This suspicion is strengthened by the observation that spiral arm-like strings of bright knots are seen to be located in the outer parts of this galaxy on direct plates.

It might, of course, be argued that NGC1275 is not an elliptical galaxy at all. The strongest argument against this view is that the core of the Perseus cluster (of which NGC1275 is the brightest member) consists only of E and SO galaxies. The brightest members of such early-type clusters are also almost invariably found to be of type E or SO.

Possibly the gas that was ejected from evolving stars in NGC1275 was not swept out of this galaxy by intra-cluster gas because NGC1275 has little or no net motion relative to the centre of mass of the Perseus cluster. This view is consistent with the observation by Chincarini and Rood (1971) that NGC1275 has a radial velocity of 5291 km sec^{-1}, which does not differ significantly from the mean cluster velocity of (5460 $\pm$ 200) km sec^{-1}. Similarly it might be argued that the gas was not swept out of NGC5128 because this object is one of the rare examples of a giant elliptical galaxy that is not situated in a cluster.

Some support for the notion that violent explosive events trigger star formation is provided by the observation that the nucleus of the post-eruptive galaxy M82 (Kronberg et al. 1972) is surrounded by a dozen young "super starclusters" each of which has $M_V \approx -15$.

According to Sandage (1972) giant elliptical galaxies that are radio sources are often bluer than are those which are radio quiet. This observation might be accounted for by recent bursts of star formation and (or) by the presence of emission lines. Possibly elliptical galaxies with relatively strong hydrogen lines in their nuclei such as NGC3379 (Spinrad 1972) are objects in which a burst of star formation took place within the last $\sim 1 \times 10^9$ years.

REFERENCES

Ardeberg, A., and de Groot, M., 1973, Astron. Ap. 26, 53.
Baade, W., 1944, Ap. J. 100, 137.
Baade, W., 1951, Publ. Univ. Michigan Obs. 10, 7.
Baldwin, J.R., Danziger, I.J., Frogel, J.A., and Persson, S.E., 1973, Ap. Letters 14, 1.
Baum, W.A., 1959, P.A.S.P. 71, 106.
Bell, R.A., and Parsons, S.B., 1972, Astrophys. Letters 12, 5.
Faber, S.M., 1972, Astron. Ap. 20, 361.
Faber, S.M., 1973, Ap. J. 179, 731.
Freeman, K.C., and Munsuk, C., 1972, Proc. Astron. Soc. Australis 2, 151.
Gascoigne, S.C.B., 1969, M.N.R.A.S. 146, 1.
Gunn, J.E. and Gott, J.R., 1972, Ap. J. 176, 1.

Hagen, G.L., and van den Bergh, S., 1974, Ap. J. Letters 189, L103.
Hartwick, F.D.A., 1970, Ap. Letters 7, 151.
Hodge, P.W., 1965, Ap. J. 142, 1390.
Hodge, P.W., 1972, P.A.S.P. 84, 365.
Hodge, P.W., 1974, P.A.S.P. (in press).
Janes, K.A., and McClure, R.D., 1972, Bull.A.A.S. 4, 241.
Kronberg, P.P., Pritchet, C.J., and van den Bergh, S., 1972, Ap.
J.Letters 173, L47.
Lloyd-Evans, T., 1971, Colloquium on Supergiant Stars, M. Hack
(ed.) (Trieste Obs.-Trieste) p. 125.
Madore, B.F., 1974, unpublished Toronto PhD thesis.
Mathews, W.G. and Baker, J.C., 1971, Ap. J. 170, 241.
McClure, R.D. and van den Bergh, S., 1968, Astron. J. 73, 1008.
Morgan, W.W., 1959, Astron. J. 64, 432.
Morgan, W.W., and Mayall, N.U., 1957, P.A.S.P. 69, 291.
Morgan, W.W., and Osterbrock, D.E., 1969, Astron. J. 74, 515.
Morton, D.C. and Thuan, T.X., 1973, Ap. J; 180, 705.
Nassau, J.J. and Blanco, V.M., 1958, Ap. J.128, 46.
Osmer, P.S., 1973, Ap. J. Letters, 184, L127.
Peimbert, M., 1968, Ap. J. 154, 33.
Peimbert, M., 1973, paper presented at IAU Symp. No. 58 in Canberra,
Australia.
Przybylski, A., 1968, M.N.R.A.S. 139, 313.
Przybylski, A., 1971, M.N.R.A.S. 152, 197.
Przybylski, A., 1972, M.N.R.A.S. 159, 155.
Rood, R.T., 1973, Ap. J. 184, 815.
Sandage, A.R., 1972, Ap. J. 178, 25.
Searle, L., 1971, Ap. J. 168, 327.
Searle, L., and Sargent, W.L.W., 1972, Ap. J. 173, 25.
Spinrad, H., 1961, P.A.S.P. 73, 336.
Spinrad, H., 1972, Ap. J. 177, 285.
Tifft, W.G., 1963, M.N.R.A.S. 125, 199.
Van Agt, S.L.T.J., 1967, Bull. Astron. Inst. Netherlands 19, 275.
Van den Bergh, S., 1958, Astron. J. 63, 492.
Van den Bergh, S., 1971, Astron. J. 76, 1082.
Van den Bergh, S., 1972, R.A.S.C. Journal 66, 237.
Van den Bergh, S., 1974, Astron. J. 79, 603.
Van den Bergh, S., 1974, Ap. J. (in preparation).
Van den Bergh, S., 1975, Stars and Stellar Systems, 9, A.R.Sandage
and J. Kristian (eds.) (Univ of Chicago Press, Chicago).
Van den Bergh, S., and Henry, R.C. 1962, David Dunlap Obs. Publ.
2, 281.
Walker, M.F., 1964, Astron. J. 69, 744.
Wares, G.W., Ross, J.E., and Aller, L.H., 1968, Astrophys. and
Space Sci. 2, 344.
Whitford, A.E., 1972, Bull. A.A.S. 4, 230.
Wolf, B., 1972, Astron. Ap. 20, 275.
Wolf, B., 1973, Astron. Ap. 28, 335.

STELLAR POPULATIONS

L. Gratton

University of Roma, Italy.

1. THE BASIC IDEA OF STELLAR POPULATIONS.

The concept of Stellar Population - one of the most useful in Astrophysics - was first introduced by Baade (1944). When trying to resolve the nucleus of the Andromeda nebula and some dwarf nearby
elliptical galaxies, he found that on red-sensitive plates the two
companions of M31, the central region of M31 itself and some other
galaxies could be resolved by the 102-inch telescope "into an un-
beleivable mass of the faintest stellar images" (fig.1). Baade concluded that the brightest stars in these objects were red stars -

Fig. 1. N.G.C. 205; Baade's plate showing resolution in stars.

G. Setti (ed.), Structure and Evolution of Galaxies, 13–43. All Rights Reserved.
Copyright © 1975 by D. Reidel Publishing Company, Dordrecht-Holland.

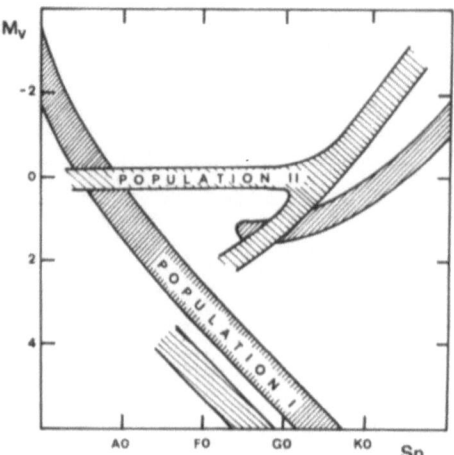

Fig.2. The two basic stellar populations in the H.R. diagram (adapted from Baade).

presumably K giants – of photographic magnitude $M_{pg} \approx -1$ ($M_V \approx -2.5$) similar to or slightly fainter than the brightest stars in globular clusters, but definitely brighter than normal field giants, whose absolute photographic magnitude is about +1.7 ($M_V \approx 0$).

From these observations Baade assumed the existence in the galaxies of <u>two kinds of stellar populations</u>, namely:

<u>population I</u>, typical of peripheral regions of spiral galaxies, as represented by galactic clusters, whose brightest members are blue stars(spectral types O-B2) with $M_V \approx -7$; red giants, if present, are normally not brighter than, say, $M_V \approx 0$ (fig.2);

<u>population II</u>, to be found in elliptical galaxies and the central regions of spirals, and typified by globular clusters, whose brightest stars are red (spectral types K2-K5) with $M_V \approx -2$; the blue stars in globular clusters – the so-called horizontal branch stars – are usually of $M_V \approx +0.5$ of fainter.

The difference between the two kinds of Stellar Populations was thought to be one of age, population II being the older. But it soon appeared both from the theory of the internal constitution of red giants and from spectroscopic evidence that the chemical composition of stellar matter was also involved and indeed should be considered as the primary factor.

Thus stars of the dwarf ellipticals and of the central regions of spirals are no longer called population II stars, but are considered as old (metal-rich) population I stars; only very metal-poor stars, like those of the globular clusters and of the galactic halo (subdwarfs) are now considered as "bona fide" population II stars.

The great fruitfulness of the concept of Stellar Populations is due to the discovery by Baade himself and many others that the

Table I

Some Properties of Stellar Populations.

Property	Population I	Population II
H.R. diagram	Galactic clusters	Globular clusters
Galactic distribution	Disk	Halo
Kinematics	Low (relative) velocities	High (relative) velocities
Galactic orbits	Plane and circular	Inclined and eccentric
Interstellar matter	Closely associated	No connection
Age	Young and old	Old
Chemical composition	Metal-rich	Metal-poor

type of stellar population is connected with a number of "a priori" apparently unrelated properties, such as the character of stellar variability, kinematical properties of stars in the vicinity of the sun,etc...The explanation of these correlations is one of the great achievements of the modern theory of galactic evolution.

Table I lists some of the main properties of Stellar Populations.

Of course, besides stellar variability – which is not mentioned in the table – one might add other properties which are very imperfectly known or can only be guessed, but might prove to be of basic importance, such as the mass-luminosity relation, the initial luminosity function, He-content and so on.

A more refined scheme (Blaauw 1969) is given in Table II.

The general idea which relates the concept of Stellar Populations to the evolution of the Galaxy is that stars originate through fragmentation and condensation of interstellar matter and thus inherit the properties (kinematical,chemical and physical) of the medium at the time and place where they have been formed; early in the history of the Galaxy only stars of population II (metal-poor) were formed, but very soon, after perhaps no more than 10^9 years or less, the formation of population II stars was arrested and during the remaining 10^{10} years of galactic life, only population I stars were born.[†]

In these lectures I shall discuss the following properties and

---

[†] This is the currently accepted view. However recent results by V. Castellani and his group (unpublished) seem to show that the ages of globular clusters in our galaxy span over a much larger interval of time, from 2 billion years for the oldest and 0.9 billion years for the youngest; during this time the He-content of clusters increased continuously from 15 to 35% , that of metals being more or less randomly distributed.

L. GRATTON

Table II

Principal Galactic Population Components

Population component	Tot. mass (10^9 $M_\odot$)	Age (10^9y)	Examples
Population I: "Extreme"	2	0.1	Gas, Associations, Supergiants, Youngest clusters, Youngest spiral population.
Population I: "Older"	5	0.1-1.5	A-stars, Strong-line stars, Young spiral population.
Disk population	} 47	1.5-8	Planetary Nebulae, Weak- line stars.
Population II: "Intermediate"		8-10	High-velocity stars ($\|z\|>30$ km / sec), Long -period variables.
Population II: "Halo"	16	10	Subdwarfs, Globular clusters, RR Lyrae variables with period ≥ 0.4 days.

their correlations to age:
(a) kinematics and space distribution,
(b) chemical composition from spectroscopic and broad-band photometric observations.

Among the various general papers on Stellar Populations, I
will quote only the book of the 1957 Vatican Conference (O'Connell
ed. 1958), Blaauw's article in "Galactic Structure" (Blaauw and
Schmidt ed.s 1965) and a more recent excellent review paper by I.R.
King (1971).

2. SOME DYNAMICAL CONCEPTS.

It seems useful to begin by recalling some basic concepts of stellar
dynamics.

The Dynamics of the Galaxy is dominated by coexistence and
partial superposition of two components with very different mechanical properties:

(a) Stars, whose mean free paths are very long (collision-time, say, equal to or longer than 10^{10} years);
(b) Interstellar matter, consisting of gas and dust clouds, whose collision-time is very short.

Only gravitational forces need to be considered, but if magnetic forces are important, obviously they will act only upon the interstellar medium.

More properly these conditions correspond to the surroundings of our sun, where the mass-ratio of stars to interstellar matter is of the order of 2:1 or 3:1. This is a peripheral region of the Galaxy, where the spiral structure is well developed; in the central region the collision-time might be quite different. In dwarf elliptical galaxies no interstellar matter is observed, the star density is much greater, but the mean free path is still long (on account of the larger relative velocities; see e.g. Woltjer 1967 and these lectures).

As a first approximation it may be assumed that the galactic system has attained a quasi-stationary state and admits an axis (z-axis) and a plane (z=0) of symmetry. The general field may thus be represented by a time-independent potential function ϕ which in cylindrical coordinates $\tilde{\omega},\theta,z$, is a function of $\tilde{\omega}$ and z, but not of θ; furthermore $\phi(\tilde{\omega},z)=\phi(\tilde{\omega},-z)$.

Under these circumstances the galactic orbit of a star depends only upon the initial position and velocity. More precisely, if we leave aside the difficulties connected with the existence and analytical form of the so-called third integral, it may be shown that a stellar orbit does not change during the whole life of a star - some 10^{10} years or more - in the sense that it is completely determined by the following three invariants of the motion (isolating integrals): the energy integral,

$$E = \tfrac{1}{2}(\dot{\tilde{\omega}}^2 + \tilde{\omega}^2\dot{\theta}^2 + \dot{z}^2) - \phi(\tilde{\omega},z),\qquad(2.1)$$

the momentum integral,

$$h = \tfrac{1}{2}\tilde{\omega}^2\dot{\theta},\qquad(2.2)$$

a third integral, which to a first approximation may be taken as

$$I_3 = \tfrac{1}{2}\dot{z}^2 - \psi(z),\qquad(2.3)$$

where ψ is a function of z which changes slowly also with $\tilde{\omega}$. Equation (2.3) is an exact expression for the third integral in the case when the potential ϕ can be separated in the sum of two terms,

$$\phi(\tilde{\omega},z) = \phi_0(\tilde{\omega}) + \psi(z),$$

which are, respectively, functions of $\tilde{\omega}$ and z only.

The following points must be noted:
(a) If, as usual, $\phi(\tilde{\omega},z)$ is taken to be equal to zero when

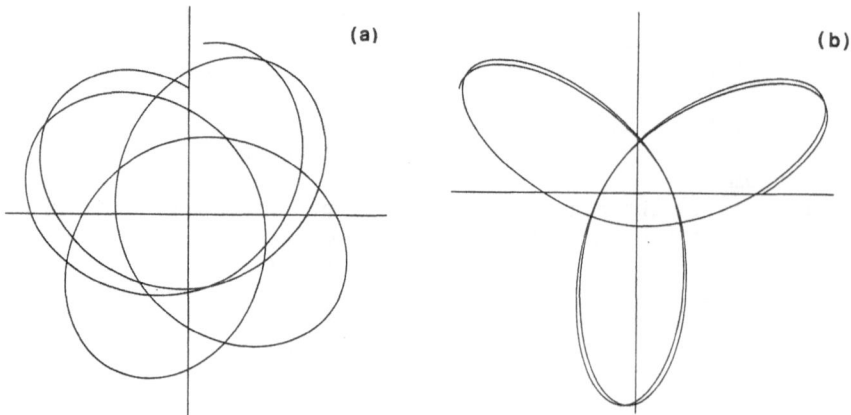

Fig.3. Galactic orbits. (a) An orbit of intermediate eccentricity.
(b) An orbit of high eccentricity.

either $\tilde{\omega}$ or z or both become infinite, positive and zero values of
E correspond to orbits which extend to infinity (like the hyperbolic
or parabolic cases of keplerian motion); negative values of E correspond to "closed" orbits (elliptic case), that is to orbits which
remain at all times at a finite distance.

(b) When E<0, the projection of the orbit on the galactic
plane may be described as a closed elliptic-shaped curve, whose
axis rotates uniformly around the center of the system (fig.3).
The perigalactic and apogalactic distances, p and a, and the
ellipticity of the orbit

$$e = \frac{a - p}{a + p} , \qquad (2.4)$$

are determined by the integrals of the motion and therefore remain
constant at all times. The motion perpendicular to the galactic
plane (z-motion) has an oscillatory character and if one neglects
the change of $\psi(z)$ with $\tilde{\omega}$, the star reaches at each oscillation the
same height above and below the galactic plane; the variation of
ψ with $\tilde{\omega}$ causes a periodic change of the amplitude of the oscillations.

(c) Another useful way to look at an orbit with E<0 is to use
a system of reference rotating at the same speed of the projected
orbit; in this system the projection of the trajectory òn the galactic plane is an oval-shaped curve (epicycle, fig.4) contained between
the pericentric and apocentric distances. If one takes into account
also the z-motion, then in general it is found that the orbit fills
an "orbit box" whose shape is invariant throughout the motion of the
star; another constant of the motion is then i, one half the maximum
angle subtended by the box at the center of the system (fig.5.).
Hence a,p,e and i are invariants of the motion which possess a simple
geometrical meaning and may be used to describe the orbits of the stars

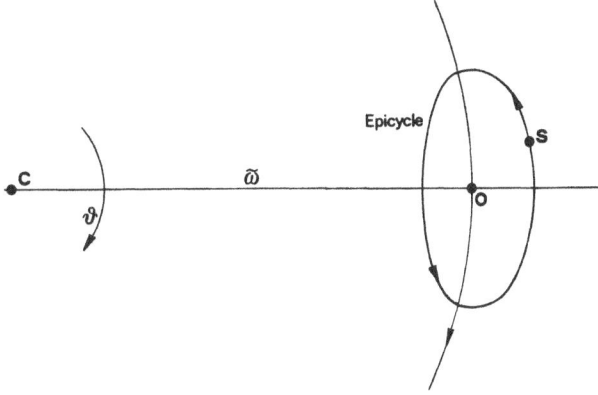

Fig.4. Epicycle.

     Consider now the motion of the interstellar clouds. It is read-
ily seen that due to the short free paths, the only general stationary pattern of motion is one in which the clouds move along circular
orbits in the galactic plane (z=0); the velocities of the clouds are
at each point perpendicular to the direction of the center and
their values are equal to the (local) <u>circular velocity</u> given by

$$\Theta_c^2 = -\tilde{\omega}\,\frac{\partial\phi}{\partial\tilde{\omega}}\ . \tag{2.5}$$

     From the observations of the 21-cm line of interstellar H it
is, thus, possible to obtain the <u>rotational curve</u> of the Galaxy,
namely the curve giving the variation of Θ_c with $\tilde{\omega}$ (in the galactic
plane). One can also use the result from the motions of the nearby

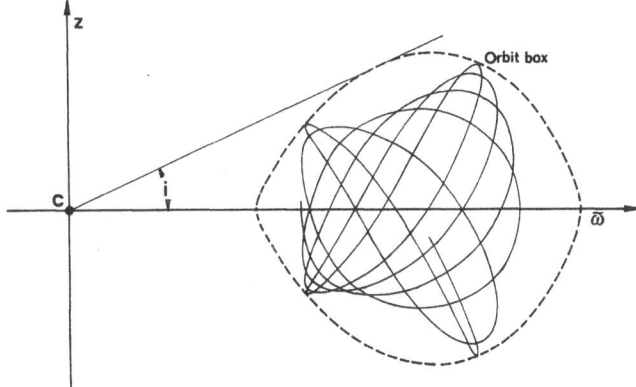

Fig.5. Orbit box in a plane through the center of the Galaxy perpendicular to the galactic plane.

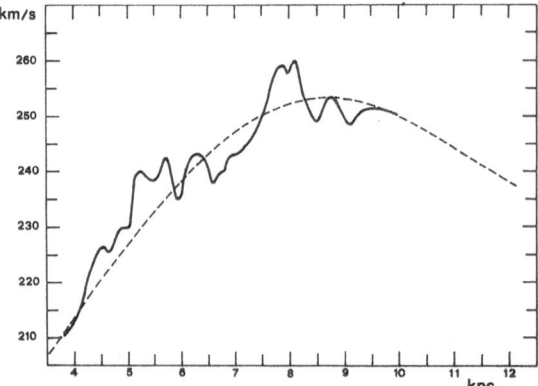

Fig.6. Rotational curve of the Galaxy (Shane and Smith).

stars to obtain the value of Θ_c at the place of the sun, 10 kpc
from the center of the Galaxy, which is found to be

$\Theta_{\odot}$= 250 km/sec.

The rotational curve obtained by Shane and Smith (1965) together
with the theoretical curve computed by Schmidt (1965) from a model
of the mass distribution inside the Galaxy is shown in fig.6.
Non-circular motions (expansion or contraction) of the interstellar medium either general or local have been often suggested (e.g.
Dixon 1967, Pismis 1960). Outwards motions of considerable amount
in the region inside 3kpc from the center do certainly exist and
may have an important evolutionary significance which is not understood at present. Outside this region non-circular motions should
be considered as small deviations from the general pattern; they
are probably related to the wavy form of the observed curve as compared with the smooth theoretical one.
A small perturbation in the interstellar medium originates a
system of density waves which propagate with the velocity of sound.
It may be shown that, due to the general (differential)rotation of
the medium corresponding to equation (2.5) a system of waves possessing a certain degree of permanence must have a spiral structure.
Since young stars inherit the structure of the medium in which they
have been formed, the spiral form of many galaxies is currently
explained as the result of a spiral pattern of density waves moving
relative to the interstellar medium; the whole spiral structure
rotates with a velocity somewhat different from Θ_c (either smaller
or greater). This interpretation of the spiral structure in terms
of density waves was first envisaged by Lindblad and later developed
by Lin (see e.g. Lin 1967, Contopoulos 1972).
There are two more important points which I would like to stress,
although they are only indirectly related with the subject of my

lectures.

(a) From Table II one can see that the objects which share the spiral pattern of the Galaxy form at most the 10% of its mass and probably no more than the 2 or 3% of it, because the A stars do not show any trace of a spiral structure. In other words, although optically the spiral shape of many galaxies is exceedingly conspicuous, the mass distribution is much more uniform. This explains why it is possible to obtain a fair representation of the general gravitational field of the Galaxy by means of a smooth potential function ϕ depending only on the coordinates $\tilde{\omega}$ and z.

(b) Only interstellar matter may evolve towards a flat disk-like configuration; due to the invariability of stellar orbits, a system containing only stars can change only very slowly its general shape. Following this view, the old conception of the evolution of galaxies along the well known Hubble's sequence - ellipticals from E0 to E7, followed by spirals Sa, Sb, Sc in order - cannot be maintained. Instead one must visualize the shape of a galaxy as the result of two competitive processes: star formation and the tendency of the interstellar medium to collapse to a flat disk; the relative speed of these two processes is determined by such physical parameters as density, total mass and rotational momentum. The shape of a galaxy becomes essentially frozen if at a certain moment all the interstellar matter has been used in the process of star formation. If this happens before a disk has been formed, the galaxy becomes an elliptical with a degree of flattening which will later remain practically constant; if there is time enough to form a disk the resulting shape will be that of a spiral (Sandage,Freeman and Stokes 1970). This may be a somewhat oversimplified view, which ignores such effects as those due to the activity of galactic nuclei or to encounters between galaxies, but it has the merit of employing only very simple and clear physical concepts and should not be too far from the truth.

3. KINEMATICS AND AGE OF NEARBY STARS.

An obvious corollary of what has been said in the preceding section is that young stars, formed after the interstellar medium has attained its present pattern of circular motion, will move along essentially plane circular orbits (i and e small); conversely, stars moving along orbits with large i and e must have been formed before the medium reached its stationary state of motion.

It is very convenient to represent the motion of the stars near the sun in a frame of reference moving with the circular velocity Θ_0 (local standard of rest). Then nearby stars moving along almost circular orbits will have very small velocities (relative to the local standard of rest).

In this frame the three components of the solar motion are:

$$U_\odot = -9, \qquad V_\odot = +12, \qquad Z_\odot = +7 \qquad (km/sec),$$

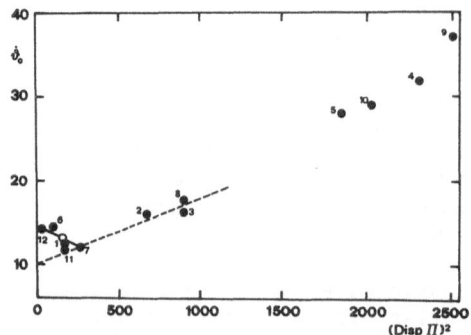

Fig. 7. Relation between the component θ_0 of the solar motion and disp Π for various groups of stars. The numbers correspond to the groups of Table III; the open circle represents the average for the 5 groups with lower θ_0.

(Delhaye 1965); here U is directed towards the anticenter of the Galaxy, V lies in the galactic plane in the direction of the general rotation of the Galaxy and Z is perpendicular to the galactic plane.

It is well known that the study of stellar velocities shows a well defined correlation of the U-component of the velocity of the centroid of a group of stars (relative to the standard of rest) to the internal dispersion of the velocities of the single stars (asymmetric drift). This is shown in Table III and the accompanying fig .7 from Delhaye's article; θ_0 is the velocity of the sun relative to the

Table III

Solar Motion (in the V-Direction) and Dispersion of the Velocities (in the U-Direction) for Some Groups of Objects

Group	$\dot{\theta}_0$ (km/sec)	disp Π (km/sec)
1. Supergiants	+12.2	13
2. gA-gF	+16.0	26
3. gG-gM	+16.2	30
4. Carbon stars	+31.8	48
5. Subgiants	+28.0	43
6. B0	+14.5	10
7. dA	+12.1	16
8. dF-dM	+17.7	30
9. White dwarfs	+37	50
10. Planetary nebulae	+29	45
11. Classical Cepheids	+12.0	13
12. Interstellar Ca II	+14.4	6

The asymmetric drift may be understood if one assumes that the different groups of objects in Table III were formed at different epochs, before the interstellar medium had attained its present (quasi-stationary) state of rotation. In the past, when the stars of the older groups were formed, the interstellar medium was still contracting from a more extended configuration towards a very flat disk of the present size; the deviations from plane circular orbits-correlated to a larger dispersion of the velocities- have been inherited by the groups of stars from the motion of the medium at the time of formation.

There are in fact two reasons why we should expect higher eccentricities in the older groups: (a) deviations of the motion of the interstellar clouds from which the stars were born from a plane circular motion (due to contraction and flattening to a disk) and (b) variation with time of the potential ϕ, due to the evolution of the system towards a stationary state. Of course during the contraction, the component of the velocity of the interstellar medium in the direction of the general rotation would increase, because of the conservation of the angular momentum and this explains why the older groups rotate more slowly than the younger.

This picture is supported by a number of observational facts. For instance, 0-B2 stars are certainly very young; their ages may be estimated in the following way. For main-sequence stars the luminosities are roughly proportional to the 4th power of the mass

$$L \propto M^4. \tag{3.1}$$

Since these stars derive their energy from H-burning, the total available energy is

$$E_H = 0.006\alpha c^2 M, \tag{3.2}$$

where α is the fraction of the total mass corresponding to H converted into He. Assuming $\alpha \approx 0.1$, the maximum age of a main-sequence star of mass M is

$$t_{max} = \frac{E_H}{L} \simeq 10^{10} M^{-3} \text{ years}, \tag{3.3}$$

if the mass is expressed in solar units.

Thus for the sun $t_{max} \approx 10^{10}$ years. For a star 0-B2 with $M=10M_\odot$, $t_{max} \approx 10^7$ years and since it is reasonable to assume that during the last ten million years the motion of the interstellar medium did not change appreciably, we must expect these stars to move along plane circular orbits with very small velocities relative to the standard of rest.

In fact the peculiar motions of the B-stars relative to their centroids are of the order of 10 km/sec (see Table III). It is also interesting to note that in 10^7 years they will have travelled only about 3×10^{15} km or 100 pc from their place of birth; if they were born inside a given spiral arm - whose cross section is of the order

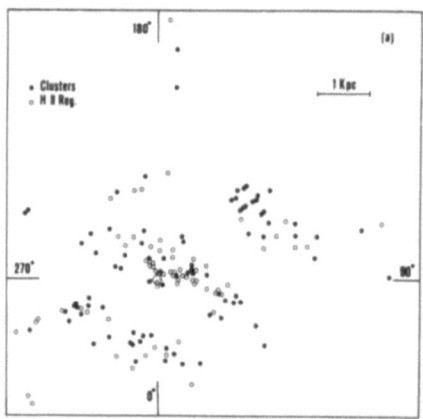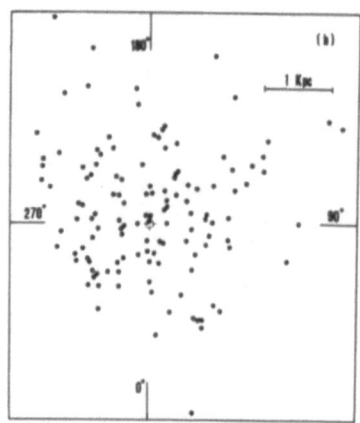

Fig. 8. (a) Distribution of young clusters and HII regions in the
galactic plane (Becker and Fenkart). (b) Distribution of old clusters
in the galactic plane (Becker and Fenkart).

of 200 pc- they must be still well inside the same arm. In other
words, by looking at the space distribution of the O-B2 stars, we
must be able to find some trace of the spiral structure of the interstellar medium from which they were born.

It is very well known that Morgan and his coworkers found evidence that clusters containing O-B2 stars inside two or three kpc
from the sun are distributed along three lanes suggesting three
distinct spiral arms (Morgan, Whitford and Code 1953 ; see figure
8(a) from Becker and Fenkart 1970).

On the other side, a star of 3 solar masses will stay on the
main sequence (as an A-star) roughly 5×10^8 years; during this time
it will have travelled quite a long way from its birth-place. In
fact the distribution of older galactic clusters whose earliest
main-sequence stars have spectral types ranging from B3 to F8 does
not show any trace of spiral structure (fig. 8 (b)) and is fairly
homogeneous.

Stars nearer than 25 pc, whose distances are known from trig-
onometric parallaxes have been investigated recently by Woolley and
his coworkers (1971). They have computed for 1286 stars the orbital
invariants e and i assuming a convenient form for the potential.
They find a marked increase of the eccentricity from types B to G;
the median e is practically the same for types G, K and M.

This is very well shown by stars of the main sequence (Table IV);
white dwarfs have a median e equal to 0.2. The box angle i shows
an analogous variation with spectral type, namely it increases from
B to G and then remains practically constant for types G, K and M.
This is to be interpreted as an increasing mean age for stars of
the different spectral types: among B and A-stars there are only
young objects; F-stars are somewhat older; G, K, and M-stars are in
the mean the oldest, although among them we find fairly young objects.

Table IV

Distribution of Eccentricity With Spectral Type
(only main-sequence stars with good parallaxes).

Range in e		B	A,Am	F0-F4	F5-F9	G	K	M	Me	Total
0	to .050		11	3	8	7	22	11	10	72
.050	.075	1	7	2	8	16	17	10	4	65
.075	.100		5	2	14	21	20	20	2	84
.100	.125			2	10	9	9	21	6	57
.125	.150			2	5	13	14	13	3	50
.150	.200				6	19	33	21	3	82
.200	.250			1	2	21	22	13	2	61
.250	.300				3	5	15	7	1	31
.300	.400				1	7	9	15	1	33
	> .400					2	1	4		7
Total		1	23	12	57	120	162	135	32	542
Median e			.0554	.0909	.0902	.1445	.1495	.1311	.1080	
Maximum e			.0936	.2036	.3485	.7857	.5280	.5442	.3881	

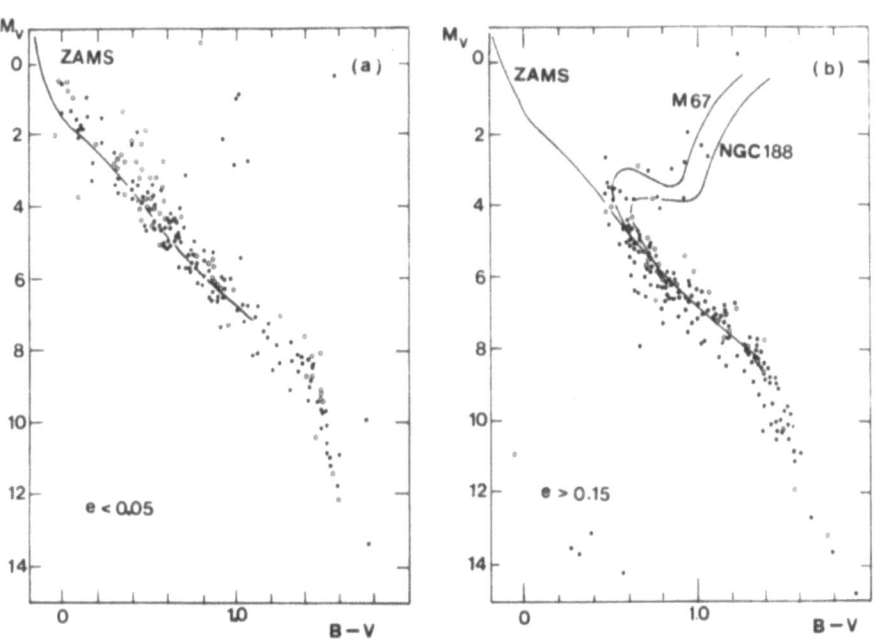

Fig. 9. Color-magnitude diagram.(a) Stars with e< 0.050(Woolley et
al.). (b) Stars with e >0.150 (Woolley et al.). Filled circles
correspond to single stars, open circles to binaries.

The color-magnitude diagrams for stars with different eccentricities are especially interesting. Figures 9(a) and (b) show the
diagrams for stars with e<0.050 and e>0.150, respectively, obtained
by combining figs. 2a, 3a, and 4a and figs. 6a, 7a, and 8a of the
paper by Woolley et al.

The diagram in fig. 9(a) corresponds to what we would expect
from a mixture of stars similar to those of rather young clusters
(like the Hyades) and stars from somewhat older clusters (contri-
buting a few G5-G8 subgiants). On the other side the diagram in
fig. 9(b) is very similar to those of rather old clusters, like
M67 and NGC 188, whose ages are evaluated around 5.5×10^9 years and
9×10^9 years from the magnitudes of the brightest main-sequence stars
(Sandage and Eggen 1969).

4. THE HIGH-VELOCITY STARS.

It has been known for a very long time that there are no stars with
space velocity relative to the sun larger than about 65 km/sec directed towards the hemisphere centered on the point with galactic coordinates $\ell=90°$ and $b=0°$, although velocities as large as 500 km/sec
and more can be found in the opposite direction. The asymmetry of
the high velocities has been especially investigated in a classical
paper by Oort (1926) who found that high-velocity stars are also
physically different in many respects from ordinary (low-velocity)
stars.

According to the present view, a high velocity means a large
deviation from circular motion, that is an orbit of large eccentricity (e$\gtrsim$0.350). In fig.10 are represented the curves of constant
eccentricity and constant apocentric distance in the U,V-plane (the
so-called Bottlinger diagram); these curves have been computed by
means of a very simple model of the Galaxy, in which the potential
in the galactic plane is given by the equation

$$\phi(\tilde{\omega}) = \frac{G\,M_g}{b^2 + (\tilde{\omega}^2 + b^2)^{\frac{1}{2}}} \tag{4.1}$$

where M_g and b are constants. In order to obtain for Oort's constants
of galactic rotation the classical values A=+15, B=-10 (km sec^{-1}
kpc^{-1}), one must assume

b = + 2.74 kpc

M_g = 2.58 x 10^{11} solar masses

(Eggen, Lynden-Bell and Sandage 1962). Although very simple, this
model may serve well for the present purpose; among other things
M_g is a fair estimate of the total mass of the Galaxy.

Fig. 11 shows the distribution of 656 stars from a catalogue
of high-velocity stars published by Eggen (1964) in the Bottlinger

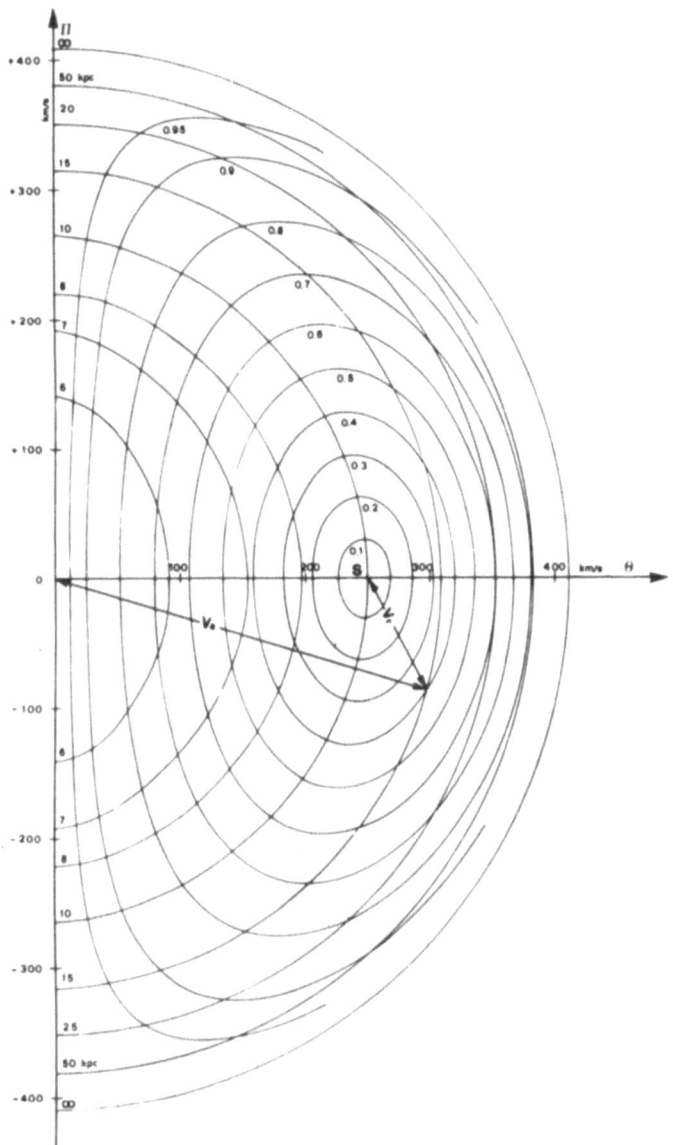

Fig. 10. Bottlinger diagram. Θ and Π are the components of the velocity of a star in the galactic plane in the direction of galactic rotation and of the anticenter, respectively; S marks the circular velocity in the solar neighborhood (the local standard of rest), V_r represents the velocity of a star relative to the local standard of rest, V_a its absolute velocity; the two families of curves correspond to constant a and e.

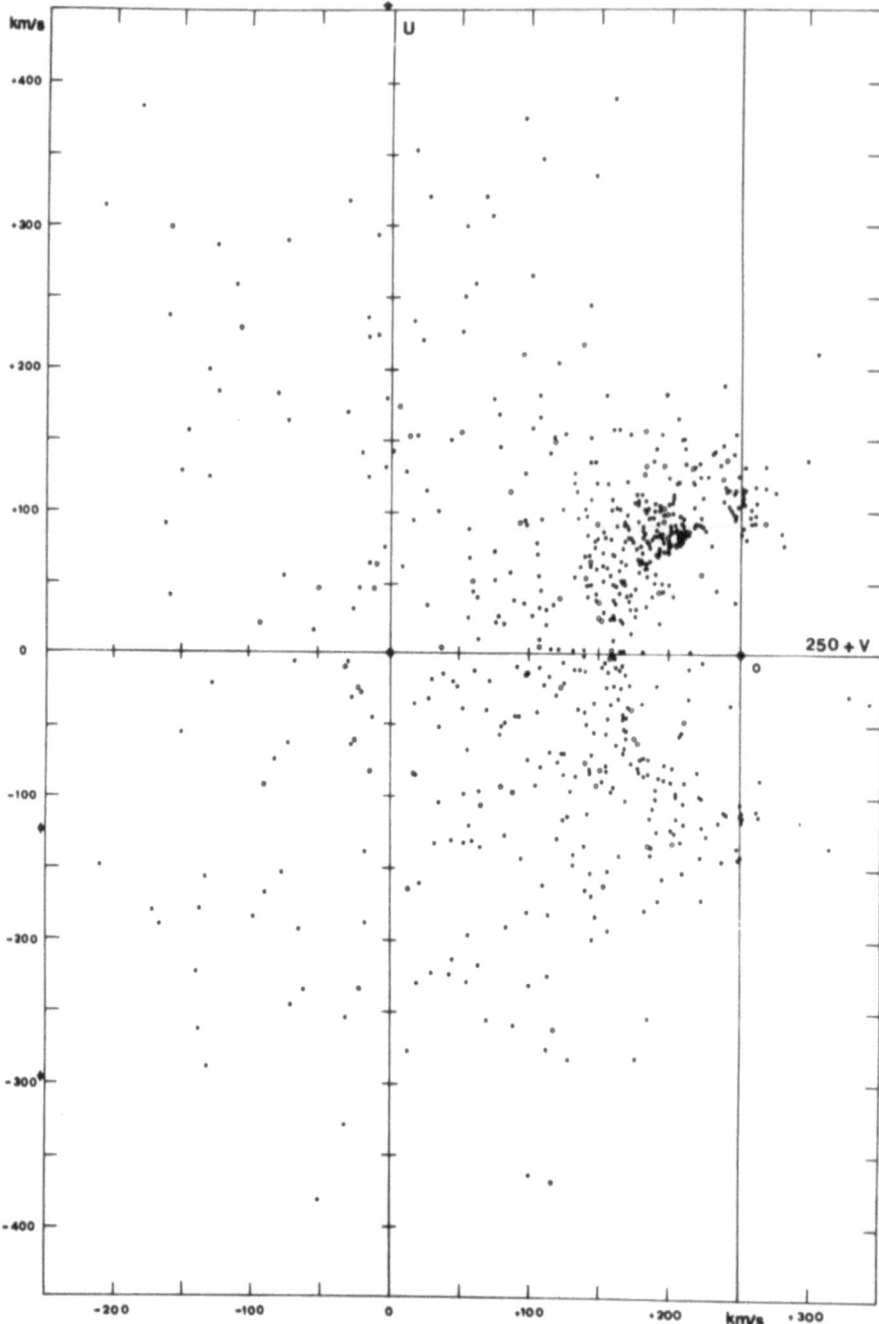

Fig. 11. The high velocity stars in the Bottlinger diagram.

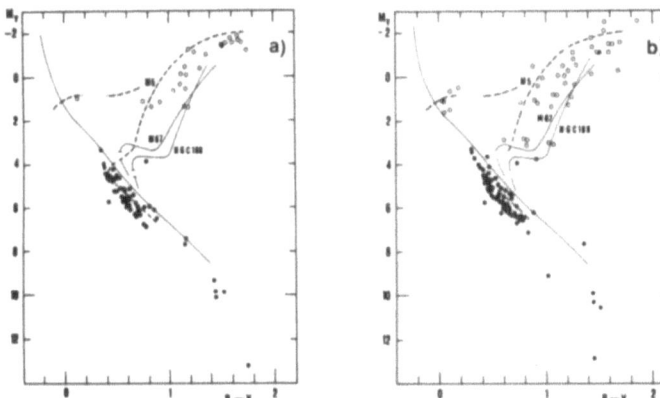

Fig.12. The color magnitude diagram. (a) For stars with a>15 kpc.
(b) For stars with |z|> 65 km/sec. Filled circles correspond to well
determined parallaxes, open circles to observations of minor weight.

diagram. By comparison with the curves of fig.10 it is clear that
all stars with a high velocity have very large eccentricities, but-
with very few exceptions, some of which may be due to errors of the
distances or, in lesser degree, of the proper motions-their velocities
are directed in such a way that the apogalactic distances are finite
and lie inside 25 or 30 kpc at most, although there are a number of
stars with retrograde motion. Notwithstanding the various uncertainties, thus, there is a strong evidence that practically no star in
the region of the sun has an apocentric distance larger than, say 30
or 40 kpc.
    The color-magnitude diagram for (high-velocity) stars with apo-
centric distance larger then 15 kpc is shown in fig 12(a); filled
circles correspond to stars with a well determined distance-modulus
m-M, mainly from trigometric parallaxes, open circles to stars
with a more uncertain distance; the data are from Eggen's catalogue.
Since the number of well determined distance-moduli of stars fainter
than M_V = 4 is large enough, uncertain moduli for these stars were
excluded.
    Typically all the faint stars in the diagram are bluer than the
main-sequence stars with the same M_V; in other words: all the stars
fainter than M_V = 3 and apocentric distance larger than 15 kpc are
not normal main-sequence stars but belong to a sequence of subdwarfs.
The converse is not true; there are subdwarfs also among high-velocity
stars with a<15 kpc, but they are outnumbered by those which lie
upon the ordinary main sequence.
    The result for giants is much less convincing because the data
are not so good; the luminosities having been taken from all available sources. Even so, fig.12(a) suggests that the giants with apocentric distance larger than 15 kpc are definitely brighter and bluer
than those of the old galactic clusters M67 and NGC188 and appear

to approach those of the globular cluster M 5. Interstellar reddening is probably not very large for the stars in Eggen's catalogue;
but anyway a correction for interstellar reddening would make the
stars even brighter and bluer.

A similar result is obtained by selecting among the high-velocity stars those with a large velocity component perpendicular to
the galactic plane, for instance, those with $|z|>65$ km/sec. In the
corresponding color-magnitude diagram (fig.12(b))we find a larger
number of giants and also a group of blue stragglers, suggesting
a blue horizontal branch. Variable stars have been omitted from the
diagrams of figs. 12(a) and (b). Note also that the two diagrams
are not independent,since there are many stars in common to both.

Of course red giants of relatively young galactic clusters like
M11 or the Hyades are known to have absolute magnitudes around 0 or
+1 and B-V colors around +1.5, which would fit with the giants of
figs. 12 (a) and (b). But the absence of main-sequence stars fainter
than 2.0, the presence in their stead of a sub-dwarf sequence and
of a number of subgiants may be considered as a sufficient evidence
that these stars belong to population II.

It is thus safe to conclude that stars in the neighborhood of
the sun with apocentric distance a>15 kpc and/or with $|z|>65$ km/sec
belong to population II. Together with RR Lyrae variables (not discussed here) and with globular clusters they form a halo of population
II stars extending several kpc above and below the galactic plane and,
in the galactic plane itself, some 10 kpc or perhaps more farther
than the usually adopted radius of the Galaxy (15 kpc). Population I
stars are not to be found more than, say, 1 kpc away from the galactic plane and farther than 15 kpc from the center.

These results are usually taken as a proof that, initially, some
10^{10} years ago, the Galaxy was a large spheroidal gaseous cloud with
an equatorial radius of 25 or 30 kpc; at that time - as it is shown
by the age of the oldest population II stars on the main (subdwarfs)
sequence - stellar formation began and the gas collapsed very rapidly
to a flat disk, leaving behind a halo of globular clusters and population II stars, moving along highly eccentric orbits. The time of
the collapse was not much longer than the time of free fall, which
for a sphere with a radius of 30 kpc and a mass of 10^{11} $M_\odot$, is considerably less than 10^9 years. Apparently, during a time as short as
a few times 10^8 years all the halo population II stars have been
formed and since then only population I stars have been born; when
the oldest stars of population I began to form, possibly the contraction of the interstellar medium was still going on, although at a
greatly reduced rate. This is necessary to explain the correlation
between the shape of the color-magnitude diagram and orbital eccentricity for stars whose orbits have e<0.350. Although this model may
be oversimplified, it seems sufficiently well supported by observations
to be used with confidence as a working hypothesis in a number of
problems.

5. CHEMICAL COMPOSITION OR AGE?

In the preceding sections it was shown that globular clusters and
such field stars as the high velocity subdwarfs (and RR Lyrae variables) form a galactic halo of population II stars, older than all
population I stars.

It seems however impossible that age alone might be the cause
of the difference in the color-magnitude diagram between the two
types of populations. For instance, no theory of stellar structure
can explain the subdwarf sequence without taking into account some
other parameter.

There are also some difficulties with red giants in globular
clusters; their presence in clusters as old as these can be understood only if they are stars of about 1 $M_{\odot}$ or less, which have spent
a long time on the main sequence and then migrated to the right in
the H.R. diagram, brightening at the same time by no less than 5
magnitudes. In this respect they differ from the red giants of the
Hyades, which have more or less the same brightness as the brightest
main-sequence stars in the same cluster and must have essentially
the same mass (from 2 to 4 $M_{\odot}$).

Now, stars on the giant branch are known to have a H-exhausted
core and to burn H in a shell (see e.g. Cox and Giuli 1968, sect.
26.4c); they possess a dense core and an expanded envelope, which is
practically fully convective. Due to this circumstance, during the
shell H-burning phase a star establishes itself upon the so-called
<u>Hayashi track</u> corresponding to its mass and moves along it in the
direction which precedes the main-sequence stage, i.e. expanding
and brightening. It is understood that only the envelope expands; while

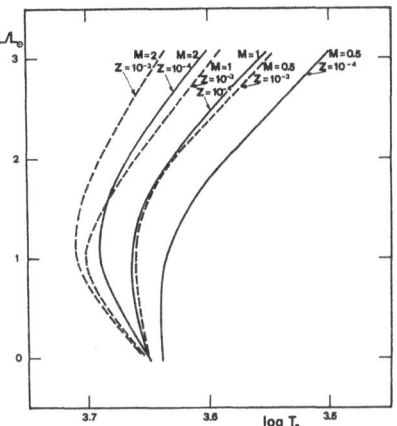

Fig.13. Hayashi tracks for stars of different mass and metal content.
Each curve is labelled with the values of the Mass (in solar masses)
and of the metal content Z (data kindly furnished by Caputo, Castellani and D'Antona).

the core becomes denser and hotter; the expansion ends when the core attains the temperature of ignition of the He-burning reactions.

For stars of a given mass the radii and the luminosities along the Hayashi track depend appreciably on their chemical composition, essentially through the opacity of the envelope; tracks of stars with greater metal content are distinctly redder and less steep than those of stars with smaller metal content (see fig. 13).

Consider, then, a group of stars all of the same age and chemical composition like those of a cluster; since the giant phase lasts only a very short time - compared with the length of the main-sequence phase - all stars on the giant branch have practically the same mass and we shall find them scattered along the Hayashi track due to the (small) differences in the time at which they left the main sequence. The differences of the Hayashi tracks due to different chemical compositions will cause the giant branches of metal-rich clusters to lie to the right of those of metal-poor clusters.

The upper limit of the brightness - for a given mass - is essentially determined by the threshold temperature of the He-burning reactions; it requires a larger radius and brightness to reach the core temperature necessary for the ignition of He-burning, when the envelope has a small opacity, that is a small metal content. These characteristics of the giant branches of galactic and globular clusters are schematically shown in figures 12(a) and (b); in these figures the color-magnitude diagrams for M 67 and NGC 188 are taken from Sandage and Eggen (1969). Incidentally, these diagrams show the tendency of the giant branches to converge (funnelling) towards an apparently single region defined by the Hayashi convective track; the branches shown in the figures 9 (a) and (b) have been copied from Woolley et al. paper (1971) and have been reproduced from older less complete results.

For all these reasons, when trying to fit the theoretical giant sequences to those of the galactic and globular clusters, it was found necessary to adopt different chemical compositions; good agreement for galactic clusters was obtained with "solar" abundances, that is with a metal content from 2 to 5%, but for globular clusters a much lower metal content, from 0.1 to 0.01%, or even smaller, had to be assumed.

It appears, therefore, that the difference among the color-magnitude diagrams of the two populations is due primarily to differences in their chemical compositions rather than in their ages. The fact that the old halo stars belong to population II while those of the disk, even those not much younger than those of the halo, belong to population I, means that the collapse of the interstellar matter was accompanied by a (sudden) increase of its metal content - whatever the cause of this change - by a factor up to one thousand.

This remarkable conclusion needs some sort of confirmation; I wish then to discuss some more direct observational evidence concerning the chemical composition of stellar populations.

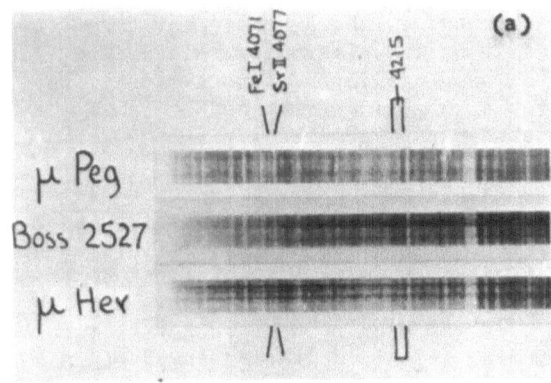

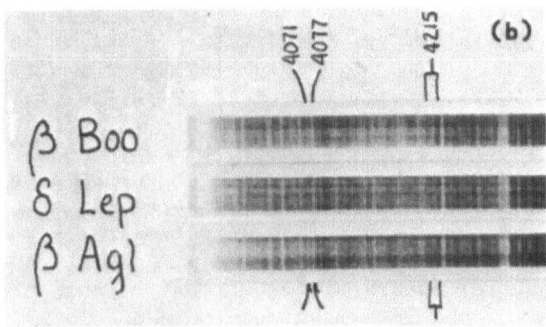

Fig. 14. The spectrum of the high-velocity stars Boss 2527 (a) and
δ Lep (b) (from M.K.K.Atlas).

6. GENERAL SPECTRAL CHARACTERISTICS OF STELLAR POPULATIONS.

It has been known for a long time that the usual luminosity criteria
often fail in the case of high-velocity stars. Figures 14(a) and (b)
are reproductions of two plates of Morgan, Keenan and Kellman "Atlas
of Stellar spectra" (1943), showing this effect for the two high-
velocity stars Boss 2527 and δ Lep; a similar effect has also been
noted for α Boo and other high-velocity giants.

In the high-velocity stars the high ratio of the line 4077 Sr II
to the adjacent lines of Fe I indicated giant characteristics, but
the break at the CN-band 4215 is as small as in subgiants or in
main-sequence stars of the same spectral type. Similar effects have
been studied in more detail by Keenan and Keller (1953).

In a series of papers Roman (1950,1952) showed that, if one
leaves apart stars with anomalous CN-bands, the remaining stars with

spectral types F5-K1 may be divided into two groups; one group,
which she called the weak-line group, is characterized by the fact
that the G-band and the line 4227 Ca I are stronger relative to all
other lines than in the other group (called the strong-line group).
She found that the stars of the strong-line group had a mean space
velocity of 26 km/sec as compared with 42 km/sec for the weak-line
group.

Similar spectral differences are especially conspicuous if one
compares stars in globular clusters with ordinary stars of the same
spectral type.   Indeed the extreme weakness of metallic lines in globular cluster giants induced the early observers to assign them to
the spectral types F (see, for instance, Baade 1952); however when
good spectra became available, it was soon evident that real spectral types-viz those corresponding to the excitation temperature-
were at least G5 or later in agreement with the colors (Deutsch
1953).

Individual stars of globular clusters, and also of galactic
clusters if one wants to observe main-sequence stars of types later
than F or G, are difficult to observe due to their faintness.   This
difficulty may be overcome by modern techniques employing very efficient spectrograph and image tubes; in the near future spectra of
stars as faint as the 10th magnitude with relatively high resolution
will become available; with lower resolution the 15th magnitude is
already attainable.   However information on stars in crowded regions
like the central parts of globular clusters or of spiral galaxies
and the main body of ellipticals can be obtained only from their
integrated spectrum.

If $f_\lambda(M_V,B-V)$ is the spectral energy distribution (the absolute
flux) corresponding to a star of absolute magnitude M_V and color
B-V, and $N(M_V,B-V)dM_V \, d(B-V)$ is the number of stars in the region observed, whose magnitude and color lie in the intervals dM_V and $d(B-V)$,
respectively, the integrated spectrum is

$$F_\lambda = \int f_\lambda(M_V,B-V)N(M_V,B-V) \, dM_V \, d(B-V), \qquad (6.1)$$

the integration being extended to all magnitudes and colors.   If it
is necessary to consider the velocities of individual stars, the
function under the integral sign must be convoluted with a function
of λ corresponding to the velocity distribution.

In practice stars may be grouped in a limited number of groups
(e.g. K-giants, A and F stars, G-dwarfs, etc....), so that equation
(6.1) may be written as

$$F_\lambda = \Sigma_i f_i(\lambda)N_i, \qquad (6.2)$$

where $f_i(\lambda)$ is the spectral energy distribution for stars of group
i and N_i their number.   From observation of F_λ, knowing the functions
$f_i(\lambda)$ some information concerning the N_i may be obtained for the

groups which contribute mostly to the observed light. In principle
it is possible also to get some information on the $f_i(\lambda)$; for instance, one can decide whether the K-giants belong to the strong-
line or to the weak-line group.

This type of analysis has been carried on by Spinrad and others
for the nuclear regions of spirals. In a somewhat less sophisticated
form it was applied by Mayall (1946) and by Morgan (1956) to the
stellar population of globular clusters. Morgan comments that in the
case of globular clusters the spectra show "a surprising absence of
compositeness"; in other words, in practice the light comes from a
single dominant group of stars. The classification of the spectra-
by comparison with ordinary standards - is however not unique.

Morgan defines three different spectral types:
the <u>fundamental</u> type, based on the intensity ratio of the G-band to
Hγ;
the "<u>hydrogen</u>" type, based on the absolute (estimated) intensities
of the hydrogen lines;
the "<u>metallic-line</u>" type based on the intensities of the strong
Fe I lines in the region λ 4200-4400.

Table V gives these types for a number of globular clusters and
for two well known subdwarfs. It is seen that in most cases the me-
tallic-line type is the earliest, while the fundamental type lies in
the middle of the other two; for the first three clusters no metallic-

Table V

Integrated Spectra of Globular Clusters (Morgan).

Cluster	Fundamental	Hydrogen	Metallic
NGC 6341 (M 92)	F2	F6	?
7078 (M 15)	F3	F6	?
5024 (M 53)	F4	F6	<FO
5904 (M 5)	F5	F8	FO
6205 (M 13)	F5	F8	FO
5272 (M 3)	F7	F8	F1
6229	F7	F8	FO
5139 (Cen)	F7	F8	FO
6522	F8	GO	G2
6356	G5	G2	G5
6637 (M 69)	G5	G8?	G2
6440	G5?	G5?	G2?
HD 140283	F2	F8	<FO
BD +2°2538	F2	F8	<FO

line type could be estimated. The spectra of these three clusters
closely match the spectrum of the subdwarfs. This shows that the
extremely numerous subdwarfs and not the bright but few K-giants
contribute almost the totality of the light of the clusters. The
last three clusters have rather strong metallic lines and the CN-
absorption may be present; in NGC 6356 the metallic abundance is
probably close to the solar value.

Perhaps the most important and surprising result from these
spectra is that although all globular clusters have essentially the
same age within 10^9 years or less – from the brightest members of
their main (or subdwarf) sequence – their metal content differ by
very large factors: from 1/1000 or 1/100 to about the solar value.

7. INFORMATION FROM HIGH-RESOLUTION SPECTRA.

Detailed information on individual elements and quantitative results
can be obtained only from high-resolution spectra. In principle these
yield also information on (excitation) temperature, electron pressure,
surface gravity and – if a good parallax is available – radii, luminosities and masses may also be obtained. To this purpose, however,
very good model atmospheres are necessary and unfortunately at present they are not available.

There are essentially two methods of handling spectroscopic
data: differential curve-of-growth or coarse analysis and the so-
called fine analysis; the latter avoids some of the pitfalls of the
coarse analysis, especially if as in many cases lines with saturated
Doppler core have been used. However if proper care is taken to employ
only weak lines, the simpler method is just as good (Gratton 1953,
Cayrel and Cayrel de Strobel 1966). Two excellent summaries of the
work done in this field have been published recently by Pagel (1970,
1973).

The information which one especially wishes to obtain, concerns
the enrichment in metals of the interstellar medium – from which
the stars have been formed – with age. Hence stars showing anomalous
effects, like Carbon stars (types R and N), S-stars, H-deficient
stars, etc.... must be avoided. The best would be of course to compare
stars of approximately the same spectral type and luminosity from
different clusters; but, with few exceptions, even K-giants in galactic clusters are too faint. Thus in general one must use field stars
and rely on such age indicators as kinematical properties as shown
in the preceding sections.

An interesting idea is to correlate abundances of elements originated by different nuclear processes; for instance, elements of the
Fe-peak with those produced by α-processes or by s- and r-processes.
If, however, abundance-ratios are considered it is not clear at first
which kind of information may be obtained by this way. Now, if we
assume that initially – that is for the first ganeration of stars
in the Galaxy – only H and He were present and all the other elements
were synthesized later in the stars at constant rates, it is obvious

that the relative abundances of the elements other than H and He
will be the same in all stars and equal to the ratios of the respec-
tive generation rates; this, at least, if the total amount of H converted in heavier elements is not too large. Conversely, if we find
that there are changes among the relative abundances of elements
produced by different processes, we must conclude that at a certain
moment the rates of the processes underwent some change. This would
be in itself a very important result, because it would point to a
very deep physical difference between stars of different populations.

Since the analysis of high dispersion spectra is extremely long
and painstaking, the number of stars for which detailed results are
available is not very large. Quite recently semiautomatic methods
of data handling employing digitized microphotometers and large computers have become available. In order to obtain comparable results
from different stars it seems essential to keep as constant as possible the observational techniques; so the spectra must be obtained
with the same spectrograph and emulsion type, must be calibrated in
the same way and be measured with the same microphotometer; the re-
ductions - especially the localization of the continuum and the choice
of the wave lengths - must be carried on in the same way and, of course
the same lines must be measured in all stars. I may mention a large
investigation of about 60 field K-giants which is in progress at
our Laboratory in Frascati on these lines.

The data available in 1973 have been discussed by Pagel (1973),
to whose paper we refer for more details; an older - but still very
valuable - paper by R. Cayrel and G. Cayrel de Strobel (1966) may
also by quoted.

The comparison of data from different sources may be misleading
due to the differences in observational techniques; results from the
same star by different authors show often greater differences than
those from different stars. When this is taken into account we may
conclude that the results on the metal-to-hydrogen ratio from high
resolution spectra confirm those obtained with low dispersion and
yield reasonably reliable quantitative data.

We consider a little more in detail the case of the K-giants.
It is usual to employ the symbol [El/H] to indicate the logarithm
of the relative abundance ratio of a certain element to hydrogen for
a certain star with respect to the sun; thus a negative value of
[El/H] corresponds to a deficiency of the element - relative to H-
in the star with respect to the sun.

The results from all available sources show that K-giants may
be divided into three groups:
Group a: Population II stars, like members of globular clusters or
very high velocity stars, going from extreme cases like HD 122563
or HD 221170 to less extreme ones, like γ Pav. In this group [Fe/H]
ranges from -0.7 to 3.0, showing a very high deficiency of metals
of the Fe-peak relative to the sun.
Group b: intermediate stars with orbits of relative high eccentricity,
such as α Boo, ζ Ret, η Cep, etc... In this group [Fe/H] ranges mostly
from -0.4 to -0.1 and shows thus a mild deficiency of metals, but

with a large dispersion of the values for the individual elements.
Group c: ordinary stars with circular orbits, like αUMa,εVir, the
Hyades etc... For these stars [Fe/H] is not far from 0, but the dispersion is very large and it is very difficult to distinguish between
real differences and those due to observational causes.

As a whole these results confirm the general decrease of the
metal content with increasing age; but this seems to be a general
trend and by no means a one to one relationship. There is a considerable overlap of the individual values obtained for stars of the
three groups.

The results from single elements are even more uncertain and
the only reasonable conclusion is that they all show essentially the
same trend; that is, when Fe is up or down for a given factor, all
the heavy elements are up or down for the same factor (relative to
the sun) inside the limit of observational errors. Perhaps there is
a suggestion that, where Fe is low, C and N are lower and an analogous behavior is hinted by the elements from Zn to Zr.

For the general problem of metal-abundance correlation with age,
the observations of subgiants are especially interesting, because
for these stars it is possible to obtain individual estimates of their
ages from their position in the H.R. diagram. For about a dozen of
field subgiants spectrum analysis are available; they lie in the range
-0.44. Since these are evolved stars, whose ages are in some cases
older than that of the sun, it must be concluded that the metal-
content in the disk did not change appreciably since the halo collapsed
10^{10} years ago (Eggen and Sandage 1969). This surprising result con-
firms that obtained from globular clusters, that the main change in
the general metal abundance took place during the first few hundreds
of million years of galactic life.

8. BROAD BAND PHOTOMETRY; BLANKETING.

In a very well known paper, Eggen, Lynden-Bell and Sandage (1962)
derived chemical differences between stars of different population
types by means of U-B and B-V colors. In short, the crowding of
absorption lines- mostly from metals and molecules- has a blocking
effect in some parts of the spectrum greater than in others; indeed
in those regions where the lines are less and weaker, the flux may
be enhanced due to a back-warming effect, because the energy blocked
in the lines must be radiated away in the spaces between the lines,
if the total flux -i.e. the effective temperature- must remain the
same.

In particular the U-region is more depressed, the B-region is
less so and the V-region is reinforced in stars of spectral type
later than G0 with strong lines. For stars later than K5 the resulting shift in the two-color diagram is nearly parallel to the main-
sequence line, but for stars between G0 and K5 one can obtain by
this way a very useful information about their chemical composition.
For more details upon this effect, generally known as line-blanketing,

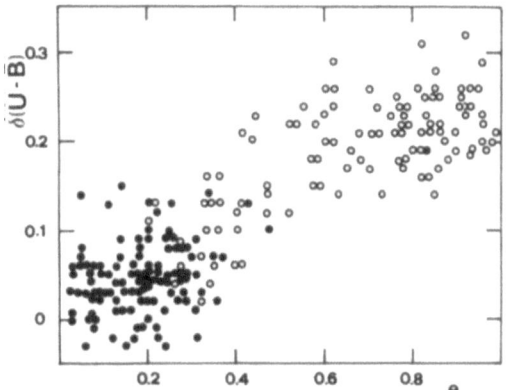

Fig. 15. Correlation between δ(U-B) and e (Eggen, Lynden-Bell and Sandage). The filled and open circles correspond to two different Catalogues by Eggen.

we refer to a paper by Wildey, Burbidge, Sandage and Burbidge (1962; see also the paper by Eggen et al. already quoted and the Varenna lectures by Sandage and Gratton 1962).

Of course from these observations one cannot obtain detailed

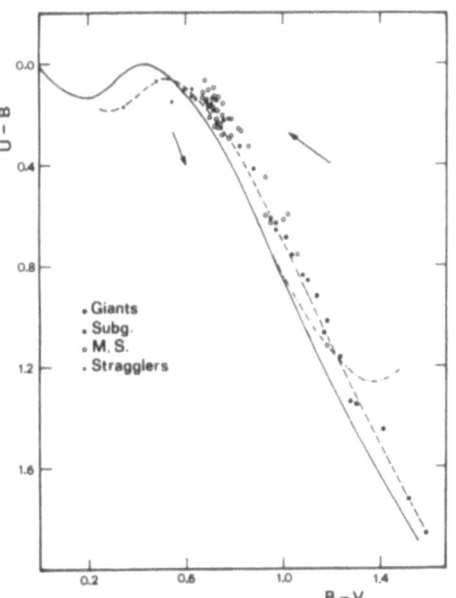

Fig. 16. Two-color diagram for N.G.C. 188 (Eggen and Sandage). For explanations see the text.

information upon chemical abundances, but only a general <u>index of</u> <u>metallicity</u>; as such one chooses the ultraviolet excess $\delta(\overline{U-B})$ rel- ative to the Hyades main-sequence line in the two-color diagram at the same B-V; the difference $\delta(U-B)$ is taken in the sense Hyades-Star, so that the excess is positive when the star lies above the Hyades line in the two-color diagram. Hence $\delta(U-B)$ is greater when the metal content is smaller. Stars with zero metal content should lie upon an envelope line, corresponding to $\delta(U-B) = 0.2$ at B-V = 0.5.

Fig. 15, from an Eggen et al. paper, shows the correlation between the ultraviolet excess and the orbital eccentricity e (see section 3) for a sample of 221 stars. Although the scatter of the points is large, the general decrease of metal content with increasing eccen- tricity is clearly shown.

The stars of this sample are relatively near objects; for more distant stars proper care must be taken to eliminate the effect of interstellar reddening. How this can be done, that is how reddening and blanketing may be separated for the stars of a distant cluster, is better explained in fig. 16 from a paper by Eggen and Sandage (1969).

In this figure the two-color diagram for the Hyades is shown as a solid line; the dashed line is obtained from the solid one by a parallel shift along the reddening vector, $\Delta(B-V) = 0.09$, $\Delta(U-B)= 0.06$, that is by the amount of reddening of the cluster NGC188. The deviations of the stars of the cluster relative to the dashed line are caused by blanketing - apart of course from observational errors.

The deviations are small, but quite evident, especially near B-V = 0.7, showing that the metal content of NGC 188 is slightly smaller than that of the Hyades (see, however, the next section).

Line blanketing affects also the position of the stars in the color-magnitude diagram; for instance, a star with a metallicity

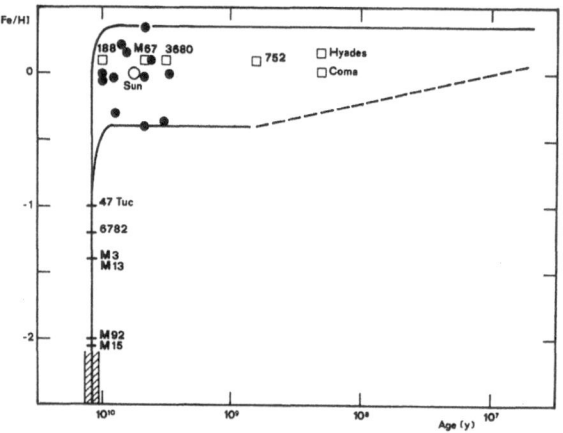

Fig. 17. Variation of metal content with age in galactic objects according to Eggen and Sandage.

corresponding to an ultraviolet excess about 0.3 may have a $\Delta(B-V)$ greater than 0.2 relative to the Hyades. This is of the same order of magnitude of the horizontal shift of the subdwarf sequence relative to the ordinary sequence (see figs 12(a) and (b); hence the color difference between subdwarfs and ordinary main-sequence stars of the same absolute magnitude is essentially an atmospheric effect connected with the difference of chemical composition.

Figure 17, from Eggen and Sandage (1969), summarizes the results concerning the change of metal content with age; since then the observational situation has not changed essentially. Clusters are indicated by their names; closed circles correspond to a dozen subgiants. This figure depicts very well the rapid increase of metal content during the collapse of the galactic halo and its essential constancy during the remaining 10^{10} years; the 2×10^{8} year hatched interval of time corresponds to the era of Population II star formation; during the rest of the time only Population I stars were formed.

9. SUPERMETALLICITY.

To conclude these lectures I wish to make a short mention of the so-called supermetallicity of some stars discovered a few years ago by Spinrad and others (see especially Spinrad and Taylor 1969) by means of a scanner technique.

This technique is essentially a refinement of the broad-band photometry, in the sense that the flux is measured at a large number of wave lengths (33 in all) from 3880 A to 10700 A; the pass-bands are chosen to contain either strong absorption features or regions as free as possible of absorption.

The colors derived from the ratio of the fluxes in two different pass-bands are used to obtain temperature indices and absorption feature strengths. Standard stars of different temperature, luminosities and metal content were used to calibrate a number of criteria from which physical results could be obtained for field and cluster stars.

The most startling result from these investigations was the high abundances of metals found in some evolved stars. The cool evolved stars of the clusters M 67 and NGC 188 were found to have in average an abundance of Ca, Mg and Na three times larger than the sun, in contradiction to the result from (ordinary) broad-band photometry.

According to the present view the stars in M 67 are as old as the sun and those in NGC 188 are considerably older; thus some doubts have been advanced concerning the validity of the scanner abundances. In principle, of course, there is the possibility that the chemical evolution had progressed at different rates in different regions of the Galaxy, so that stars formed in different places at the same time need not to have the same metal content. It is also possible that – especially in some clusters – the atmospheric composition does not reflect the initial composition of old stars due to accretion of nuclearly evolved matter from more massive stars. But the

existence of super metal-rich stars is in itself a very interesting
fact and deserves a confirmation.

On the other side, the ultraviolet excess of the stars in NGC
188 and M 67 according to Eggen and Sandage (1964, 1969) and Racine
(1971) yields for these clusters a metal-content (slightly) smaller
than that of the Hyades. This is in contradiction with the observations by Spinrad, Greenstein, Taylor and King (1970) who find from
the spectra of dwarfs in M 67 and NGC 188 that the metallic lines
are stronger than in the Hyades dwarfs; they note a conspicuous
discrepancy between the spectral types derived from H-lines and those
from metallic lines in 7 stars in M 67, the spectra from metallic
lines being considerably later. They take this as an indication that
the cluster is super metal-rich.

However a reexamination of the same plates by Morgan and Abt
(1973) showed that the spectral types estimated by Spinrad et al.
were incorrect, possibly due to some observational effect. According
to Morgan and Abt no systematic difference exists between the spectra
determined from metallic and H-lines. The spectra resemble very
closely those of the Hyades.

The question is thus still controversial. It seems that while
some super metal-rich stars may (and do in fact) exist, the metal
content of the old galactic clusters is essentially equal or, perhaps,
somewhat smaller than that of the Hyades. On the whole the conclusion
reached in the previous sections, viz. that the differences in metal
content among Population I stars are rather small, amounting to a
factor 4 at most, can be regarded with confidence; the variation with
age, as indicated by spectroscopic and color observations is very
slight and may be masked by observational and other factors. This
contrasts with the very low average metal content as clearly shown
by extreme Population II stars (globular clusters and subdwarfs);
besides in Population II stars we observe very large differences in
chemical abundances amounting to a factor 100 or even 1000 between
the metal content of different stars.

REFERENCES

Baade, W., 1944, Ap. J. 100, 137.
Baade, W., 1952, Trans. I.A.U. vol. 8 , 682.
Becker, W. and Fenkart, R., 1970, The Spiral Structure of our Galaxy,
I.A.U. Symp. n° 38 (Becker and Contopoulos eds.) page 205.
Blaauw, A., 1969, Structure and Evolution of the Galaxy (Mavridis ed.)
page 295.
Cayrel, R. and Cayrel de Strobel, G., 1966, Ann. Rev. of Astr. and
Astroph. 4,1.
Contopoulos, G., 1972, The dynamics of Spiral Structure, Univ. of
Maryland.
Cox, J.P. and Giuli, R.T., 1968, Stellar Structure.
Delhaye, J., 1965, Solar Motion and Velocity Distribution of Common
Stars in "Galactic Structure" (Blaauw and Schmidt eds.).

Deutsch, A.J. 1953, Principes fondamentaux de Classification Stell-
aire (Centre Nat. Rec. Sc., Paris), page 25.
Dixon, M.E. 1967, A.J. 72, 429.
Eggen, O.J. 1964, Roy. Obs. Bull. n° 84.
Eggen, O.J., Lynden-Bell, D. and Sandage, A.R. 1962, Ap.J. 136,748.
Eggen, O.J. and Sandage, A.R. 1964, Ap. J. 140, 130.
Eggen, O.J. and Sandage, A.R. 1969, Ap. J. 158, 669.
Gratton, L. 1953, Mem. Soc. Roy. Sci. Liège, 14,419.
Keenan, P.C. and Keller, G. 1953, Ap.J. 117, 241.
King, I.R. 1971, Publ. A.S.P. 83, 377.
Lin, C.C. 1967, Stellar Dynamical Theory of Normal Spirals in "Rel-
ativity theory and Astrophysics; 2. Galactic Structure" (Ehlers ed.).
Mayall, N.U. 1946, Ap. J. 104, 290.
Morgan, W.W. 1956, Publ. A.S.P. 68, 509.
Morgan, W.W. and Abt, H.A. 1973, A.J. 78, 386.
Morgan, W.W. , Keenan, P.C. and Kellman, E. 1943, An Atlas of Stellar
Spectra.
Morgan, W.W., Whitford, A.E. and Code, A.D. 1953, Ap.J. 118, 318.
O'Connell, D.J.K. (ed.) 1957, Stellar Populations.
Oort, J.H. 1926, Groningen Publ. n° 40.
Pagel, B.E.J. 1970, Quart.J.R.A.S.11, 172.
Pagel, B.E.J. 1973, Space Science Reviews, 15, 1.
Pismis, P. 1960, Bol. Tonantzintla and Tacubaya Obs. 19, 3.
Racine, R. 1971, Ap.J. 168, 393.
Roman, N. 1950, Ap.J. 112, 554.
Roman, N. 1952, Ap.J. 116, 122.
Roman, N. 1965, High Velocity stars in "Galactic Structure" (Blaauw
and Schmidt eds.).
Sandage, A.R. and Eggen, O.J. 1969, Ap.J. 158, 685.
Sandage, A.R.,and Freeman, K.C. and Stokes, N.R. 1970, Ap.J. 160,
831.
Sandage, A.R. and Gratton, L. 1962, Observational approach to Stellar
Evolution in "Star Evolution" (Gratton Ed.).
Schmidt, M. 1965, Rotation Parameters and Distribution of Mass in
the Galaxy in "Galactic Structure" (Blaauw and Schmidt eds.).
Shane, W.W. and Bieger-Smith, G.P. 1965, Bull. Astr. Inst. Nether-
lands, 18, 263.
Spinrad, H., Greenstein, J.L., Taylor, B.J. and King, I.R. 1970,
Ap.J. 162, 891.
Spinrad, H. and Taylor, B.J. 1969, Ap.J. 157, 1279.
Wildey, R.L., Burbidge, E.M., Sandage, A.R. and Burbidge, G. 1962,
Ap.J. 135, 94.
Woolley, R., Pocock, S.B., Epps, E.A. and Flinn, R. 1971, Roy. Obs.
Bull. n° 166;
Woltjer, L. 1967, Structure and Dynamics of Galaxies in "Relativity
Theory and Astrophysics; 2. Galactic Structure"(Ehlers ed.).

STELLAR DYNAMICS

L. Woltjer

Department of Astronomy, Columbia University,
New York, U.S.A.

1. INTRODUCTION

A typical major galaxy consists of 10^{11} - 10^{12} stars. In addition
some gas may be present but we shall neglect this for the moment.
The stars move in the gravitational field of the galaxy and the
gravitational field is generated by the stars and gas. If we
characterize the gravitational field by a potential Φ, the orbits
of the stars are given by

$$\frac{d^2 r}{dt^2} = - \nabla \Phi \tag{1}$$

The gravitational potential follows from Poisson's equation

$$\nabla^2 \Phi = 4\pi G\rho , \tag{2}$$

with G the gravitational constant and ρ the mass density. The
equations in this form are not particularly useful. There are
10^{11} equations (1) for 10^{11} stars, while ρ in equation (2) consists of 10^{11} δ-functions if the stars are treated as mass points.
  In order to make further progress we shall write

$$\Phi = < \Phi > + \delta\Phi \tag{3}$$

with Φ a suitably averaged potential -- averaged perhaps over 10^5
stars in a realistic galaxy. Similarly taking $<\rho>$ to be the density averaged over similar regions we have

$$\nabla^2 < \Phi > = 4\pi G < \rho > \tag{4}$$

We shall now first demonstrate that in typical stellar systems the instantaneous dynamical effects of the $\delta\Phi$ are negligible, although in some cases their secular effect may be important.

2. THE TIME OF RELAXATION

Suppose we have a star in a region in which $\nabla< \Phi > = 0$. This star whose mass we take to be m^* will move in a straight line orbit if the effects of $\delta\Phi$ are disregarded. Suppose that somewhere along its trajectory there is a star of mass m and that the impact parameter (the smallest distance between the unperturbed orbit of m^* and m, the latter supposed to be at rest) is D. If the velocity of m^* is V then during a time of order $2D/V$ the moving star experiences an acceleration of order Gm/D^2 transverse to the unperturbed orbit. As a result the star acquires a velocity $\Delta v_\perp$ transverse to its original velocity vector given by

$$\Delta v_\perp \approx \frac{2D}{V} \frac{Gm}{D^2} \approx \frac{2Gm}{DV} \tag{5}$$

In almost all relevant cases $\Delta v_\perp \ll V$. If we now consider the effects of N subsequent encounters with all the same magnitude of the impact parameter and if we assume that subsequent encounters are statistically independent it is clear that $<\Delta v_\perp>$ on the average vanishes (because the directions of $\Delta v_\perp$ in the plane transverse to $\underset{\sim}{V}$ are random). However $<(\Delta v_\perp)^2>$ does not vanish and for the average $v_\perp^2$ after N encounters we have

$$<v_\perp^2> = N(\Delta v_\perp)^2 \tag{6}$$

The number of encounters with impact parameters between D and $D + dD$ in a time T is equal to $2\pi DdDTVn$ with n the number density of perturbing stars along the trajectory. Inserting this into eq.(6) and integrating over all impact parameters between D_{max} and D_{min} we have

$$<v_\perp^2> = \int_{D_{min}}^{D_{max}} 2\pi TVn \frac{4G^2m^2}{D^2V^2} DdD = \frac{8\pi G^2m^2Tn}{V} \ln \left(\frac{D_{max}}{D_{min}}\right) \tag{7}$$

The integral diverges at both ends. If we had integrated the equations for the two body encounter exactly, equation (5) would be slightly modified and the divergence at the smaller impact parameters would be eliminated. In fact the correct expression is obtained by replacing D_{min} in equation (7) by $G(m + m^*)/V^2$. The divergence at the upper end is real and associated with the infinite range of gravitational forces. However at large impact parameters the whole theory breaks down because the star m^* now interacts not just with one other star at a time, but with several. Some qualitative reasons may be given for believing that the correct result is obtained by taking D_{max} equal to the dimension of

a stellar system. In practice the choice of D_{max} is not very important because it occurs in a logarithm which has a value in excess of about 10 in typical cases. Defining now the relaxation
time T_D(with D for deflection) as the time it takes for $<v_\perp^2>^{1/2}$ to
become equal to V we have

$$T_D = \frac{V^3}{8\pi G^2 m^2 n \ln \Lambda} = \frac{V^3}{8\pi G^2 m \rho \ln \Lambda} \approx \frac{10^8 V^3 (km/sec)}{m^2(\odot)n(pc^{-3})}$$ (8)

with $\Lambda = V^2 D_{max}/[G(m + m^*)]$, and ρ the mass density. After a time
T_D the "memory" of a star as to its unperturbed orbit has largely
become lost. Note that in equation (8) the product $m\rho$ occurs.
Thus if in a region of fixed mass density we distribute the matter
in larger clumps the relaxation process occurs faster: A galaxy
made of globular clusters relaxes much more quickly than a galaxy
of the same mass composed of single stars.

In a realistic treatment of course we should take into account
the fact that not only m^* but also all the other stars move. This
does not usually lead to large numerical changes in T_D. In addition to T_D several other relaxation times may be introduced, like
the equipartition time, T_{eq}, which indicates how long it takes two
distributions of stars with different mean energies to reach equal
mean energies, T_E the time on which the energy of a star is appreciably changed and T_f the time for dynamical friction -- the loss
of momentum in the direction of the unperturbed trajectory. If
all stars have comparable masses all these relaxation times tend
to be of the same order. If not, important differences occur.
For example $T_f \approx 2m/(m + m^*)T_D$. Hence a massive globular cluster
will be slowed down rather fast. On the basis of this fact
Ostriker has recently argued that the semistellar nucleus of the
Andromeda Nebula may have resulted from the accumulation of massive clusters near the center.

Some typical values for n, V and T_D are given in the following table where the mass m has been taken equal to that of the sun.

Region	$n(pc^{-3})$	V(km/sec)	T_D(years)
Solar Neighborhood	0.1	20	10^{13}
Elliptical galaxy	10	200	10^{14}
Inner part globular cluster	1000	10	10^8
Galactic (= open) cluster	10	0.5	10^6

It is seen from the table that in galaxies the effects of relaxation are unlikely to be very important at the present time.
More violent collective relaxation processes in the early phases
of a galaxy will be discussed by Dr. Prendergast.

If we take a spherical stellar system in equilibrium the virial theorem states that

$$zE_k + E_g = 0 \quad , \tag{9}$$

where E_k is the kinetic energy of the system $(1/2M < V^2 >)$ and E_g the gravitational potential energy $(-3/5GM/R$ for a uniform sphere). Here M and R are the mass and radius of the configuration. We therefore have

$$< v^2 > = \frac{3}{5} \frac{GM}{R} \tag{10}$$

For a nonhomogeneous distribution of matter the coefficient 3/5 has to be changed. Inserting this into the expression for T_D and noting that $M \propto R^3 n$ we find

$$T_D \propto \frac{V^3}{n} \propto M^{1/2} R^{3/2} \tag{11}$$

Hence for a system of given mass the relaxation time decreases rapidly with decreasing radius.

If we have a system in which the relaxation time is short energy exchanges between the stars will cause some to obtain enough energy to escape. Because the energy changes in each encounter are rather small, most stars escaping from a cluster will have just barely enough energy to leave the cluster: they arrive at infinity with essentially zero velocity and therefore have zero energy. This implies that the total energy E_{tot} of the cluster remains constant. We clearly have

$$E_{tot} = E_k + E_g \tag{12}$$

or with equation (9)

$$E_{tot} = - E_k = \frac{1}{2} E_g \tag{13}$$

Making use of the same assumptions as went into equation (10) we have

$$E_{tot} = \frac{1}{2} M <v^2> = - \frac{3}{10} \frac{GM^2}{R} \tag{14}$$

As stars evaporate from a cluster, M decreases and consequently the cluster contracts while the velocity dispersion increases. From equation (11) it is then seen that $T_D \propto M^{7/2}$. Consequently once evaporation of stars begins, the process rapidly accelerates itself, a fact made use of in certain current quasar theories.

3. THE BOLTZMANN EQUATION

Neglecting the effects of stellar encounters we now shall make a
statistical description of a stellar system by introducing the
distribution function $f(\underset{\sim}{r}, \underset{\sim}{v}, t)$, defined as the number of stars
with coordinates in a unit interval around $\underset{\sim}{r}$, and velocities in a
unit interval around $\underset{\sim}{v}$ at the time t. Introducing a six dimen-
sional phase space with x, y, z, v_x, v_y, and v_z as independent
coordinates, we evaluate the change in the number of stars in a
box located at x, y, z, v_x, v_y, v_z and with sides dx, dy, dz,
dv_x, dv_y, dv_z. Applying Gauss's theorem to this box we immediately find

$$\frac{\partial f}{\partial t} + \nabla_6 \cdot (f \underset{\sim}{V}_6) = 0 \tag{15}$$

with $\nabla_6 \cdot$ and $\underset{\sim}{V}_6$ the divergence and velocity in the six dimensional
phase space. The components of $\underset{\sim}{V}_6$ are $\dot{x}$, $\dot{y}$, $\dot{z}$, $\dot{v}_x$, $\dot{v}_y$, $\dot{v}_z$ or making use of equation (1) v_x, v_y, v_z, $-\partial\Phi/\partial x$, $-\partial\Phi/\partial y$, $-\partial\Phi/\partial z$. Noting that $\underset{\sim}{r}$ and $\underset{\sim}{v}$ are independent coordinates in phase space and
that Φ does not depend on $\underset{\sim}{v}$ we find that $\nabla_6 \cdot \underset{\sim}{V}_6 = 0$ and therefore
obtain

$$\frac{\partial f}{\partial t} + v_x\frac{\partial f}{\partial x} + v_y\frac{\partial f}{\partial y} + \frac{1}{2}\frac{\partial f}{\partial z} - \frac{\partial \Phi}{\partial x}\frac{\partial f}{\partial v_x} - \frac{\partial \Phi}{\partial y}\frac{\partial f}{\partial v_y} - \frac{\partial \Phi}{\partial z}\frac{\partial f}{\partial v_z} = 0 \tag{16}$$

which is the collision free Boltzmann equation (Vlasov equation
in plasma physics). From the definition of f it follows that
$\rho = m \int f \, d\underset{\sim}{v}$ with $d\underset{\sim}{v}$ representing the integration over the velocity
coordinates. Hence the Poisson equation (4) becomes

$$\nabla^2\Phi = 4\pi Gm \int f \, d\underset{\sim}{v} \tag{17}$$

where from equation (16) onward Φ represents the smoothed potential. Any function f that is positive definite and satisfies
equations (16) and (17) represents a possible stellar system (although not necessarily a stable one).

The general solution of equation (16) may be written as

$$f = f(I_1, I_2, I_3, I_4, I_5, I_6) \tag{18}$$

where $I_1 - I_6$ are the integrals of the equations of motion of a
single particle ($v_x = dx/dt........$, $dv_x/dt = -\partial\Phi/\partial x........$).
For suppose that there were just one such integral I_1. Then the
orbit of a particle in phase space would be located on a hyper-
surface I_1 = constant. If no other integrals existed, the orbit
would have to fill that surface ergodically. Writing equation
(15) as Df/Dt = 0 -- with D/Dt the time derivative following the
motion in phase space -- we see that f must be constant along a
trajectory in phase space and therefore on a surface covered
ergodically by such a trajectory, f must be a constant. But along

different trajectories, that is on different surfaces I, the
value f may be different. Hence we would have $f = f(I_1)$. With
six integrals the reasoning proceeds in an entirely analogous
manner leading to equation (18). If the system is time independent the time may be eliminated between the six integrals, reducing their number to five. Sometimes integral surfaces fill
the relevant part of the phase space ergodically. Such integrals
are called "non isolating". Only the isolating integrals are
relevant in equation (18).

To illustrate the theory let us consider a spherically symmetrical time independent potential and see which of the five
integrals are isolating. The integrals in such a potential are
easily enumerated by starting out with a 1/r potential. In such
a potential the particle would move in an elliptical orbit and
the integrals would be the energy E, the three components of the
angular momentum J and an integral specifying the orientation of
the major axis of the ellipse. All these integrals would be
isolating. If the potential is not of the 1/r type, then generally
the motion still may be described in terms of an elliptical orbit,
but now in a rotating frame of reference (precession of the peri-
helion). As a result the major axis of the orbit may now point in
any direction (orthogonal to J) and the corresponding integral is
no longer isolating. Therefore the general solution for the
spherical case will be given by

$$f = f(E, J_1, J_2, J_3) \tag{19}$$

with $J_1 - J_3$ the three components of J.

We construct a simple solution by choosing f to be given by

$$f = Qe^{-2\alpha^2 E} = Qe^{-\alpha^2(v^2+2\Phi)} \tag{20}$$

which clearly is a special case of equation (19). Poisson's
equation now becomes

$$\nabla^2\Phi = 4\pi Gm \int_0^\infty f4\pi v^2 dv = 16\pi^2 GmQe^{-2\alpha^2\Phi} \int_0^\infty e^{-\alpha^2 v^2} v^2 dv$$

$$= \frac{4\pi^{5/2} GmQ}{\alpha^3} e^{-2\alpha^2\Phi} \tag{21}$$

which is just the equation for the isothermal gas sphere. Solving
this equation with the appropriate boundary conditions one obtains
$\Phi(x,y,z)$ and upon inserting this into equation (20), one finds the
explicit dependence of f on r and v.

In the more general case of a system that has only axial

symmetry the situation is less clear. Certainly the energy
$E = \frac{1}{2} v^2 + \Phi$ and the angular momentum component $J_z = \overline{\omega} v_\phi$ (in
polar coordinates $\overline{\omega}$, ϕ, z with z the axis of symmetry) remain
isolating integrals. In most potentials of this type also a
"Third integral" appears to be isolating, but this is by no means
always the case. A simple case is provided by a separable poten-
tial $\Phi(\overline{\omega}, z) = \Phi_1(\overline{\omega}) + \Phi_2(z)$. In this case the energy associated
with the z motion is separately conserved and we may write
$I_3 = \frac{1}{2} v_z^2 + \Phi_2(z)$. Such a separable potential is a reasonable
representation to the potential in the solar neighborhood. For
the local stars we therefore should have

$$f = f(E, J_z, I_3) = f[v_{\overline{\omega}}^2 + v_\phi^2 + v_z^2 + 2\Phi_1, \overline{\omega} v_\phi, v_z^2 + 2\Phi_2(z)] \quad (22)$$

If I_3 were non isolating, f would be symmetrical between $v_{\overline{\omega}}$
and v_z and hence the velocity dispersions $\langle v_{\overline{\omega}}^2 \rangle$ and $\langle v_z^2 \rangle$ would be
equal. The observed situation is $\langle v_{\overline{\omega}}^2 \rangle \approx 4 \langle v_z^2 \rangle$ and as noted long
ago by Jeans this shows the importance of I_3.

If we take the velocity distribution in the z direction to
be Gaussian and independent of the velocities in the plane we
may write

$$f(E, J_z, I_3) = f_1(E - I_3, J_z) e^{-\alpha^2 I_3} = f_1 e^{-\alpha^2 v_z^2} e^{-2\alpha^2 \Phi_2(z)} \quad (23)$$

from which we see (since f_1 is independent of z and v_z) that

$$\langle v_z^2 \rangle = \frac{\int f v_z^2 \, d\underline{v}}{\int f \, d\underline{v}} = \frac{1}{2\alpha^2} \quad (24)$$

is a constant independent of z. For the number density as a
function of z we have

$$\frac{n(z)}{n(0)} = e^{-2\alpha^2 [\Phi_2(z) - \Phi_2(0)]} \quad (25)$$

Taking the log and differentiating twice we have

$$\frac{\partial^2 \Phi}{\partial z^2} = -\langle v_z^2 \rangle \frac{\partial^2 \ln n(z)}{\partial z^2} \quad (26)$$

The gas in the disk of our galaxy and stars with low random velocities revolve around the galactic center approximately in centrifugal equilibrium. Denoting the corresponding "circular velocity " by θ we have

$$\frac{\theta}{\overline{\omega}} = \frac{\partial \Phi}{\partial \overline{\omega}} \quad (27)$$

and hence

$$\frac{1}{\varpi} \frac{\partial}{\partial \varpi} (\varpi \frac{\partial \Phi}{\partial \varpi}) = 2 \frac{\theta}{\varpi} \frac{\partial \theta}{\partial \varpi} \tag{28}$$

Adding equations (26) and (28) we obtain the value of $\nabla^2 \Phi$ (equation 28 making a rather small contribution to the total) and therefore the total mass density near the sun from equation (4). The result of Oort and Hill is $\rho = 0.15 \ M_\odot/pc^3$. It is interesting to compare this with the total mass density obtained from direct observation. Various contributions are as follows:

Gas {	H, H$^+$, corresponding He	0.025	$\odot/pc^3$
	H$_2$ corresponding He	0.015	
Main sequence stars		0.050	
White Dwarfs		0.010	
		0.100	$\odot/pc^3$

For some time it appeared that this apparent deficiency might be real and that a "missing mass" problem existed. However recent investigations by Sanduleak and Murray and by Weistrop have revealed the presence of an abundant population of low velocity M-dwarfs (not included under the heading "Main sequence stars" in the table). While their precise number is still uncertain, they will appreciably increase the "main sequence stars" total and thereby probably account for most of the "missing mass".

One other result from the Boltzmann equation is frequently useful. Writing the equation in cylindrical coordinates, multiplying with v_ϖ and integrating over all velocities we obtain

$$\theta^2 - <v_\phi>^2 = <(v_\phi - <v_\phi>)^2> - \frac{1}{n} \frac{\partial}{\partial \varpi} (\varpi \ n \ <v_\varpi^2> \tag{29}$$

which relates the difference between the circular velocity (equation 27) and the average velocity of revolution of a stellar population with the velocity dispersions and density gradients. On this basis the fact may be understood that low velocity stars revolve nearly with the circular velocity, while the high velocity stars lag behind.

4. EPICYCLE ORBITS

In cylindrical coordinates ϖ, ϕ, z the equations of motion for a star in an axisymmetric gravitational field may be written as

$$\ddot{\varpi} = \varpi \dot{\phi}^2 - \partial \Phi/\partial \varpi$$

$$\varpi \ddot{\phi} = - 2 \dot{\varpi} \dot{\phi} \tag{30}$$

$$\ddot{z} = - \partial \Phi/\partial z$$

The second equation may be integrated immediately to $J = \varpi^2\dot{\phi}$, with J the angular momentum integral. Eliminating $\dot{\phi}$ with this relation we have

$$\ddot{\varpi} = \frac{J^2}{\varpi^3} - \frac{\partial\Phi}{\partial\varpi} = -\frac{\partial\Psi}{\partial\varpi}$$

$$\ddot{z} = -\partial\Phi/\partial z = -\frac{\partial\Psi}{\partial z} \tag{31}$$

with $\Psi = \Phi + \frac{1}{2} J^2/\varpi^2$. Hence the integration of orbits in an axi-symmetric three dimensional potential is reduced to a two dimen-sional problem. We now refer the orbit of a star to a not too different circular reference orbit situated in the plane of sym-metry of the potential. We therefore write $\varpi = \varpi_0 + \delta\varpi$ and $z = 0 + \delta z$, with ϖ_0 a constant. Under the assumption that $\delta\varpi$ and δz are small compared to ϖ_0 and to the characteristic scale of the potential, we linearize the equations to obtain

$$\delta\ddot{\varpi} = \frac{J^2}{\varpi_0^3}\left(1 - 3\frac{\delta\varpi}{\varpi_0}\right) - \left(\frac{\partial\Phi}{\partial\varpi_0}\right)_0 - \left(\frac{\partial^2\Phi}{\partial\varpi^2}\right)_0 \delta\varpi$$

$$\delta\ddot{z} = -\left(\frac{\partial^2\Phi}{\partial z^2}\right)_0 \delta z \tag{32}$$

We choose the reference orbit such that $J = \varpi_0\theta$, that is such that it has the same angular momentum as the star being considered. We then find

$$\delta\ddot{\varpi} + \left[\frac{3}{\varpi_0}\left(\frac{\partial\Phi}{\partial\varpi}\right)_0 + \left(\frac{\partial^2\Phi}{\partial\varpi^2}\right)_0\right]\delta\varpi = 0$$

$$\text{or}\quad \delta\ddot{\varpi} + \kappa^2\delta\varpi = 0 \tag{33}$$

Integrating we have

$$\delta\varpi = a \sin[\kappa(t - t_1)]$$

$$\delta z = b \sin[(\partial^2\Phi/\partial z^2)_0^{\frac{1}{2}} (t - t_2)] \tag{34}$$

Similarly referring the phase ϕ to that of a particle moving uniformly in a circular orbit we have

$$\varpi_0^2\dot{\phi}_0 = \varpi_0\theta = J = \varpi^2\dot{\phi} = \varpi_0^2\dot{\phi}_0 + \varpi_0\dot{\phi}_0\delta\varpi + \varpi_0^2\delta\dot{\phi} \tag{35}$$

which upon integration yields

$$\delta\phi = \frac{2\,\theta\,a}{\varpi_0^2\,\kappa} \cos[\kappa(t - t_1)] \tag{36}$$

Hence the relative orbit (with respect to the reference point) in

the galactic plane is an ellipse -- the so-called "epicycle" and
the angular frequency κ is called the "epicyclic frequency".
Introducing the epicyclic period T_{ep}, the period associated with
the z-motion T_z and the period of a particle in a circular orbit
around the galactic center T_{circ}, we have near the sun

$$T_{circ} = 250 \times 10^6 \text{ years}$$

$$T_{ep} = 200 \times 10^6$$

$$T_z = 70 \times 10^6$$

Applications of epicyclic orbits will be made in later lectures.

5. MASS AND MASS DISTRIBUTION IN OUR GALAXY

According to equation (29) a population of stars with very small
random velocities will revolve around the galactic center essentially with the circular velocity. Of course at different distances from the sun and at different galactic longitudes ℓ the
circular velocity will be slightly different. Expanding the circular velocity field θ near the sun in a Taylor series as follows

$$\theta(r) = \theta(o) + (r \cdot \nabla)\theta \tag{37}$$

We may easily derive simple equations for the radial V_r and tangential V_t velocities observed at the sun from a region in the
galactic plane whose radius vector from the sun is r. The result is

$$V_r = A r \sin(2\ell)$$
$$V_t = Br + Ar \cos(2\ell) \tag{38}$$

with

$$A = \tfrac{1}{2} \left[\frac{\theta_o}{R_o} - \left(\frac{\partial\theta}{\partial\varpi}\right)_o \right]$$
$$\text{and} \quad B = -\tfrac{1}{2} \left[\frac{\theta}{R_o} + \left(\frac{\partial\theta}{\partial\varpi}\right)_o \right] \tag{39}$$

Here A and B are the "Oort constants", R_o is the distance of the
sun to the galactic center. Note that

$$(A - B)R_o = \theta_o \tag{40}$$

To construct a model of our galaxy it is essential to know θ_o and
R_o and therefore the accurate determination of A and B is of much
importance.

To determine A is comparitively straightforward. Measuring
average stellar radial velocities at various distances from the
sun and at different longitudes we obtain A from the first of
equations (38). In practice complications do occur in particular
because of uncertainties in the interstellar absorption entering
the determination of r. The determination of B is more difficult.
The transverse motions are to be deduced from the proper motions
of stars. Because the angular proper motion is proportional to
V_t/r, the effect of B is to give all stars the same proper motion
in longitude. The effect is only 0''.002/year and it is difficult
to evaluate accurately because of the systematic uncertainties in
all proper motions. The values which are generally used nowadays
for the Oort constants are as follows:

$$A = 15(km/sec)/kpc$$
$$B = -10(km/sec)/kpc \tag{41}$$

The uncertainty in A is probably ±3(km/sec)/kpc that in B is
probably larger. Extra uncertainty is added because the spiral
structure in the galaxy may cause deviations from circular
motions which have not been taken into account in equations (38).
The value of R_0 is equally uncertain. The best determinations
are based on estimates of the distances of the RR Lyrae stars
near the galactic center. Absorption corrections and uncertainties
in the absolute magnitudes of the RR Lyrae stars cause difficulties. The commonly accepted value is $R_0 = 10$ kpc, but the
uncertainty should be at least ±2 kpc. Recent analyses by van
den Bergh and by Oort and Plaut suggest that the 10 kpc value may
be somewhat too large.

Adopting for A and B the values of equation (41) and for R_0
a value of 9 kpc we have $\theta_0 = 225$ km/sec. Assuming the galactic
mass distribution to be spherically symmetrical, the mass outside
the solar orbit does not affect the gravitational forces at the
sun's location. From equation (27) we then have

$$\theta_0^2 = \frac{GM(\varpi < R_0)}{R_0} \tag{42}$$

With the parameters chosen we obtain $M(< R_0) = 1.1 \times 10^{11} M_\odot$. If
alternatively we assume that the equidensity surfaces are very
much flattened spheroids, all of the same eccentricity an interior
mass of $0.7 \times 10^{11} M_\odot$ is obtained.

Measuring radial velocities of the interstellar gas in more
distant regions of the galaxy (by making use of the 21-cm spin
flip line of atomic hydrogen) we can derive the run of θ between
the sun and the galactic center. Although small variations occur
the value of θ does not appear to change by more than 20% between

the sun and distances of a few hundred parsec from the center.
From equation (42) we conclude that $M(< R) \propto R$ corresponding to
$\rho(r) \propto r^{-2}$.

The question as to whether most of the mass is in a flattened
disk or a spherical halo has not yet been conclusively answered.
From the observed mass density near the sun it follows that the
galactic disk within the solar orbit would have a mass of about
$2 \times 10^{10} M_{\odot}$, even if the density would not increase towards the
center. Therefore at least one third of the total gravitational
field would be due to the disk, leaving a maximum interior mass in
the halo of $7 \times 10^{10} M_{\odot}$.

Direct evidence on the mass of the halo is quite fragmentary.
An estimate of the extreme halo population may be made as follows.
The total luminosity of the metal poor globular clusters in the
halo interior to the sun is about $5 \times 10^{6} L_{\odot}$. Mass to Light ratios
of about 1-2 in solar units appear applicable to these clusters,
and hence the total interior mass of the globular clusters is
around $7 \times 10^{6} M_{\odot}$. For every RR Lyrae star in a globular cluster
there appear to be about 100 in the surrounding halo, which would
then lead to an interior halo mass of $7 \times 10^{8} M_{\odot}$. However the low
mass to light ratios of the globular clusters may be caused by
the preferential escape of low mass stars as a consequence of relaxation and a resulting tendency towards equipartition. If so
the halo mass estimate might have to be increased by a substantial
factor to a value of several times $10^{9} M_{\odot}$. An earlier estimate by
Oort, based on high velocity M subdwarfs leads to a similar value.

So far we have only considered the extreme halo population.
Somewhat more metal rich globular clusters and related stars might
also make a substantial contribution and it cannot be entirely
excluded that the total interior halo mass is as large as $7 \times 10^{10} M_{\odot}$.
Ostriker and Peebles have given some reasons for believing that a
disk galaxy without a massive halo might be unstable. It might
therefore be attractive to make the halo at least as massive as
the disk.

The mass outside of the solar orbit is harder to estimate.
Of the metal poor halo globular clusters about 60% is located
outside the solar orbit. Since this is the least concentrated
halo population we know, an upper limit to the total halo mass is
probably obtained by multiplying the interior maximum halo mass
by 2.5, leading to a total halo mass of $1.8 \times 10^{11} M_{\odot}$. Including
also the disk it would seem that the total mass of our galaxy
might be as large as $2 \times 10^{11} M_{\odot}$, but probably not much more than
that. If almost all of the matter were in a very flattened disk
distribution a mass somewhat in excess of $1 \times 10^{11} M_{\odot}$ would seem
likely.

REFERENCES

1.   A. Blaauw and M. Schmidt (Ed.) Stars and Stellar Systems.
     Vol. 5: Galactic Structure, Univ. of Chicago Press, 1965.
2.   L. Woltjer, Lectures in Applied Mathematics, Vol.9: Rela-
     tivity Theory and Astrophysics, Part 2: Galactic Structure,
     p.1, Am. Math. Soc., 1967.

MODELS OF GALAXIES[*]

K.H. Prendergast

Columbia University, New York

1.1 Introduction

We would like to make a physical description of the various types
of galaxies. We know that galaxies can be classified into morphological types in the Hubble sequence

$$EO - E6 - SO \begin{cases} Sa - Sb - Sc \\ SBa - SBb - SBc \end{cases} Irregulars$$

together with all other galaxies loosely described as peculiar
galaxies. It should be noted that the SO galaxies have less simi-
larities amongst themselves (in terms of gas content, or colour,
for instance) than the other types on the Hubble sequence.

We know certain general features about galaxies that we would
like to reproduce theoretically:

(i) Neutral hydrogen is a negligible contributor to the mass
in ellipticals ($<10^{-4}$ of the mass). It is present in some SO galaxies,
and in increasing proportion along the sequence Sa to Sc.

(ii) The colours of the galaxies are redder towards the ellipticals and get bluer along the spiral sequence. There is little
colour variation over the surface of an elliptical, whereas the
arms of spirals are bluer than the rest of the galaxy.

(iii) There is little individuality among elliptical galaxies,
whereas spirals tend to each have characteristics which make them
immediately recognisable to experienced astronomers. This means
we must carefully define what we hope to explain by a general
theory of spiral galaxies.

---

[*]Notes taken by C.R. Alcock, P.C. Crane, and E.L. Turner.

G. Setti (ed.), Structure and Evolution of Galaxies, 59–70. All Rights Reserved.
Copyright © 1975 by D. Reidel Publishing Company, Dordrecht-Holland.

1.2 Elliptical Galaxies

We will do some classical stellar dynamics in terms of the distribution function $f(\bar{r}, \bar{u}; t; \log T_e, M_{bol})$ where $\bar{r}, \bar{u}$ are position and velocity vectors, t is the time, and $\log T_e$ and M_{bol} describe the type of star considered. From now on we will consider only one type of star, so that $\log T_e$ and M_{bol} will be forgotten.

The collisionless Boltzmann equation is:

$$\frac{\partial f}{\partial t} + u_i \frac{\partial f}{\partial x_i} - \frac{\partial \psi}{\partial x_j} \frac{\partial f}{\partial u_j} = 0 \qquad (1.1)$$

where ψ is the gravitational potential, and the Einstein summation convention has been used.

Consider a time independent solution (i.e. no evolution). Then the distribution function can be a function of the integrals of the particle motion only. We will consider an axisymmetric system and the integrals are:

$E = \frac{1}{2}|\bar{u}|^2 + \psi$ = total energy of particle

$J_z = (\bar{r} \times \bar{u}) \cdot \hat{k}$ = z component of particle angular momentum.

There may be a third integral of the motion, but unless it is required by observational evidence we need not consider it.

The gravitational potential is given by the relation

$$\rho = \int <m> f\{\tfrac{1}{2}|\bar{u}|^2 + \psi(\bar{r}); (\bar{r} \times \bar{u}) \cdot \hat{k}\} d^3u \qquad (1.2)$$

together with the Poisson's equation

$$\nabla^2 \psi = +4\pi G\rho \ , \qquad (1.3)$$

where ρ is the density and $<m>$ the average star mass.

This prodedure is very straightforward when we have spherical symmetry, but rather difficult when we consider flattened systems. The primary reason for the difficulty in solving Poisson's equation is that, with the form of the distribution function used, ρ is a functional of ψ. In practice the form of ψ, and hence ρ, is computed using an iterative technique, and this could not be done until the advent of large computers. We guess a reasonable form for ψ initially, say $\psi^{(0)}(\bar{r})$. Then we compute $\rho^{(0)}(\bar{r})$ and solve Poisson's equation to find $\psi^{(1)}(\bar{r})$. This tentative procedure is repeated until a stable form $\psi^{(n)}(\bar{r})$ is formed. Using Legendre Polynomials we find that if

$$\rho^{(n)}(r,\mu) = \sum_{k=o}^{N} \rho_k^{(n)}(r) \ P_k(\mu) \qquad (1.4)$$

$$\psi^{(n+1)}(r,\mu) = \sum_{k=o}^{N} \psi_k^{(n+1)}(r) \ P_k(\mu) \qquad (1.5)$$

where $\mu = \cos\theta$, then

$$\psi_k^{(n+1)}(r) = \frac{-C^{(n)}}{2k+1} \{ r^k \int_r^\infty \rho_k^{(n)}(s)\, s^{1-k}\, ds +$$
$$+ r^{-(k+1)} \int_0^r \rho_k^{(n)}(s)\, s^{k+2}\, ds \} , \qquad (1.6)$$

where $C^{(n)}$ is a constant.

This procedure generally diverges because $\psi^{(n+1)}(\bar{r})$ is everywhere less than $\psi^{(n)}(\bar{r})$, with the result that $\rho^{(n+1)}(\bar{r})$ is everywhere greater than $\rho^{(n)}(\bar{r})$. But this procedure does not conserve mass, and so by normalizing our total mass at each step we obtain a convergent sequence.

We will consider a distribution function:

$$f = \begin{cases} A\, e^{-\alpha E + \beta J z} & \text{for } \psi_o \leqslant E \leqslant E_o < 0 \\ 0 & \text{elsewhere,} \end{cases}$$

where ψ_o is the energy of a star at rest in the center of the potential well. We must truncate the function at $E_o < 0$, or else the galaxy will extend out indefinitely in space.

In a computational procedure we start with $\beta = 0$, which gives a spherical system. As β is increased from zero progressively more flattened systems result. We can match this sequence of models against the Hubble sequence E0 --- E6.

There are three specific comparisons we can make:

(i) The model predicts a form of intensity as a function of radius:

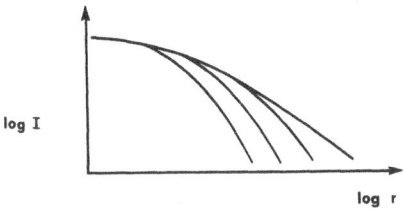

This does mimic the observations of log(I) versus log(r) rather well.

(ii) This distribution function predicts that all the isophotes have roughly the same ellipticity. As far as the observations can show, this is approximately true for elliptical galaxies.

(iii) The sequence of elliptical galaxies finishes at E6. We would like to predict this by increasing β until we produce an E6 galaxy, and then finding that a further increase in β admits no solution. This does in fact happen as β is increased beyond an E6 galaxy. The distribution function attempts to form a doughnut shaped distribution in space. An indication of incipient breakdown is that the outer isophotes become nearly rectangular.

Another distribution function that has been used is

$$f(E,Jz) = e^{-\alpha E} -(1-\alpha E)e^{\beta J_z+\gamma J_z^2}.$$

This approaches f=o smoothly for $E \rightarrow 0$.

A very general problem that has not been discussed above is
which distribution function to use. Any function of the integrals
of motion automatically satisfies the collisionless Boltzmann
equation, and so our procedure cannot predict the functional form
of f. Observationally we cannot establish that f is unique, as
all we see are various integral moments of f. For instance: (i)
Surface brightness; (ii) Line of sight velocities; (iii) Line of
sight velocity dispersions. However, it appears that elliptical
galaxies form a good one-parameter family, and we would like to
find a unique one-parameter form for f.

In conventional gas dynamics, for instance, the ambiguity is
resolved by the collision term. But the two-body relaxation time
in stellar systems is much greater than the Hubble time and hence
two-body relaxation cannot lead to a unique f. Two other possibilities for removing the ambiguities are specifying initial conditions, or discovering some "violent relaxation" processes.

2. VIOLENT RELAXATION

"Violent relaxation" occurs because the potential $\psi(\bar{r})$ is a function
of the particle density $\rho(\bar{r})$. The nature of this relation can be
seen in the following simple example: On the one hand, consider an
externally applied non-harmonic potential $\psi(x)$. Between x and x+dx,
place some particles initially at rest. The particles will move
independently in $\psi(x)$. Supposing that the particles start in phase
at a time t=o in phase space, at a time t later all phase information
is lost, although the trajectories of the individual particles are
known. On the other hand, collective relaxation occurs because
$\psi(x)$ depends on the positions of the particles themselves: If there
is a dip in $\psi(x)$, it is because $\rho(x)$ is high there. Consequently,
the energy E of a particle is no longer a constant of its motion.
In addition to the loss of phase information, the trajectories of
the particles themselves are "messed up", and the orbits lose all
the information about the initial conditions.

Lynden-Bell (1967) has used techniques of statistical mechanics
to derive an exclusion principle in phase space-the area in phase
space is preserved. This leads to an equilibrium distribution f,
but does not determine how long it takes to achieve it.

Another avenue of investigation which has been used is to make
numerical calculations of the motions of particles in their collective
potential. For the collision-free Boltzmann equation, the density
that a particle sees in phase space is not time-dependent. However
two-body interactions, if they were important, would very quickly
smear f out. Numerical errors, which increase the "entropy", will

mimic collisions in this regard, but smear f much faster. To insure
that the numerical method is reliable, we have required that, if
there are no collisions, the calculations must be time-reversible.
This does not mean that the true force equation is obeyed; a systematic error in the technique is still possible.

     The simplest such calculation is for one-dimensional self-
gravitating slabs (stars) moving in the x-direction.

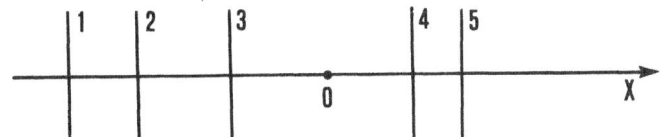

The magnitude of the force exerted by such a slab is independent of
distance. Thus, for instance, the force on slab 3 is zero, and that
on slab 5 is minus four in arbitrary units. We can integrate the
trajectories analytically. When a collision occurs, we can regard
the two slabs as either bouncing off or passing through each other.
There will be a discontinuity in the force seen by a slab at the
time of a collision. This graininess in the force-field will emphasize the collective effects. Let us consider a set of symmetric
initial conditions

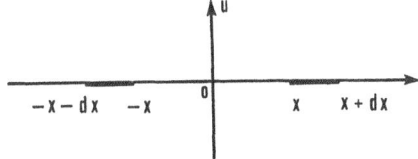

A ring in phase space develops after about seven crossing times.
If we look at the density distribution at very great t, we see what
looks like a gaussian with a thin population of high-energy particles
and with low-amplitude wiggles.

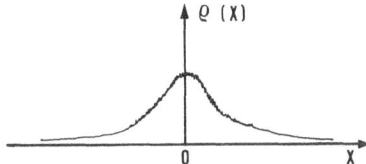

These wiggles contain all the information about the initial conditions, but the gross properties of the distribution do not depend
on the initial conditions. The long-term behavior is unknown but
it may evolve slowly.

     Let us now consider a more realistic three-dimensional planar
distribution. This calculation-done with R.H. Miller (1968)-is a
time-reversible calculation for ∿120,000 self-gravitating mass points.
We find that f(x,y,u,v,t) quickly evolves from absurd initial conditions to a "steady state". As an example, consider an initial

uniform circular density distribution in (x,y). The particles at
each point (x,y) are each given a fixed fraction of the circular
velocity at (x,y) plus a random additional velocity component.
The results of this calculation do not significantly depend upon
whether the random-velocity distribution is projected from one
quadrant into the other three, or whether it is random everywhere.
In units of the dynamical time $t_D=(G\rho)^{-1/2}$, and looking along the
x>o axis, we get the results shown in the figures on this page.
The gross distribution is very smooth but superimposed are many
small wiggles, retaining information about the initial conditions,
which are not of particular interest. Apparently any system designed
to mimic a stellar-dynamical distribution quickly evolves to a
quasi-static distribution. And, in fact, if we take an initial
circular distribution, with somewhat more angular momentum, we get
a bar-rotating, stable, and with a 2/1 axial ratio. But in every
case, we get a system that is <u>smooth</u> and <u>hot</u> (most motion is random).
These results correspond to the purely classical stellar-dynamical
distributions; they give no individuality!

To get more interesting results, we need a system that is not
too hot, that is to say not pressure dominated. To produce a gravitational instability, we need at least a subpopulation of gas, or
stars, that is cold. As a first attempt to get a cold system, we
can take the simple-minded approach of instantaneously cooling an
evolved system. However, this just throws the system into violent
oscillations again; it becomes hotter and denser. Or we can cool an
evolved system a little bit at each time step. This gives totally
different results: The system never has a chance to become organized,
and produces a "mess" of lumps. However, it is important to find a
physical system like this, i.e. a self-gravitating system whose
random velocities are decreased, as in inelastic collisions.

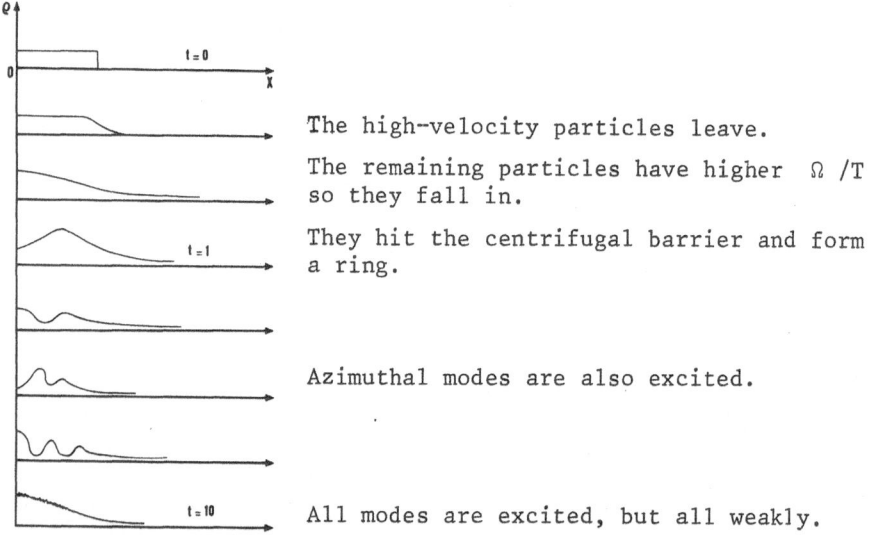

The high-velocity particles leave.

The remaining particles have higher Ω /T
so they fall in.

They hit the centrifugal barrier and form
a ring.

Azimuthal modes are also excited.

All modes are excited, but all weakly.

3. NUMERICAL MODELS WITH GAS AND STAR FORMATION

As we saw in the previous sections, models with no collisions
(stellar) produce smooth and hot systems resembling E galaxies,
while models with inelastic collisions simulated by artificial
cooling after each time step (gaseous) produce irregular messes.
In this section two-component (gas and stars) models will be considered, in which the gaseous component is transformed into a stellar
component to simulate star formation.

It will be seen that this two-component model can produce very
different systems from either of the one-component models. Essentially the introduction of star formation adds a new time scale to
the problem. Previously we had only the gravitational time scale
$t_D = (G\rho)^{-1/2}$ and the two-body relaxation time scale, which however
is too long to be of any interest. If we introduce a star-formation
rate law

$$\frac{dN_S}{dt} = KN_G^2 \qquad\qquad (3.1)$$

where N_S and N_G are the local star and gas densities, respectively,
and K is a constant, a new time scale

$$t_{SF} = \frac{N_S}{KN_G^2} \qquad\qquad (3.2)$$

is introduced. The character of the model is now controlled by the
ratio $R = t_{SF}/t_D$ of the two interesting time scales.
For R<<1: Nearly the same as purely stellar models.
For R>>1: Nearly the same as purely gaseous models.
Intermediate R values: Spiral type systems?
The important result is that one can reproduce the Hubble sequence
from early to late types by increasing the angular momentum (E types)
and R values (S types) of the models. This would also allow one to
explain the correlation of late morphological type with high gas
content.

Now we specifically consider a model with initial conditions
like those described in section 2, and $R \gtrsim 1$, $N_G \neq 0$, $N_S = 0$, that is
star formation on an essentially gravitational time scale and all
gas (no stars) initially. This model neglects any star-to-gas
transformations, any gas heating (by supernovae, etc.), a stellar-
luminosity function (and lifetime distribution), and is no longer
time-reversible. As an aside, about four hours of 360/75 computer
time is required to integrate the equations of motion for this
120,000 body system.

Having performed the numerical calculations and arrived at the
"answer", the even more difficult problem of extracting useful information from the computer results remains. In a certain sense,
one must "observe" this simulated galaxy.

The first step is to generate a motion picture showing the
appearance of the gas and star surface distributions as the system

evolves. Generally, the system resembles a two-armed spiral after
the initial violent oscillations subside into a relatively stable
state. The appearance is not, however, particularly stable with the
arms being much more visible at some times than others. The following
points are also apparent from a visual inspection:
1) Many orbits are highly non-circular; thus a bumpy rotation curve
   is expected.
2) The spiral arms are density waves, with points overtaking them
   and passing through them.
3) The spiral arms are trailing.
4) The spiral pattern and other features are quite apparent in the
   gas distribution, but are at best barely visible in the stars
   which appear hot and smooth.
Further information can be obtained numerically in a direct way:
1) rotation curves (gas and stars)
2) Oort A and B constants
3) epicyclic frequencies
4) epicyclic axial ratios
5) line-of-sight velocity distribution
A Fourier analysis of the gas pattern around a circle at various
radii from the center gives the following information:
1) The dominant mode is m=2 (two-armed spiral), but there is also
   some power in the m=1 mode, i.e. the nucleus is not exactly
   centered, and there is considerable power in the other modes,
   i.e. the noise level is high.
2) The pattern speed is slowly decreasing with time.
3) The arm spacing and arm-disk density contrast both secularly
   decrease with time.
The last two points are open to some suspicion as all secular trends
in numerical simulations must be.
     A similar Fourier analysis of the star pattern reveals that
the spiral pattern is present but very weak. It is enhanced if the
hottest 90% of the stars are left out. It is not clear whether this
is because the low velocity stars remember that they were formed in
spiral arms, or whether they are simply more affected by the potential
perturbation of the pattern.
     It is interesting to note that the spiral pattern is not visible
beyond the corotation point.
     In his thesis Quirk (1970) has carried out a number of experi-
ments to determine which aspects of the model are required to produce
a spiral pattern. These are summarized below.
1) A system with no self-gravitation did not produce spiral structure
   even when stirred by a central, rotating bar-like potential perturbation. The "background" potential used for these experiments
   is obtained by smoothing and symmetrizing the field of a relaxed
   system.
2) In a relaxed system with an artificial "background" potential
   such as described above, spiral structure will appear (with or
   without stirring), if self-gravitation is turned on. Thus, a
   cooled gaseous system appears to be unstable to spiral pertur-

bation if, and only if, it is self-gravitating.
3) If a "burst" of star formation is simulated by changing all remaining gas to stars, the stellar configuration changes rapidly
   ($t \sim t_D/4$). In particular, the central gaseous disk rapidly forms
   a long thin rotating bar of stars which is stable for at least 1
   rotational period.
4) Rotating bar-like potentials quickly sweep out two regions at
   right angles to the bar, even in the absence of self-gravitation.
   In conclusion we note that numerical experiments like these
cannot replace the mathematical analyses such as those of Lin, Shu,
Toomre, and others. They are rather intended to supplement them.

4. QUALITATIVE DYNAMICS

Suppose we observe the motion of a bound star (E<0) in an axisymmetric
potential. Viewed from an inertial frame, it will describe a rosetta
orbit, which in general does not close on itself.
   If we view the motion from a star with the same angular momentum,
but in circular motion, the star describes a Lindblad epicycle.

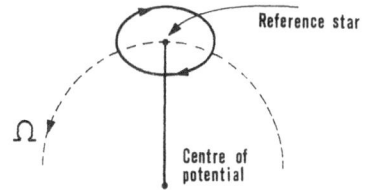

Now consider the general motion of stars in a rigidly rotating
frame, angular velocity Ω. We must solve the equation

$$\frac{d^2\bar{r}}{dt^2} + 2\Omega\hat{k} \times \frac{d\bar{r}}{dt} = -\bar{\nabla}(\psi_{grav} + \psi_{cent}) = -\bar{\nabla}\psi_{total} \qquad (4.1)$$

where $2\Omega\hat{k} \times \frac{d\bar{r}}{dt}$ is the Coriolis force and $\psi_{cent} = -\frac{1}{2}\Omega^2 r^2$.

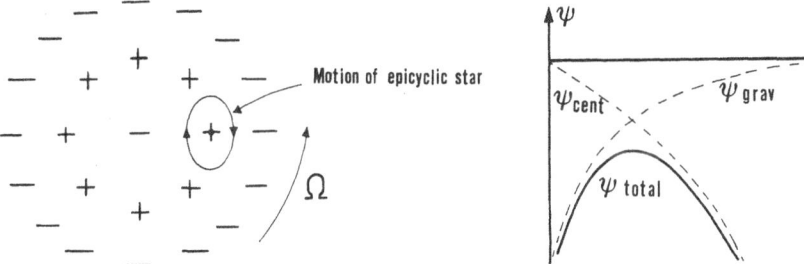

Suppose we consider a star that is not on the potential maximum; instead of moving radially outwards or inwards, the Coriolis
forces will curve the moving star around and the resulting motion
will be

Note that it is possible to roll uphill in a rotating system. We
can draw a number of useful analogies from plasma physics here-
for instance, the motion of an electron in a uniform B field parallel
to the rotation, and an E field lying in the x-y plane.

Now, let us destroy the axial symmetry. A perturbation due
to a rotating bar produces a potential with two discrete maxima
and two saddle points. Since this is topologically different from
the previous case, we cannot expect continuous changes in orbital
motion as we turn on the perturbation-a totally new phenomenon,
shown below, represents a star "escaping" through the saddle point.

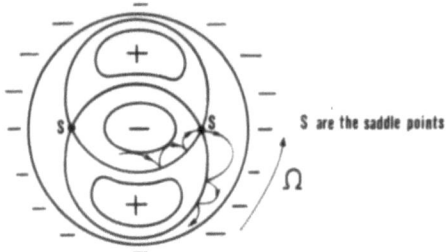

S are the saddle points

Orbits which do not approach the saddle points are qualitatively
the same. In all these orbits for low-velocity stars the motion
"minimizes" the time average of $d^2\bar{r}/dt^2$ by balancing Coriolis forces
against $\bar{\nabla}\psi_{total}$.

Consider a high velocity star in a bar-like potential. The
trajectory looks like

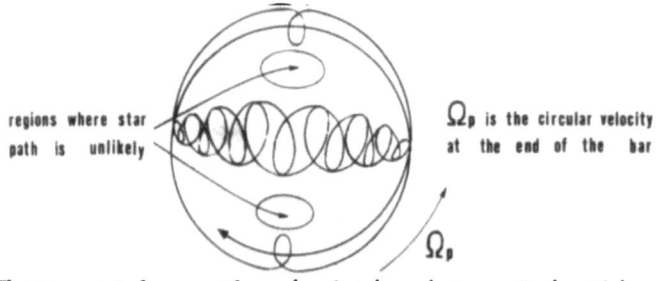

regions where star
path is unlikely

Ω_p is the circular velocity
at the end of the bar

There must be another isolating integral in this problem which
removes large regions of phase space, in order to prevent high
velocity stars from escaping.

A low-velocity star might be able to "escape" through a saddle

point and become a "runaway star", as shown below in the rotating
and non-rotating frames.

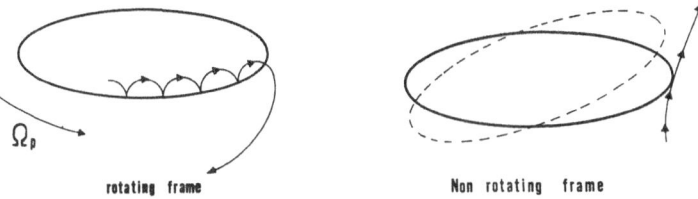

These results qualitatively explain many features of barred spirals:

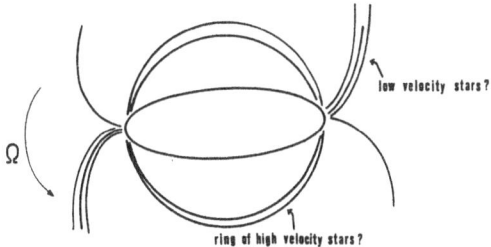

None of these pictures has internal consistency. Freeman (1966a,b,c)
has models with internal consistency which indicate that as the bar
loses low velocity stars from its edge, it becomes progressively
fatter.

There are spiral galaxies in the Hubble Atlas which show evidence of a ring of stars—perhaps this indicates stirring by a faint
bar at the center.

The rotation curve of barred spirals is not well known; usually
all we have is

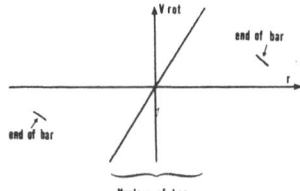

In one case (NGC 7479) the rotation curve is a straight line
up to the end of the bar.

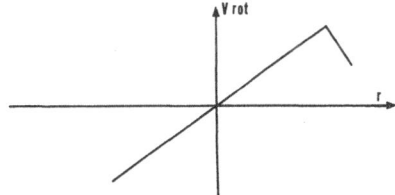

Note: Gas dynamics has not been discussed here. It is possible
to predict shock waves where dust lanes are seen in barred spirals,

by following the motion of the gas.

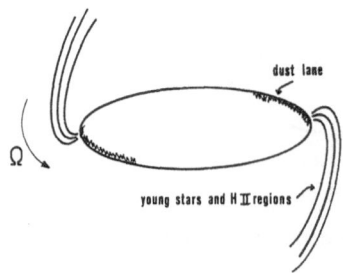

REFERENCES

Freeman, K.C., 1966a, M.N.R.A.S., 133, 47.
Freeman, K.C., 1966b, M.N.R.A.S., 134, 1.
Freeman, K.C., 1966c, M.N.R.A.S., 134, 15.
Lynden-Bell, D., 1967, M.N.R.A.S., 136, 101.
Miller, R.H., Prendergast, K.H., 1968, Ap. J., 151, 699.
Miller, R.H., Prendergast, K.H., Quirk, W.J., 1970, I.A.U. Symposium No. 38, p.365.
Prendergast, K.H., Tomer, E., 1970, Astron. J., 75, 674.
Quirk, W.J., 1970, unpublished Ph.D. thesis, Columbia University.

MASS DETERMINATIONS IN SPIRAL GALAXIES; NON-CIRCULAR MOTIONS

E. Margaret Burbidge

Department of Physics, University of California
La Jolla, California 92037 U.S.A.

I. INTRODUCTION

The most direct method of determining masses of spiral galaxies is to measure their rotation from line-of-sight velocity measurements. One can use either ionized gas (H II regions or generally-distributed ionized gas), atomic hydrogen (21-cm measures), or stars. The basic methods are described in a review by Perek (1962) and by Burbidge and Burbidge (1975). The basic principle is that one considers the acceleration of the gas or stars under the action of the gravitational field being exerted by the object under study. Since accelerations cannot be directly observed, they must be deduced from observations of the velocity field at one instant of time. Thus certain assumptions are implicit:

1) The velocity field is independent of time, i. e. the galaxy is in a steady state.
2) Since only line-of-sight velocities are measurable, symmetry of some sort must be assumed -- generally axial symmetry.
3) Newtonian gravitation is the dominant force, e. g. the velocity field must not be perturbed by energy input from explosive events in the nuclei of galaxies.

From the point of view of the optical observer, one can make some general observational comments. Firstly, galaxies with plenty of ionized gas distributed throughout the luminous body are the easiest to study. Emission lines produced in such gas are usually fairly narrow and easy to measure, unlike absorption lines produced by the stellar component of a galaxy, which are usually diffuse and must be measured against a continuum which will be faint in the outer parts of a galaxy. Further, ionized gas in a galaxy may be assumed to lie close to the equatorial plane, so that as long as one is not observing a spiral galaxy edge-on, a given velocity measurement can be tied to a unique point in the galaxy. This is not the case in observing absorption lines from the stars in a galaxy without gas or dust, when the line of sight integrates light from all stars in a pencil beam going right through the galaxy (stars do not shadow one another). Incidentally, spiral galaxies with a fairly large orientation of the rotation axis to the line of sight will be the best objects to study, as long as this angle is not too near 90°, because correction of the measured velocities for projection will not be too large.

As regards 21-cm measures, the situation until recently has been that the velocity resolution was very much better than for optical measures, while the spatial resolution within a galaxy was much poorer. There has been a convergence here from both sides: the radio telescopes at Westerbork and Bonn can give good spatial resolution and optical image-tube techniques have made possible the use of higher spectral resolution.

II. MASS DETERMINATIONS

In general, an ideal rotation curve for a galaxy corrected for inclination looks like this:

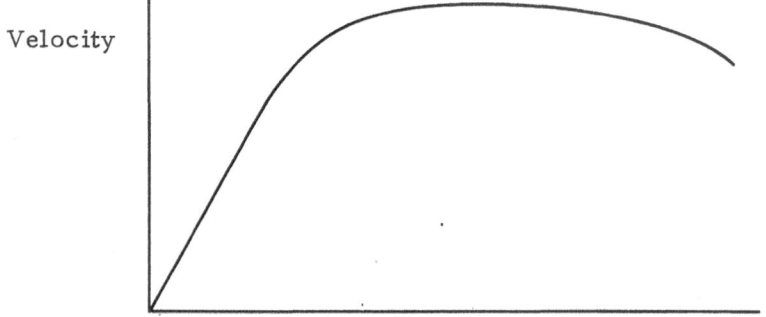

Distance from center

A curve like this will yield the mass distribution within a galaxy with the following restrictions:

1) A model for the galaxy must be postulated, since we seek three-dimensional information from one-dimensional measures (the only galaxy for which some three-dimensional observational information is available is our own).

2) We can derive mass information only out to the last observed point. Any mass further out which provides too little light to be observed optically, or which for any reason cannot be observed in 21-cm radiation, remains undetected.

3) The velocity field may be complicated near the center by non-circular motions, as we shall discuss later, and a true mass distribution may be unobtainable here.

We shall discuss various mass models, which have been applied to spiral galaxies mostly of type Sc, with some Sb and very few Sa. Because of their lack of axial symmetry, barred spirals present special problems and little has been done on their masses. We shall pretend that we have an ideal smooth curve like Figure 1, in which velocities have reflected exactly about a well-determined central velocity that corresponds to a point which is both the obvious optical center and the mass center (noting that several galaxies, including the Large Magellanic Cloud, apparently have different mass and optical centers).

In an axisymmetric galaxy, in cylindrical coordinates $(\tilde{\omega}, \theta, z)$, we have:

Attractive force per unit mass = $Q_{\tilde{\omega}}$

Rotational velocity = u_θ

Gravitational potential = $\tilde{\Phi}$

$$Q_{\tilde{\omega}} = \frac{\partial \tilde{\Phi}}{\partial \tilde{\omega}}$$

$$Q_{\tilde{\omega}} \, \tilde{\omega} = - u_\theta^2 (\tilde{\omega}) \tag{1}$$

i.e. in general

$$M(\widetilde{\omega}) = \frac{k}{G} u_\theta^2 (\widetilde{\omega}) \tag{2}$$

(see Perek 1962, Burbidge and Burbidge 1975).

Let us consider a few mass models.

1)   The simplest model is a mass point -- the total interior mass is derived from a single rotational velocity at a point near the outside of a galaxy:

$$M = \frac{\widetilde{\omega} u_\theta^2}{G} \, . \tag{3}$$

This generally gives a larger mass than the more detailed methods, because it assumes the greatest central concentration.

2)   Another simple model is the homogeneous spheroid. Let the axial ratio be c/a, and let the uniform density be ρ. Then:

$$Q_{\widetilde{\omega}} = \frac{3}{2} \frac{GM}{a^2 - c^2} \left[\frac{a}{(a^2 - c^2)^{1/2}} \cos^{-1}\left(\frac{c}{a}\right) - \frac{c}{a} \right] \, . \tag{4}$$

Let the quantity in square brackets = α/a^2. Then equation (1) gives (since $\rho = 3M/4\pi a^2 c$)

$$M_{total} = \frac{a u_\theta^2}{G\alpha} \, , \tag{5}$$

where α is of order unity.

At any point interior to the point $\widetilde{\omega}$,

$$M(\widetilde{\omega}) = \frac{4\pi\rho}{3} \frac{c}{a} \widetilde{\omega}^3 \, ,$$

so that

$$u_\theta = \left[\frac{4\pi G\rho\alpha}{3} \frac{c}{a} \right]^{1/2} \widetilde{\omega} \, , \tag{6}$$

i.e., the rotation curve is linear with a slope proportional to $\rho^{1/2}$. This mimics solid body rotation.

As $c/a \to 1$, $\alpha \to 1$, so for a spherical mass, this reduces to the mass point equation.

One can combine a mass point and one, two, or more homogeneous spheroids.

3) <u>Non-homogeneous spheroids</u> give a wider range of possible models. Perek (1962) discussed confocal spheroids; Burbidge, Burbidge, and Prendergast (1959) used similar spheroids. Some analytic expression must be assumed for the variable density. As before, circular velocities in the equatorial plane are given by:

$$u_\theta^2 = Q_{\tilde{\omega}} \, \tilde{\omega}$$

and if we now let <u>a</u> be the variable and take the integration out to <u>each observed point</u> $\tilde{\omega}$, we have

$$u_\theta^2 = 4\pi G (1 - e^2)^{1/2} \int_0^{\tilde{\omega}} \frac{\rho(a) \, a^2 \, da}{(\tilde{\omega}^2 - a^2 e^2)^{1/2}} \, , \qquad (7)$$

where the eccentricity $e = (1 - c^2/a^2)^{1/2}$.

Schmidt (1956) used $\rho = p + (q/a)$ for the density law in a model for the Galaxy. He used four separate spheroids for different populations in the Galaxy, plus 9 homogeneous spheroids.

Burbidge, Burbidge, and Prendergast used a polynomial

$$\rho(a) = \sum_{n=0}^{\infty} \rho_n a^n \, , \qquad (8)$$

which is very convenient for computation. First one fits a polynomial in $\tilde{\omega}$ to the <u>observed rotation curve</u> by least squares. Square this and write

$$u_\theta^2(\tilde{\omega}) = \tilde{\omega}^2 \sum_{\ell=0}^{\infty} v_\ell \, \tilde{\omega}^\ell \, . \qquad (9)$$

The coefficients v_ℓ are thus determined from the observations, and one then equates coefficients of $\tilde{\omega}$, $\tilde{\omega}^2$ ... in equation (7) to get the unknown coefficients ρ_n.

In practice, the polynomial (8) should contain only a few terms because the observations do not warrant more; if more than 5 or 6 terms are used, the resulting density curve has unrealistic kinks in it.

One can apply this method for a range of c/a -- say 1/5 to 1/15. The total mass within the last observed point is then obtained by integrating the density curve $\rho(\tilde{\omega})$. It proves to be fairly insensitive to the number of terms and to the axial ratio. The central density, however, is sensitive to both and is poorly determined. But in any case, as we shall see, the velocity curve in the center may include substantial non-circular motions.

Schmidt has pointed out that one may add terms in a^{-1} and a^{-2} to the expansion (8), so that terms in $1/\tilde{\omega}$ and $1/\tilde{\omega}^2$ appear in the equation for $u_\theta(\tilde{\omega})$. This simplifies the computations.

4)   Disk Models. A galaxy may be assumed to be a thin cylindrical disk of variable density. Such models have been considered by Wyse and Mayall, Aller, and Schwarzschild, and full references may be found in Burbidge and Burbidge (1975). Although at first sight we might suppose that such models would be simpler than non-homogeneous spheroids, they are actually more complicated analytically, and we shall not discuss them here.

5)   Bottlinger's Model. For our Galaxy, as discussed in the lectures by Woltjer, one has 3-dimensional velocities and a distribution function for various stellar populations. Bottlinger discussed a force law

$$Q_{\tilde{\omega}} = -\frac{a\tilde{\omega}}{1 + b\tilde{\omega}^3} \tag{10}$$

for the attraction in the radial direction, this derives from a general law

$$Q_{\tilde{\omega}} = -\frac{a_1\tilde{\omega} + a_2\tilde{\omega}^2 + \ldots}{1 + b_1\tilde{\omega}^3 + b_2\tilde{\omega}^4 + \ldots}$$

At large distance $\tilde{\omega}$ from the center equation (10) gives

$$Q \approx - \frac{a}{b\tilde{\omega}^2} \qquad (11)$$

which represents physical reality. Since at large $\tilde{\omega}$, $Q \to -GM/\tilde{\omega}^2$, we have

$$M = \frac{a}{bG} \qquad . \qquad (12)$$

The rotational velocity u_θ is given by $u_\theta^2/\tilde{\omega} = -\dot{Q}$, so if the rotation curve is observed, a and b can be found by a least squares fit:

$$u_\theta = \tilde{\omega} \left[\frac{a}{1 + b\tilde{\omega}^3} \right]^{1/2} \qquad (13)$$

to the observed curve. Thus M can be derived. For a simpler expression, one can differentiate equation (13) w.r.t. $\tilde{\omega}$ and obtain $(u_\theta)_{max}$ and $\tilde{\omega}_{max}$ where it occurs, in terms of a and b:

$$(u_\theta)_{max} = \left(\frac{a}{3}\right)^{1/2} \tilde{\omega}_{max}$$

$$\tilde{\omega}_{max} = \left(\frac{2}{b}\right)^{1/3} \qquad .$$

Thus,

$$M = \frac{3}{2} \frac{(u_\theta)^2_{max} \tilde{\omega}_{max}}{G} \qquad (14)$$

from the simply observed quantities $(u_\theta)_{max}$ and $\tilde{\omega}_{max}$.

This method has been used for 21-cm data of low resolution, where $(u_\theta)_{max}$ and $\tilde{\omega}_{max}$ can be obtained from 21-cm observations in which no details of the form of the rotation curve can be discerned.

6) Flat Disk Approximation of Brandt and Belton. This is a method for extrapolating from the formulation of Burbidge, Burbidge, and Prendergast in order to take account of mass lying exterior to the last observed point.

Let $c/a \to 0$. Use a surface density $\sigma(a)$ instead of $\rho(a)$. Then equation (7) becomes

$$u_\theta^2(\tilde{\omega}) = G \int_0^{\tilde{\omega}} \sigma(a) \; \frac{a \, da}{(\tilde{\omega}^2 - a^2)^{1/2}} \; . \tag{15}$$

This has a well-known form -- Abel's integral equation -- and it can be inverted to give σ as an integral of u_θ^2. Thus we have

$$M(\tilde{\omega}) = \frac{2}{G\pi} \int_0^{\tilde{\omega}} \frac{u_\theta^2(a) \, a \, da}{(\tilde{\omega}^2 - a^2)^{1/2}} \; . \tag{16}$$

Now we may assume some functional form for $u_\theta(\tilde{\omega})$ which gives Q that is a generalization of Bottlinger's force law:

$$u_\theta(\tilde{\omega}) = \frac{A\omega}{(1 + B^n \tilde{\omega}^n)^{3/2n}} \; . \tag{17}$$

Note that the validity of this formula depends on how good a fit it is to the observations.

For large $\tilde{\omega}$, we must have

$$u_\theta^2 = \frac{GM_{total}}{\tilde{\omega}} \; .$$

Thus in the limit of large $\tilde{\omega}$,

$$M_{total} = \frac{A^2}{G B^3}$$

$$= \left(\frac{3}{2}\right)^{3/n} \frac{(u_\theta)^2_{max} \, \tilde{\omega}_{max}}{G} \; . \tag{18}$$

The Bottlinger-Lohman form of this approximation uses $n = 3$; Brandt and Belton (1962) used $n = 3/2$. They found that this extrapolation resulted in a 50% increase in the mass of NGC 5055.

Toomre (1963) developed a refinement of this disk model, in which the potential Φ is represented by an expression that satisfies Poisson's equation and involves Bessel functions.

III. OBSERVATIONS OF NEUTRAL HYDROGEN

In some galaxies, neutral hydrogen can be detected by 21-cm measures further out than the optical observations. Three examples are given by Roberts and Rots (1973), namely M31 (Sb), M101 (Scd), and M81 (Sab). Observations reach out to some 30 kpc from the center, and lead to increased masses over those given by the last observed optical points. They found:

	M81	M31	M101
Mass within 30 kpc $(M_{\odot})$	1.4×10^{11}	2.3×10^{11}	1.8×10^{11}
Mass integrated to infinity $(M_{\odot})$	2.0×10^{11}	3.5×10^{11}	3.1×10^{11}

These may be compared with the range mentioned by Woltjer for our Galaxy of 1.2 to 4×10^{11} $M_{\odot}$.

IV. MASS-TO-LIGHT RATIOS

The data on masses can be combined with luminosities to give a parameter that is useful in consideration of the evolutionary stage of galaxies -- the mass-to-light ratio. Because of the observational limitations discussed -- that the best mass determinations are available for galaxies with the most uncondensed gas -- results are heavily weighted toward the Sc galaxies. Mass determinations for Sa, SO, and E galaxies have mostly to be determined by other means, discussed by other lecturers at this summer school -- from double galaxies, and from virial theorem.

The results tabulated by Burbidge and Burbidge (1975) give the following average values (out to the last observed point):

Sc $\langle M \rangle = 5.4 \times 10^{10} \, M_\odot$

            $\langle M/L \rangle = 5 \, M_\odot / L_\odot$.

Sb $\langle M \rangle = 9 \times 10^{10} \, M_\odot$

            $\langle M/L \rangle = 5 \, M_\odot / L_\odot$.

Sa $\langle M \rangle \sim 11 \times 10^{11} \, M_\odot$ from only two objects.

These are to be compared with values for giant ellipticals of $\sim 10^{12} \, M_\odot$ and anywhere from 0.3-$5 \times 10^{10} \, M_\odot$ for dwarf ellipticals, and M/L ratios of 10-30, from virial theorem determinations. Page's double galaxy results, discussed by Sargent elsewhere in these lecture notes, agree fairly well for spiral galaxies but give higher values for SO and E galaxies.

Roberts (1975) points out that the ratio of the mass of neutral hydrogen (from 21-cm work) to total masses is a monotonically increasing function of morphological type, average values for M_{HI}/M_{total} being:

    Sab-Sbc: 0.01
    Sc-Scd : 0.06
    Sd-Sm : 0.10
    Irr I   : 0.19

Total masses of galaxies from 21-cm work generally come out larger than those determined optically, because the neutral H extends further out.

However, there has been some questions raised as to how far the outermost 21-cm measurements in M31, the best studied object, do indeed refer to circular motion. Emerson and Baldwin (1973) believe that the extensive extrapolations for M31 may not be valid, and they find that it can be fitted with moderately low mass-to-light ratios of 5 for the disk component and 10 for the spheroidal component.

Finally, we should mention recent suggestions by Ostriker and others that spiral galaxies may be surrounded by very extended halos of stars of very low luminosity, i. e. high M/L, which increase the observed mass by factors ~ 10. Such mass would, of course, be unobservable by rotation methods, and all that can be said at present is that it is a postulate without observational evidence.

V.   CORRECTION FOR GAS PRESSURE

If there are substantial random gas motions in the center of a galaxy, which would be detected by measuring the width of emission lines there, the gas will exert a pressure which gives partial support against gravitation for the mass and the rotation curve will give too low a mass density in the central region.   The correction which must be applied is analogous to the correction for rotation which must be applied in virial theorem determinations for masses of E galaxies.   We have:

$$-\frac{u_\theta^2}{\tilde{\omega}} = \frac{\partial \tilde{\Phi}}{\partial \tilde{\omega}} - \frac{1}{\rho}\frac{\partial p}{\partial \tilde{\omega}} \ . \tag{19}$$

If $\langle u^2 \rangle$ is the mean-square random velocity,

$$p = \frac{1}{3}\ \rho \langle u^2 \rangle \ .$$

If u_θ is approximately linear with $\tilde{\omega}$ near the center, with slope k, and if $\langle u^2 \rangle$ is constant over the region of interest, equation (19) integrates to give

$$\rho = \rho_o \exp\left[\frac{3\tilde{\Phi} - \tilde{\Phi}_o}{\langle u^2 \rangle} + \frac{3k^2\tilde{\omega}^2}{2\langle u^2 \rangle} \right] \ . \tag{20}$$

Near the center, the potential

$$\tilde{\Phi} \approx -\frac{4}{3}\ \pi G \rho* \tilde{\omega}^2$$

where $\rho*$ is the average total density.   Then the scale height λ

of the gas in the $\tilde{\omega}$ direction is given by

$$\frac{1}{\lambda^2} = \frac{8\pi G\rho*}{\langle u^2 \rangle} - \frac{3k^2}{\langle u^2 \rangle}$$

or

$$\rho* = \frac{1}{8\pi G} \left[3k^2 + \frac{\langle u^2 \rangle}{\lambda^2} \right] \qquad .$$

$\langle u^2 \rangle$ can be obtained from the observed line broadening, and λ is half the distance over which the broadening occurs.

VI. NON-CIRCULAR MOTIONS

Activity in the nuclear regions can give rise to large non-circular motions which may persist out to some distance from the center. General discussion of this activity is covered in the lectures by G. Burbidge, but we note that it may occur on a small scale in the majority of spiral galaxies. It certainly occurs in M31; Münch was the first to detect it there, and a full discussion of very detailed and accurate image-tube observations can be found in Rubin and Ford (1971) and Rubin et al. (1973). The velocity plots given in the latter paper should be studied by anyone who does not realize how complicated things can be in the nuclear region of such a well-ordered galaxy as M31!

It should be noted that, in addition to gas motions, there will be gas-star interactions so stellar motions may also be peculiar, but a large part of the peculiar stellar motions may be due to the fact that there is dust obscuration and one may be looking to different distances above the equatorial plane over quite small regions.

We have heard at this summer school of the very interesting velocity field in NGC 2903, described by Dr. Simkin. This type of phenomenon has been studied in many other galaxies, e.g. NGC 253, NGC 4736, and we have also heard here of the beautiful radio observations in NGC 4258, described by Prof. Oort. It may be presumed that this phenomenon, to larger or smaller extent, is very common in spiral galaxies.

Non-circular motions, of course, occur also in barred spiral galaxies, and have been described here by Dr. Prendergast. It is these flow patterns in the non-axisymmetrical mass distribution of barred spirals which makes it so difficult to make any mass estimates in this class of galaxy.

REFERENCES

Brandt, J. C., and Belton, M. J. 1962, Ap. J. 136, 352.

Burbidge, E. M., and Burbidge, G. R. 1975, chapter in Stars and Stellar Systems, Vol. 9, ed. A. and M. Sandage.

Burbidge, E. M., Burbidge, G. R., and Prendergast, K. H. 1959, Ap. J. 130, 739.

Emerson, D. T., and Baldwin, J. E. 1973, Mon. Not. Roy. Astron. Soc. 165, 9P.

Perek, L. 1962, Adv. Astron. Astrophys. 1, 165.

Roberts, M. S. 1975, chapter in Stars and Stellar Systems, Vol. 9, ed. A. and M. Sandage.

Roberts, M. S., and Rots, A. H. 1973, Astron. Ap. 26, 483.

Rubin, V. C., and Ford, W. K. 1971, Ap. J. 170, 25.

Rubin, V. C., Ford, W. K., and Kumar, C. K. 1973, Ap. J. 181, 61.

Schmidt, M. 1956, B. A. N. 13, 15.

Toomre, A. 1963, Ap. J. 138, 385.

PHENOMENOLOGY OF SPIRAL GALAXIES

J.H. Oort

Sterrewacht, Leiden, the Netherlands

Chapter I

Classification and Populations

Galaxies have probably been formed from fluctuations in
density and motion in the universe. Many cosmologists have suggested
that the separation into individual units took place around the
time of decoupling of matter and radiation, when the radius of the
universe was about 1/1000th of the present radius. In this case
the protogalaxies must have obtained their angular momentum at a
later stage in their evolution, through interaction with other
protogalaxies or with galaxy clusters. For when the radius of the
universe was 1/1000th of its present value, and the density about
10^{-20}g cm^{-3}, a mass of, say, 10^{11} M$_\odot$ would have had a radius of 10^{21}
cm. A mass with such a radius cannot contain anything like the
angular momentum contained in a galaxy like the Milky Way system.
The regions of slightly higher than average density or
slightly slower than average expansion which separated out in this
early stage of the universe will expand to a maximum radius and
will become galaxies after collapse from this maximum to a size
where the density becomes sufficient to form stars. The angular
momentum contained in the galaxies may have been imparted to them
in the course of this evolution. But, it is still uncertain whether,
under the circumstances such as we observe in the universe, enough
angular momentum can be imparted in this way. If not, the angular
momentum must have been there from the beginning, and this could
only be the case if rotating protogalaxies had detached themselves
much later than the decoupling era, viz. at a time when the average

G. Setti (ed.), Structure and Evolution of Galaxies, 85–117. All Rights Reserved.
Copyright © 1975 by D. Reidel Publishing Company, Dordrecht-Holland.

density in the universe had fallen to about 1/10,000th of that at
the time of decoupling. But there must then have been a mechanism
which maintained in the universe a high degree of turbulence up
to that epoch.

An alternative, and perhaps more attractive possibility, is
that the regions of excess-density or less rapid expansion were in
general considerably larger than galaxies, and that the galaxies
became individual units during the collapse of these large regions.
The angular momentum could then accrue from internal streams in
these large units. In this case the fundamental turbulent elements
must have had masses equal to clusters of galaxies rather than
to individual galaxies. Unfortunately our present knowledge of the
motions and densities inside "developing" clusters, like the Virgo
cluster and its appendages, is insufficient to calculate backward
to the birth time of the galaxies, and to obtain sufficient knowledge of the conditions at that time. The actual birth process of
galaxies is therefore still uncertain.

We should even keep an open mind about the possibility that
they originated in an entirely different manner. Ambartsumian has
on several occasions put forward the view that something quite
extraordinary was, and perhaps still is, contained in the nuclei
of galaxies. He has even intimated that galaxies may have come from
embryos as small as these nuclei. I may return to these ideas at
a later time.

For the moment let us accept as a working hypothesis that
protogalaxies were in some way formed from turbulent elements in
the universe, and that they obtained the major part of their mass
and angular momentum at the time they first became separated from
the surrounding universe (cf. Fig. 1).

The properties of the completed galaxies will be determined
mainly by three things: the protogalaxy's mass, its angular momentum,
and its random internal currents.

If no stars were formed during the collapse of a protogalaxy,
and all the kinetic energy gained in the collapse would be radiated
away, the final product would be a thin gas disk concentrated toward
the centre. The radius of the disk would be determined by the
initial angular momentum, the strength of its central concentration
by the measure in which the radial velocities connected with the
collapse preponderated over the random stream velocities.

In reality the evolution of a galaxy differed from this, because
a considerable part of the gas must have been transformed into
stars before it collapsed into a disk. Stars formed at considerable
distance from the equatorial plane can never become disk stars.
They will continue to reveal in their distribution and motions some
information on the shape and dimension of the protogalaxy at the
time when it was formed. Their motions and distribution may, however,
have been randomnized by the irregularities in the protogalaxy's
gravitational field in a so-called "violent relaxation", so that
their present distribution may have become rather regular.

The stars formed in the earliest stages of contraction, when

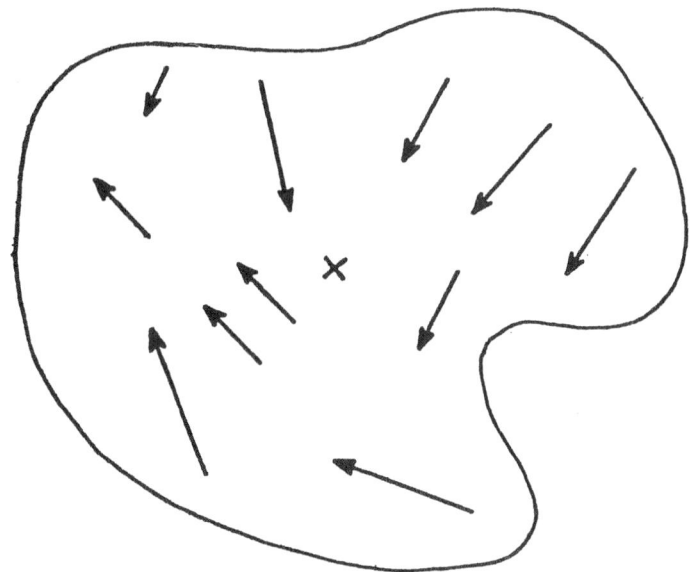

Fig. 1.

the detached lump of gas could not yet have developed a sensible
flattening, will tend to assume an approximately spherically
symmetrical distribution, with considerable concentration toward
the centre. These stars are designated as halo stars, or halo
population II. In our Galaxy, globular clusters, RR Lyrae variables
of low metal content and subdwarfs are objects that belong to this
population. They have velocity dispersions in the radial (Π)
direction of about 200 km/s and a very low average velocity of
rotation, averaging about 50 km/s. This is between 1/4 and 1/5
of the circular velocity. Although it was expected that these
halo stars would have a smaller average momentum than disk stars,
the difference mentioned is rather larger than expected. So far
as I know no entirely satisfactory explanation of this has yet
been found.

As the random currents in the protogalaxy gradually died out,
but before the protogalaxy finally collapsed into a thin disk,
other stars were formed. We observe these as an intermediate
population II. Easily recognizable stars belonging in this category
are long-period variables and RR Lyrae variables of higher metal
content. Their average distance from the galactic plane inferred
from the velocity dispersion of about 45 km/s in the z-direction
is roughly 1000 pc; this is approximately half that of the extreme
halo objects, and three or four times that of the older disk
populations. The intermediate population II has considerable
rotation, averaging some 75 or 80% of the circular velocity.

All population II stars are relatively rare in the general

neighbourhood of the Sun; the most frequent types, the subdwarfs,
are intrinsically faint, and are very incompletely known. As a
consequence we cannot from direct observation obtain a meaningful
estimate of the mass density in the halo, or, for that matter, of
the population II in general. On general dynamical grounds Ostriker
and Peebles (1) have come to the conclusion that the total mass
of very high-velocity population II stars in a spiral galaxy must
be at least equal to that of the stars in the disk. They showed
that in a stable system the ratio of the energy of rotation to
the gravitational energy must be 0.14 ± 0.02. This abundant
population II must consist overwhelmingly of intrinsically faint
stars, with a M/L ratio between 50 and 70 according to their
estimates.

There is one fact about the halo that should receive special
notice, namely that it does not consist wholly of stars that are
very poor in metals, such as would be the case if the halo stars
had formed from the primeval gas. Side by side with globular
clusters that are extremely metal-poor (with "metal" abundance
of about 1/100th of that of the Sun), there are also clusters with
metal contents that approach those of typical disk stars. This
seems to indicate that already during the halo stage the gas in
the protogalaxy must have been greatly enriched in metals,
presumably by a fairly rapid formation of a great many stars of
large mass and their subsequent explosion.

All this must have happened during a small fraction of the
present age of the galaxies. For our Galaxy the time between the
first beginning of star formation and the end of the violent
relaxation when it collapsed into a disk can hardly have been
much longer than the time of revolution of about 200 million years.
For the much denser giant ellipticals this will be still considerably less.

An alternative point of view concerning the metal enrichment
has been put forward by Unsöld. He suggests that all heavy elements
originated in the nucleus which was formed by the initial collapse
of the protogalaxy and that they were carried into the then still
largely gaseous halo by a sort of secondary big bang of this nucleus.
He bases this view on the remarkable similarity in the relative
abundances of the heavy elements in all stars in which those
abundances have been determined. Though the ratio of hydrogen
to metals can vary greatly the various heavy elements and isotopes
seem always to come in the same proportions, which he finds
difficult to understand unless they were all formed in the same
event.

In the proto spiral galaxies an intense period of star
formation probably occurred in the last stages of their evolution,
when the medium finally collapsed into the thin disk which is the
characteristic feature of spiral galaxies.

The stars and clusters formed in this earliest disk stage can
be distinguished from more recently formed stars by their mostly
lower metal content and especially by their strong concentration

towards the centre of the galaxy. In our own Galaxy planetary
nebulae and novae are conspicuous representatives of this old disk
population. Both can be observed to large distances and exhibit
clearly a strong concentration in the central region. The disk
clusters, the so-called galactic or open clusters, are particularly valuable tools in the investigation of the evolution of our
Galaxy, because their ages can be fairly well determined from the
turn-off points on their main sequence. The oldest have ages
comparable to those of globular clusters. Most galactic clusters
have much lower masses than the globulars. We cannot tell whether
this reflects a difference in birth function, or whether it is
just a consequence of difference in age. The less massive clusters
have shorter life times; it may be, therefore, that they have
originally been quite numerous in the halo, but have now all
disintegrated.

After the initial strong burst of star formation in the
collapsing disk the stellar birth rate in spiral galaxies must
have become much slower, and have continued to slow down as the
interstellar medium was gradually depleted. But it evidently still
takes place at present. It occurs mainly, or wholly, in places
where over large volumes the density exceeds the total average
disk density by several orders of magnitude. If O stars or early
B stars are formed in these concentrations the medium becomes
ionized and forms emission nebulae. These are the conspicuous
indicators of star formation; they can be observed in external
galaxies, and give, therefore, direct information on the rates of
star formation in various galaxy types. The emission regions
always lie in spiral arms, and are the extreme representatives of
what Baade has termed population I. At the Vatican conference
on Stellar Populations in 1957 (2) it was proposed to call all
stars younger than about a hundred million years "extreme
population I", those with ages between hundred and thousand million
years "older population I", and to refer to all older disk stars
simply as disk population. The notion of "populations" had been
introduced by Baade in 1947 after he had succeeded in observing
in the outer parts of the Andromeda nebula and its companions a
dense population of red stars. These contrasted strongly with
the blue giants, δ Cephei variables and emission patches that were
known previously. They had reddish colours, and showed a remarkably
smooth distribution, their density increasing strongly in the
direction of the centre. Baade concluded that they were probably
of the same type as the red giants found in globular clusters, but
it has since become clear from investigations by Morgan and Van
den Bergh that they resemble ordinary K giants, and are relatively
metal rich. Some stellar systems, like the companions to M31,
consist entirely of this type of population, which he called
population II. The other extreme is found in the Irregular galaxies,
like the Magellanic Clouds, in which type I appears to preponderate.

Population I is a typical spiral-arm population. It stands out
by its small-scale clumpiness as well as by large-scale irregular-

ities in its distribution, both in sharp contrast to the smooth
and regular distribution of population II.

There is a large class of "amphibious" galaxies which have
central "bulges" consisting of population II as well as extended
population-I spiral arms.

We now turn to the general problem of the classification of
galaxies. The most striking criterion for arranging galaxies in
a classification scheme is the contrast between the patchy and
the smooth components, and their relative importance.

The clumpy appearance is always connected with interstellar
gas and dust; the classification may therefore also be considered
as one according to gas content. However, we shall see that this
is not a sensitive criterion for distinguishing different types
of spirals.

There is a large class of galaxies in which the clumpy
population appears to be entirely absent. These "smooth" galaxies
may be divided into two main families, the elliptical and the SO
galaxies. In the "pure" ellipticals the axial ratio varies from
about 2 to 1; the surface density decreases smoothly with increasing distance r from the centre, and can be well represented
by $I = I_o/(1+r/a)^2$, as first indicated by Hubble. They are denoted
as E0 to E5, the number being equal to $10(1-b/a)$, where b and a
are the projected semi-minor and semi-major axes. Hubble and
Sandage have also an E6 and E7 class, but these may perhaps better
be described as composite, the main part being an E5 galaxy with
a superimposed smooth disk of much greater flattening.

The ellipticals cover a large range of masses, from the cD
galaxies which are often found as the central galaxies in large
clusters and are often fairly strong radio sources, and the more
normal giant E's like M87 and NGC 4472 in the Virgo cluster, down
to the dwarf ellipticals observed in the Local Group. The masses
of the former may exceed 10^{12} $M_\odot$ while those of the latter are
probably about 10^6 $M_\odot$.

The masses of the smallest elliptical dwarfs, like the Ursa
Minor and Draco systems, are of the same order as those of globular
clusters; also, they resemble these in the types of stars they
contain,and in their integrated colours. But they differ radically
in size, the clusters being of the order of 100 times smaller than
the dwarf ellipticals. The circumstances of their formation must
therefore have been quite different.

Dwarf galaxies are very numerous. Judging from the number
within 100 kpc there may be several hundred in the Local Group.
However, their masses are so small that their contribution to the
total mass of the Local Group is negligible.

As we pass from the dwarfs through the intermediate ellipticals
like NGC 147 and 185 in the Local Group and the closer companions
of M31, NGC 205 and M32, to the giant systems the integrated
colour-indices increase regularly. The population types in giant
E systems differ rather radically from those observed in clusters.
According to Morgan the principal contribution to the luminosity

in the violet region comes from giant K stars. In his classification he classifies these as "k". The amorphous central regions
of spiral galaxies have the same spectrum.

The other family of smooth galaxies, the SO systems, are
characterized by very strong flattening, comparable to that of
the population I in a spiral galaxy. A second property in which
they differ from the ellipticals is the step-like variation of
their surface brightness, indicating in some instances ring
structures in the space density of the light. In this respect they
bear some resemblance to early-type spirals. But they lack entirely
the strong spheroidal central bulge typical of these early spirals;
they may therefore actually be more related to Sc spirals than to
Sa's.

The colours of SO galaxies seem to be the same as those of
large elliptical galaxies, and it appears probable that like these
they consist largely of old stars.

While the classification of the amorphous galaxies is simple,
except that the SO galaxies when seen nearly face-on may not always
be easily distinguishable from E galaxies, the classification of
spirals is complicated. Each spiral has its own individuality.

A useful classification is nevertheless possible. The system
in general use is that introduced by Hubble (3), in which the
classification is made on the basis of the relative strength of
the amorphous central region and the degree of openness of the
arms. Sa galaxies have a strong central region and nearly circular
arms, which are usually tightly wound. The other extreme are the
Sc systems, with inconspicuous nuclei and wide-open, irregular
arms. The Sb's have intermediate properties. The systems with the
most irregular arms have afterwards been classed as Sd; they form
a transition to the irregular galaxies of the Magellanic Cloud
type.

As is well known there are two parallel series of spiral
galaxies, viz. those having a distinct bar structure and those
without obvious bars. The barred, SB, spirals display the same
progression in central concentration and tilt of arms as the non-
barred systems. About one in three spirals has a bar.

It should be remarked that the spiral galaxies do not actually
fit into a one-dimensional classification. For instance, among the
galaxies described as Sc by Hubble and Sandage, mostly on the
basis of the arm structure, there is a wide range in the relative
brightness of the nucleus.

An extension to the above classification was introduced by
Morgan (4). Morgan and Mayall (5) had observed that in the violet
region of the spectrum the luminosity of large elliptical galaxies
as well as of the amorphous central regions of spirals is mainly
due to K giants; on the other hand in very late spirals and irregular systems it comes mainly from A stars. Intermediate spectral
types were found for the central regions of Sb and Sc galaxies.
Morgan found a strong correlation between "the relative luminosity
of the nuclear region to the main body" and the spectral type as

defined above, and he classified the galaxies in a one-dimensional
system according to this strength of the nuclear region. He calls
this a classification of the forms of galaxies according to their
stellar population, and designates his classes by the letters
a, f, g, k, corresponding with average spectral types A, F, G and K.
The relation with the Hubble classes is roughly:

Yerkes (Morgan) aI-afS-fS-fgS-gS-gkS-kS-E

Hubble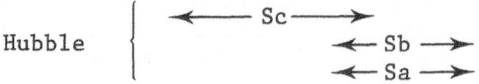

Note the large range in Morgan classes within, for instance,
Hubble's Sc class.

No classification system is entirely satisfactory, due to the
great diversity of the spirals. Several proposals have been made
to introduce additional criteria, in particular by G. and A. de
Vaucouleurs (6), who added the "ring" structure criterion.

The different forms of galaxies may be studied in the Hubble
Atlas of Galaxies prepared by A. Sandage (3). Special attention
may be drawn to the extreme flatness and step-like intensity
distribution in the So galaxies NGC 4111 and 4762; to the striking
phenomenon of dark rings in some Sa galaxies, such as NGC 5866,
and in particular NGC 4594, where a continuous ring extends from
about 12 to 28 kpc in the equatorial plane and causes a maximum
absorption of between 3^m and 4^m (cf. van Houten (7)).

In Sa spirals the arms are sometimes broad and amorphous;
an example is NGC 4826.

Chapter II

Interstellar Gas and Galaxy Types. Formation of Stars

In the preceding chapter reference has been made to the manner
in which galaxies may have formed. The difference between ellipticals and spirals is probably mainly a consequence of the amount
of angular momentum per unit mass which was contained in the mass
of primordial gas from which the galaxy formed. In the E galaxies,
with small specific angular momentum, the contracting gas was
apparently transformed completely into stars before an important
fraction could collapse into a thin disk. In spirals on the other
hand part of the initial gas remained gaseous during the initial
collapse; this was mainly the gas with high angular momentum which
prevented it from concentrating strongly towards the centre.

Soon after the gas had collected in the disk, star formation
apparently slowed down considerably, and in many of the galaxies
possessing a disk a considerable quantity of interstellar gas
remained up to the present time. The presence of such gas, (if its
mass is more than about 5% of the total mass), seems always to
lead to the formation of a large-scale spiral structure. This
structure can have a wide variety of forms, as we have seen in the
preceding chapter. In a few cases it is so irregular that it can
hardly, or not at all, be recognized as a spiral; galaxies in this
category are classified as Irregular, or as Magellanic Cloud type.

The spiral structures are always characterized by a striking
patchiness, and by irregularly distributed dark matter. The more
gas there is in a galaxy the greater its patchiness. By the presence
or absence of patchiness the galaxies can be divided into two main
groups: the smooth galaxies (E and So), and the spirals, which all
have a patchy structure. No appreciable amount of gas has so far
been detected in elliptical galaxies, except for the small amounts
of ionized gas which are sometimes observed near the nuclei. In the
later-type So galaxies some dark matter is seen, so that there must
be gas in these, though apparently not enough to make spiral arms,
or to lead to patchiness.

It seems that gas is an indispensable condition for the
formation of spiral arms. The existence of a flat disk, as found
in So galaxies, is apparently not sufficient.

How did the So galaxies originate? They have no pronounced
central bulges like the Sa type. Apparently their protogalaxies had
a high specific angular momentum like the protogalaxies of late-
type spirals, and a major part of the mass must have remained
gaseous up to the time it collapsed into a disk. Why then is there
so little gas left at present, and no spiral structure? The
appearances are that the So systems have lost their interstellar
matter by some process other than star formation. In this connection
it is suggestive that the number of So-galaxies relative to that
of spirals is very much higher in dense clusters of galaxies than

in the general field or in loose clusters like that in Virgo. The difference is striking, and one is greatly tempted to believe that the absence of gas is in some way caused by the cluster. The simplest explanation - suggested by Baade and Spitzer (8) - is that after having reached the disk stage the So galaxies lost their gas through a collision with a spiral (an intergalactic "wind" could hardly have had sufficient strength to remove the gas). I must mention that objections have been raised to this theory; the most serious one being that there exist So galaxies outside clusters. For this reason alternative possibilities for losing interstellar matter have also been considered.

Though So galaxies have no spiral features they do show a step-like distribution in surface brightness, which in some cases suggests a ring structure in the space distribution. It may be that these are vestiges of a spiral structure which existed before they lost their gas.

As mentioned in the preceding chapter nothing is known about the exact process by which population II stars are found. However, we do have some indications of how population I stars are born. The most massive of these, which appear as 0-, or early B-types, are always formed in so-called "associations", situated in regions where the gas has collected in dense clouds. These are ionized by the uv radiation of the newly born stars, and show as bright HII regions. Such emission patches are particularly abundant in Sc spirals. They are arranged in narrow chains along the spiral arms, thus showing that star formation takes place in the arms, presumably as a consequence of the compression of the interstellar medium at its passage through the spiral wave. Direct evidence for this is indicated for instance by observations of the radio continuum in the Sc spiral M51. The continuum is due to synchrotron radiation. This is enhanced by the compression in the wave. The ridge lines of the synchrotron emission follow the run of the optical arms, but do not co-incide with them. They lie shifted to the inside; the shift in position angle in the spiral plane is about 18° at $R = 5$ kpc (R being the distance from the centre) and about 35° at $R = 2.5$ kpc (cf. Fig. 2). Adopting a pattern speed equal to the rotation velocity in the outer arms these shifts correspond to a time difference of roughly 15 million years between the compression and the formation of stars. This is a reasonable time; a cloud of density 12 H-atoms per cm^3 would need this time to collapse in free fall under its own gravitation.

If star formation takes place in the spiral wave its rate should be proportional to the frequency at which the gas passes the wave. As the angular velocity of the gas increases towards the centre the rate of star formation must likewise increase.

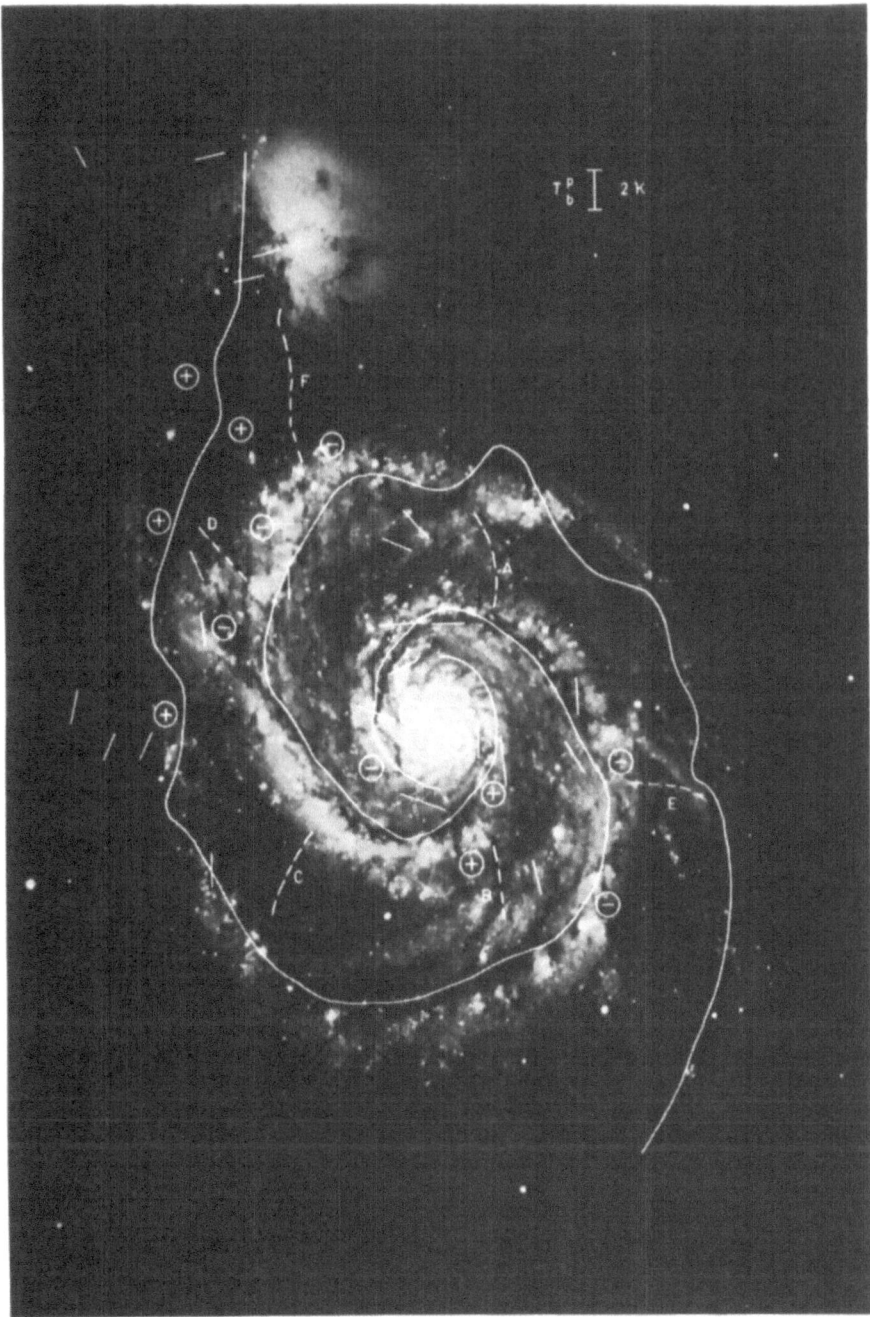

Fig. 2. Radio ridges and optical structure in M51 (Mathewson, van der Kruit and Brouw (9)).

Consequently the gas should be more and more depleted as we proceed
towards the centre. Observations of neutral hydrogen indicate that
in most spiral galaxies the HI column density does indeed decrease
as one follows the spiral arms inwards; one also observes that at
large distances from the centre it remains high out to regions
where the arms are hardly distinguishable optically. At all values
of R the gas is strongly concentrated in the spiral arms. As an
example, Fig. 3 shows the hydrogen distribution in M81.

Let ρ_i be the gas density at the epoch when stars started to
be formed in the spiral wave, and let $\omega_c - \Omega$ be the difference
in angular velocity between gas and spiral pattern, and T the
time in units of 10^9 years during which the spiral arms have
existed, then the number of times the gas has passed through a
spiral arm in a two-armed spiral is

$$n = (\omega_c - \Omega)T/\pi.$$

If α is the fraction of the gas which becomes permanently condensed in stars in one passage through the wave, and if, for a
preliminary, rough estimate, we take α to be independent of R,
the present density ρ is given by

$$\rho = \rho_i e^{-n\alpha}.$$

Taking T = 10 we then get the following results for M81:

R	$\rho_{obs.}$	$\rho(R)/\rho(9)$		
			computed	
kpc	10^{20} at/cm^2	obs.	$\alpha=0.022$	$\alpha=0.033$
9.0	16.5			
4.8	12.0	0.73	0.54	0.22
3.3	6.0	0.36	0.35	0.07
2.	3::	0.2::	0.27	0.03

The observed numbers in the last line are no more than a rough
guess. The initial density ρ_i was assumed to vary as R^{-2}. Had we,
instead, taken ρ_i to be independent of R the best representation
would have been obtained with α = 0.0065 instead of 0.022 as in
the table. These values of α are averages over the whole period
in which formation of stars in the spiral waves has taken place.

We see thus that on the average between 1 and 2% of the gas
has been used up per passage through the wave. As α is likely
to decrease as the overall density decreases the present value of
α is probably considerably lower . It should be stressed that the
uncertainty of α may be larger than indicated by the table because,
(a) α will not be independent of R, (b) the equatorial space
density corresponding to a given (observed) column density will

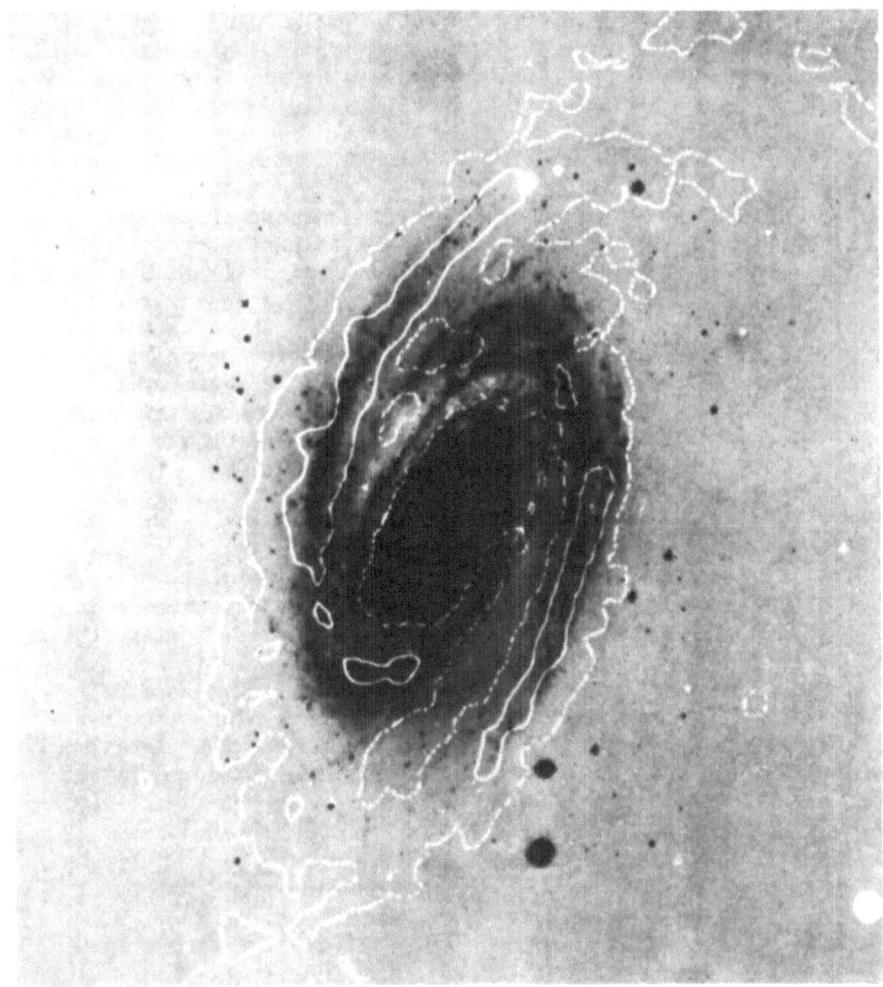

Fig. 3. Column density of neutral hydrogen in M81 (Rots and Shane (10)). The dashed contour corresponds to 3×10^{20} atom cm^{-2}, the full contour to 10×10^{20} atom cm^{-2}.

be higher at small R, (c) there might be inflow of intergalactic
gas replenishing the central regions. The low value of α to which
the above theory of star formation appears to lead gives a problem
when it is applied to the Galactic System. The run of gas density
and the rotation curve being not very different from those in M81
the value of α must also be similar. With a plausible estimate of
Ω, and therefore of n, we can then find the ratio of the present
gas density in the vicinity of the Sun to the initial gas density
ρ_i. The present density being known we can therefore derive ρ_i.
The value so found is far smaller than the observed total <u>mass</u>-
column density of the disk, and would thus indicate that not only
the halo and intermediate populations were formed during the
initial collapse, but that also the majority of the disk was
born in the early evolution of the Galaxy, before the spiral phase.
At present there are too many uncertainties in its derivation
to attach much value to this result. I have mentioned it in order
to indicate how data of this kind may ultimately contribute to
our knowledge of the evolution of the Galaxy.

Let us now return once more to the different classes of
spirals. The difference between Sa and Sc galaxies appears
principally to be connected with the amount of gas left after the
formation of the population II stars during the collapse stage.
In the Sa's the relative amount of gas left for the formation of
the disk was presumably small; no very prominent spiral structures
were formed, and, as shown by the near absence of large emission
regions, the present rate of star formation is slow. The Sc's,
on the other hand are full of emission patches and have strong
arm structure. The general picture is confirmed by the observations
of neutral hydrogen which indicate that in Sc's the interstellar
gas makes up a larger fraction of the total mass than is the case
in the early-type spirals (cf. Balkowski (11)). It should be
stressed that the Sa galaxies are by no means devoid of interstellar matter. This is shown, for instance, by the conspicuous
dust rings seen in several galaxies of this type. An interesting
example is the "sombrero" galaxy NGC 4594 where a dust ring is
observed to extend from about 12 to 28 kpc. The spheroidal halo
extends considerably beyond the latter distance (cf. van Houten
(7)). Though the Sc spirals and Irregulars show no conspicuous
amorphous central bulges they do contain halo populations, as
is indicated by the presence of numerous RR Lyrae variables and
old globular clusters in the Magellanic Clouds. It is even probable,
as was mentioned in my preceding lecture, that the halo populations
of all spirals contain a considerable fraction of the total mass.

So far as I know there is as yet no plausible explanation why
some spirals have <u>bars</u> and others have not. There appear to be no
conspicuous differences in mass, luminosity or hydrogen content
between the two classes. The origin of the bar must, like that
of the spiral arms, probably be connected with the gas. Judging
from the few available 21-cm observations it appears that there
is at present relatively <u>little</u> gas in the bars; the presence of

dark lanes in most well developed bars proves, however, that they contain certainly <u>some</u> intergalactic material.

During the last few years it has become possible to observe some external galaxies in the 21-cm line with sufficiently small beams to resolve the spiral structure. These observations will contribute greatly to our understanding of the structure, dynamics and evolution of spiral galaxies. The first, mostly still provisional, results have already shown:

1. That the gas is strongly concentrated in the spiral arms. Preliminary estimates for M81 give a ratio of the peak arm density to the interarm density of more than 3. In M101 the contrast is still higher. This important datum for the wave theory will shortly be greatly improved, and become determinable also for other spirals, like, for instance, M51.
2. The great abundance of gas in the outermost arms of spirals and the small density in the inner arms.
3. The presence of wave-like motions around the arms in at least some spirals (M81, M51).
4. The possibility of extending our knowledge of the rotation curves to much larger values of R.
5. The existence of large-scale asymmetries in the distribution of HI. Bottinelli (12) found considerable asymmetries to be present in 30% of all Sb and Sc galaxies. These asymmetries are also reflected in the rotation curves; a striking example is shown by M81 (Rots (13)).
6. The occurrence of large-scale deviations from normal rotation. As an example I mention a striking case found by Bajaja in M31. In a large region at about 5 kpc from the centre the hydrogen is seen to move at velocities that deviate by an amount of the order of 100 km/s from the normal rotation velocity. The mass of gas involved is quite large.
7. Large diffuse regions of HI surrounding some galaxies, such as M81. These appear as a sort of bridges between M81 and the companions M82 and NGC 3077, but also as huge detached clouds without optical counterparts (cf. studies by Davies, Rogstad, Rots, Guélin). Similar extended regions have been reported by Davies in the vicinity of M31. The study of this interesting domain is still in a beginning stage.

Investigations of the general "geography" and motions in our own galaxy are much more difficult owing to our situation in the central plane of the disk, so that the various arms are seen projected on top of each other. Nevertheless interesting information on the spiral motions has been derived. Prof. Lin will report on this.

Particular interest attaches to the study of the <u>nucleus</u> of the Galactic System, in which much smaller details can be observed than in external galaxies. This subject will be treated in the following chapter.

Another subject about which the Galaxy yields information
that is virtually unobtainable in other spirals is the stellar
velocity distribution. Of particular interest is the velocity
distribution of the halo stars, and the measure in which radial
motions preponderate in this population. However, the scope of
this lecture does not permit the discussion of these data.

I will now turn for a few minutes to a different subject,
namely that of the high-velocity hydrogen in our Galaxy, and the
question of whether or not the observations give evidence for a
continuous accretion of gas from surrounding space. The subject
is complex, and there is considerable controversy about the interpretation.

During a search with the Dwingeloo telescope, in 1963, for
neutral hydrogen in the halo we found some concentrated clouds at
high latitudes, moving at high velocities. They were very unevenly
distributed, with a preponderance in the north-galactic hemisphere.
All velocities, v, higher than 100 km/s relative to the local
standard of rest were negative. Though in subsequent surveys some
"clouds" with high positive velocities were discovered, there is,
in the completely surveyed area above b = +15°, a preponderance
of about 20 to 1 of the amount of gas with v < -100 over that with
v > +100 km/s.

The high-latitude high-velocity complexes generally have
features suggesting an interaction between streams of gas impinging
upon each other at high velocity. Such features are: large and
erratic velocity differences between different parts of the same
complex, sharp edges in some places, and, sometimes, queer shapes.
A plausible interpretation is that what we observe is interaction
between gas falling in from outside with irregularly distributed
halo gas. The somewhat complicated model suggested is that the
gaseous halo consists of gas expelled from the disk in local jets
or streams and that part of the expelled gas is later accelerated
to high downward velocities by a possibly continuous "wind" falling
in from surrounding space (14, 15). This picture led to an estimate
of the total inflow of intergalactic gas of roughly 2×10^{17} H atoms
per cm^2 of the galactic plane per million years, corresponding to
an increase of about 1% of the mass of the Galaxy per 10^9 years.

In lower latitudes the interpretation of the observations of
high-velocity gas is more complicated. Davies (16) and Verschuur
(17) have independently argued that in the outer regions of the
Galaxy the spiral arms extend to quite large distances from the
galactic plane and that they might extend far enough to explain
the high-velocity complexes. However, it is hard to interpret the
clouds at very high latitudes in this way, and in particular also
the features indicating that they are due to an interaction between
two streams.

Another possibility that should be kept in mind is that the high-velocity features might not be due to truly intergalactic matter, but to gas initially detached from the Magellanic Clouds and now falling into the galactic halo.

Chapter III

The central region of our Galaxy

        Although the centre of our Galaxy is hidden from our view at
optical wavelengths by the obscuring clouds in the galactic disk,
a great deal of information has been obtained during the past
twenty years by observations at radio- and infrared wavelengths,
information which for a large part cannot be obtained from other
galaxies.

        In the radio domain the first observations were made in the
continuum. They showed the presence of a strong source – one of
the strongest in the sky – in the galactic centre. The source is
extended along the galactic equator, with a number of condensations,
part of which are non-thermal and probably emit mostly synchrotron
radiation, while others are dense HII regions. The condensations
seem to be embedded in a more or less smooth medium of HII
radiation. I shall not discuss this interesting but complicated
structure, as it will be dealt with by Dr. Mezger.

        When observations of the 21-cm line became available they
showed some unexpected phenomena. In the first place they indicated
that within about 20° from the longitude of the galactic centre
the hydrogen-line profiles started to show long low-intensity
wings extending beyond what were considered to be the "tangential
velocities". The wings are caused by large non-circular motions.
These appeared to be mainly connected with features moving away
from the centre: the negative-velocity features showed up in
absorption in the direction of Sgr A, while in general the high
positive velocities showed no absorption, indicating that they
were situated beyond Sgr A, and therefore moving away from it.

        There are two main features: the "3-kpc arm", thus called
because it appears to lie at a distance of roughly 3 kpc from the
centre (as judged from its becoming "tangential" around 338°
longitude), and the "expanding arm at +135 km/s" which lies on
the opposite side of the centre, and presumably rather closer to
it than the 3-kpc arm. It is not so well-defined as the latter
(cf. Fig.s 5 and 6). Both "arms" can be followed through about
90° galactocentric longitude, and contain in those sectors an
HI mass of about 2×10^{7} $M_{\odot}$. Fig. 4, taken from an early publication
by Rougoor and Oort (18), indicates their possible location. At
the points where they cross the line through the Sun and the centre
the arms have radial motions away from the centre of 53 and 135 km/s
respectively.

        The most important problem in connection with these two features
is whether they are caused by gas expelled from the galactic nucleus
or whether they are parts of general dynamical phenomena in the
Galaxy. If they are due to explosions from the nucleus, very large
masses must have been involved. According to the computations by
Van der Kruit (19) a mass of between 5 and 10 million solar masses

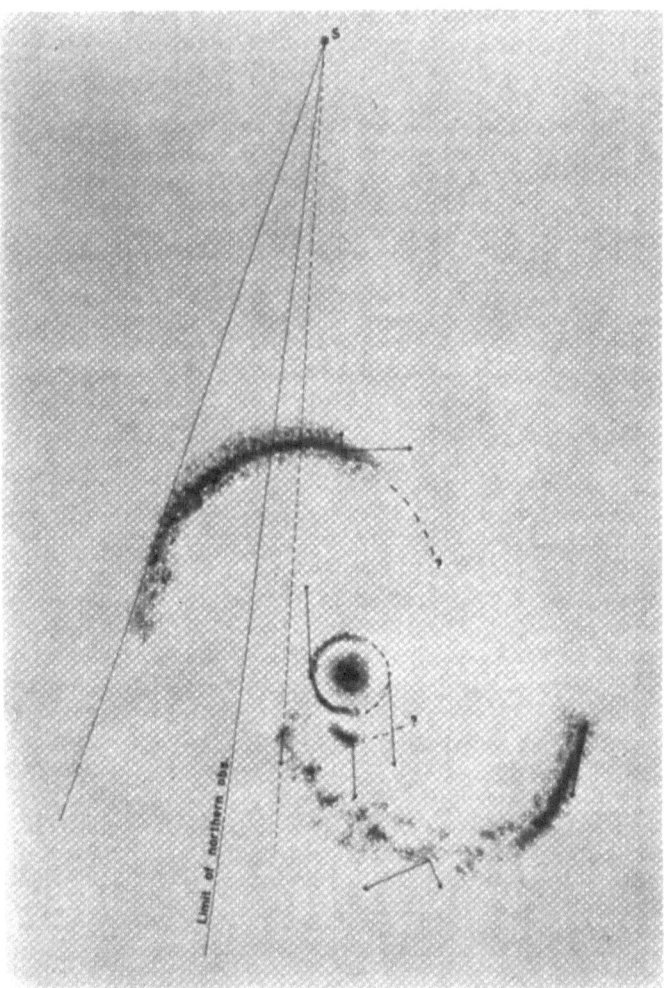

Fig. 4. Sketch of the hydrogen distribution and its motions in the
central region as proposed by Rougoor and Oort (18).

would have had to have been thrown out about 12 million years ago at
velocities around 600 km/s in order to explain the outward motion
of the 3-kpc arm. In his model the expulsion took place in two
opposite directions, at angles of 25⁰-30⁰ with the galactic plane.

    The alternative interpretation has been considered in some
detail by Simonson and Mader (20), who suggest that the 3-kpc arm
may be part of a "dispersion ring" of which the sectors moving away
from the centre happen to lie in the directions towards and
opposite the Sun, while the motions towards the centre, in the
sectors perpendicular to the former, cannot be seen. This interpretation is attractive, because it obviates the need for a recent

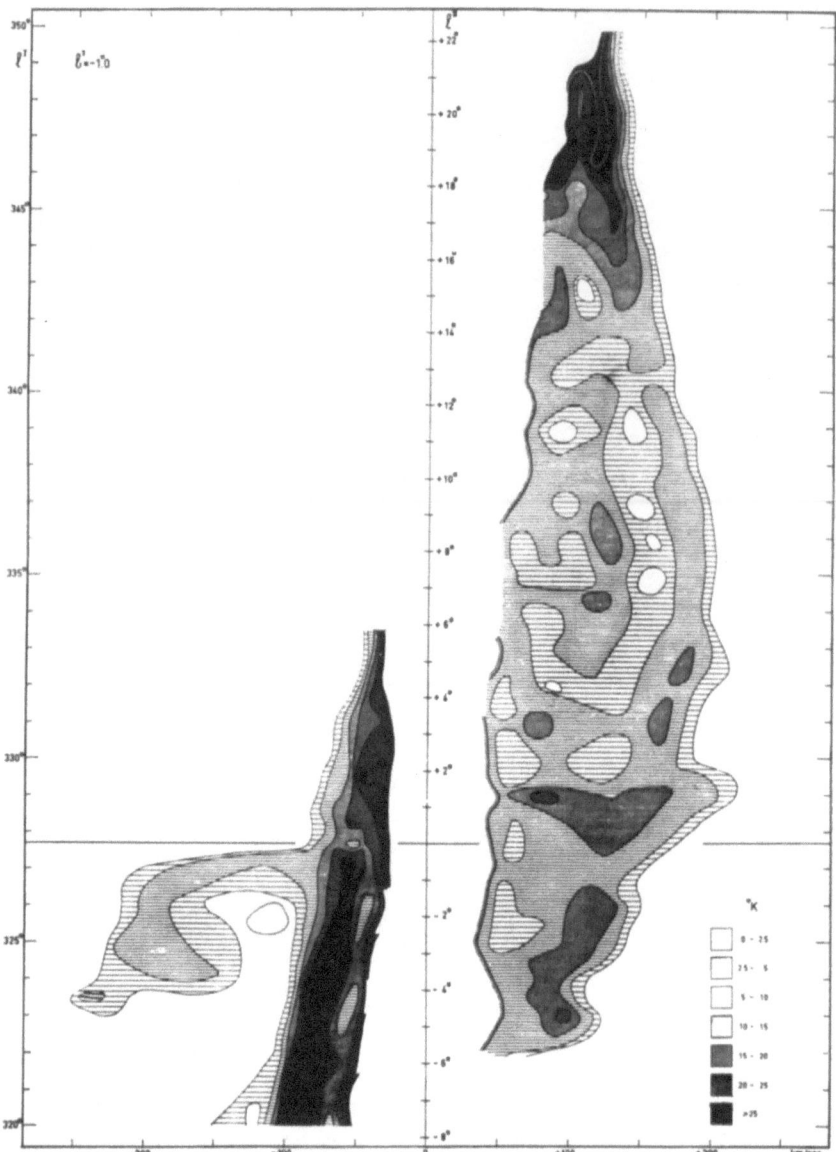

Fig. 5. Velocity-longitude contours of HI at b^{II}= -0°.5 (Rougoor
(21)).

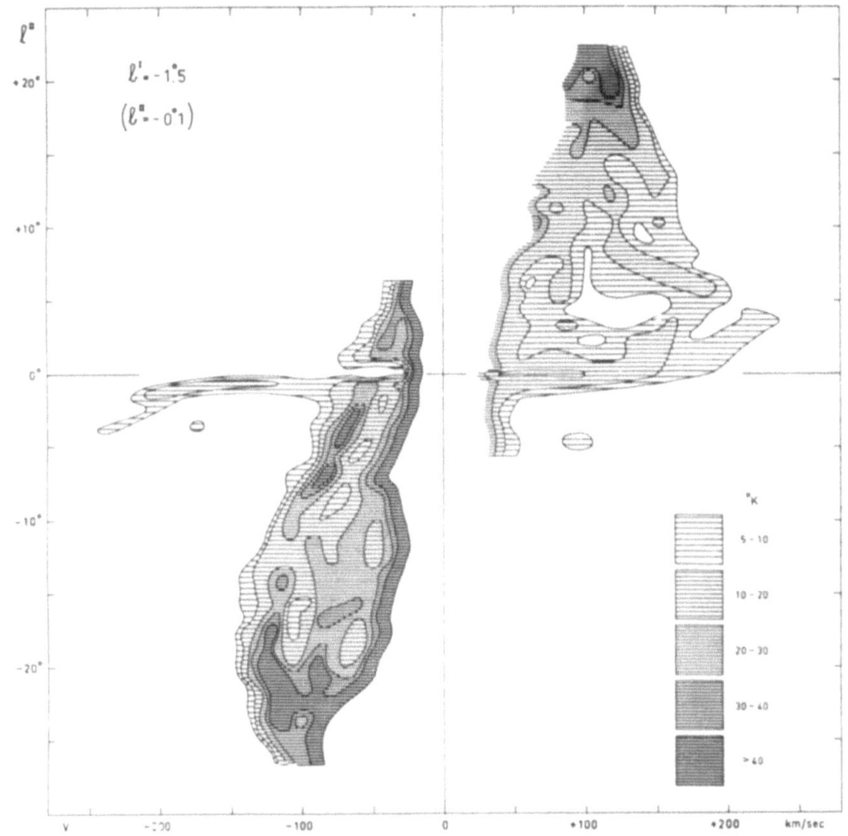

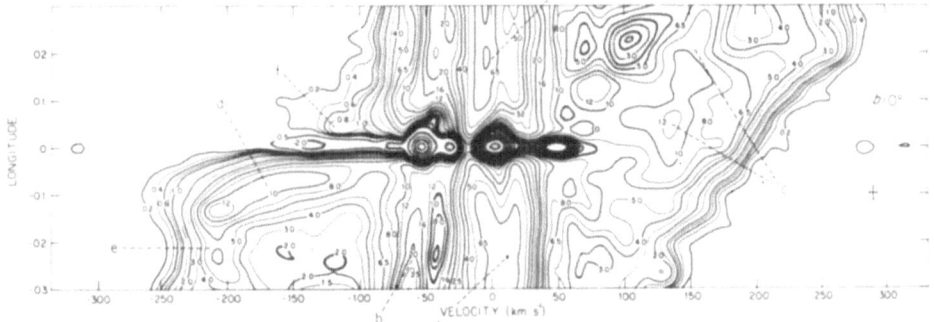

Fig. 6. HI velocity-longitude diagram at $b^{II}= -0^\circ.1$. The part between -8° and $+22^\circ$ longitude is based on Dwingeloo observations, that for the lower longitudes on observations by Burke and Tuve at the Dep. of Terr. Magnetism (from Rougoor (21)).

Fig. 7. HI velocity-longitude diagram at $b^{II}= 0^\circ$, showing part of the nuclear disk (Sanders and Wrixon (23)).

expulsion of enormous masses. But there are also objections, the
most direct being the existence of large HI masses <u>outside</u> the
galactic disk which also seem to be moving away from the centre
at considerable velocities. The most prominent one is Van der
Kruit's "Feature XII", extending from about -4° to $+8^o$ in longitude
and from -2° to -6° in latitude and containing an HI mass of
around 10^5 $M_\odot$. There are also the phenomena observed closer to the
centre, in particular those displayed by the molecular clouds,
which clearly indicate expulsion from the nucleus. In the region
within a few hundred parsecs dense molecular clouds are found to
be moving away from the nucleus at speeds of the order of 200 km/s.
Their total mass has been estimated to be at least 10^6 $M_\odot$.

In apparent contrast with these phenomena most of the atomic
hydrogen within 700 pc from the centre seems to move in a remarkably regular manner in circular orbits, forming a thin disk and
ring, which I shall refer to as the "nuclear disk". It contains
about 4×10^6 $M_\odot$ in the form of HI. The uniqueness of the feature
on the negative-velocity side is well illustrated by the "wing"
between 0° and -4° longitude in Fig. 6 (cf. also Fig.s 5 and 7).
The opposite part of this disk (at positive velocities and
longitudes) appears to be disturbed in its inner region.

The velocities of rotation in the disk agree closely with
the circular velocities corresponding to the mass density distribution inferred from the distribution of the 2.2-μ radiation
in the nuclear region and a comparison with this distribution in
the Andromeda Nebula. In the latter the distribution of the infrared emission has been shown to be identical with that in the
photographic region. By assuming that the mass-to-light ratio for
the latter is the same as that found for the main body of M31,
and that the ratio between mass and infrared luminosity is the
same as in our Galaxy we can derive the central mass distribution
in the latter. A good agreement is found with the masses derived
from the rotations (22 - 24). The following table shows the
masses within various radii as found from these observations. In
addition it gives the circular velocities, times of revolution and
the local mass densities. The data are based on an assumed
distance of 10 kpc for the centre.

R (pc)	M(R) ($10^8 M_\odot$)	Θ_c (km/s)	T (10^6y.)	$\rho(R)$ ($M_\odot/pc^3$)	
1	0.04	140	0.05	400 000	from as-
10	0.7	170	0.4	7 000	sumed M/L
100	8	180	3.	100	from
1000	150	250	25.		rotation

Inside 100 pc the data are based on an assumed M/L ratio.
They do not therefore preclude the possibility of the existence
of a large invisible mass, but an upper limit of about 10^8 $M_\odot$
is set to such a mass by the rotational velocities observed
further out. Additional information on the nuclear mass might be
obtainable from radio observations of planetary nebulae.

I now return to the molecular clouds. Most, if not all, are
moving away from the centre, and have therefore probably been
expelled from the nucleus. They appear to be confined to a very
thin disk, considerably thinner than the HI disk. This is remarkable, especially in view of the fact that much of the HI
expulsion seems to occur at large angles with the galactic plane.
It may be that the dense molecular clouds originate in some way
through an interaction between outgoing streams and the disk.

Dense clouds of molecules are observed from about $-1^\circ.5$ to
nearly $+4^\circ$ longitude. Because no dense clouds have been found
below $l = -1^\circ.3$ it is probable that a large fraction lies within
250 pc from the centre. At $l = 0^\circ$ the clouds in front of Sgr A
are moving towards us at a velocity of 140 km/s, and should thus
have been expelled during the past one or two million years. The
observations of these molecular clouds may well be considered as
the most direct evidence that matter is expelled from the galactic
nucleus, and that this occurs at a high rate; for the total mass
of these clouds must be of the order of a million solar masses.
The negative-velocity clouds on our side of the centre, between
$l = -1^\circ.2$ and $l = +0^\circ.8$, appear to form a more or less continuous
structure which looks like a sector of an expanding ring, such as
proposed by Scoville.

Most of the molecular clouds appear to lie embedded in the
disk. However, they have not been stopped in their outward motion
by the disk gas, nor carried along by its rotation. In fact, their
present column densities are probably so high that they can move
practically unimpeded through the disk. Many molecular clouds in
the central region show molecular emission for which densities
of at least 10^3, in some cases even up to 10^5 H_2 molecules per
cm^3 are required; with diameters of several parsecs their column
densities must be of the order of 10^{22} cm^{-2} or more.

There is a very strong molecular stream quite close to the
centre. It was first observed as a deep and extremely broad OH
absorption band in Sgr A at an average velocity of +50 km/s
(Fig. 8), and is therefore often referred to as the +50 km/s
absorption feature. It was originally thought to be moving towards
the nucleus. High-resolution synthesis maps of Sgr A have shown,
however, that it is not seen in absorption against the main, small
component of Sgr A ("Sagittarius A West") which is probably the
true galactic nucleus, but against a more extended, strongly non-
thermal component roughly 2' East of the former. This makes it
possible to interpret the +50 km/s band as caused by gas ejected
from the galactic nucleus and seen in absorption against Sgr A East,

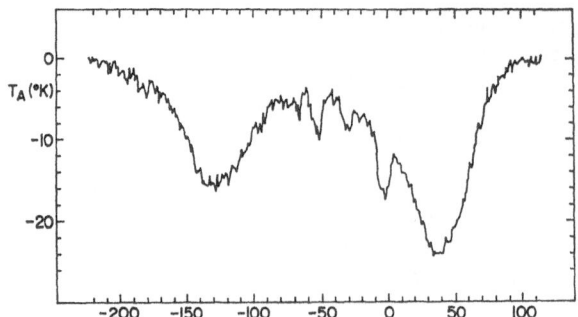

Fig. 8. OH-absorption in Sgr A (from Sandqvist (25)).

which may well be centered some 5 pc beyond the nucleus (see Fig. 9).
The proposed configuration receives support from the observations
of CO emission around Sgr A. If this interpretation is correct the
+50 km/s feature shows that expulsion of molecular gas has occurred
quite recently, and may well be continuing at present.

The true nature of the galactic nucleus is still entirely
unknown. Some evidence has, however, become available in recent
years which appears to indicate its position with fair accuracy.
This evidence has come on one hand from observations at wavelengths
of 1 and 2 microns, which indicate a strong concentration of
presumably stellar light in a small nucleus (Becklin and Neugebauer (26)). Remarkable information has been obtained from observations at wavelengths between 5 and 21 μ (cf. Rieke and Low
(27)). These revealed the presence of at least five discrete
sources within a disk of about 16", or 0.8 pc, diameter (Fig.10).
One of the sources co-incides with the stellar core discovered by
Becklin and Neugebauer. This lies close to the centre of the
"infrared disk". Equally close lies the unresolved radio core of
Sgr A West as determined by recent high-resolution observations
at 6 cm by Downes, Rogstad, Ekers, Goss and Schwarz. The disk has
radio fine structure on a scale similar to that in the infrared.
There is an unresolved central source which Balick and Brown (32)
have recently estimated to be smaller than 0".1 and to have a flux
density of 0.8 flux units at 8.1 MHz.

In connection with the nature of the nucleus mention should
be made of Mezger's suggestion that a supermassive object might
be required to explain the ionization of the extended HII region
around the centre (cf. Mezger"s discussion in this volume).

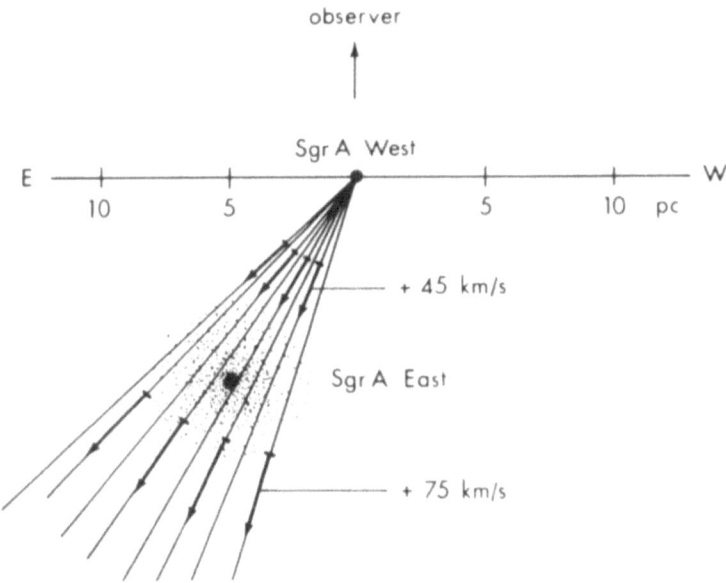

Fig. 9. Possible spatial structure of the +50 km/s stream.

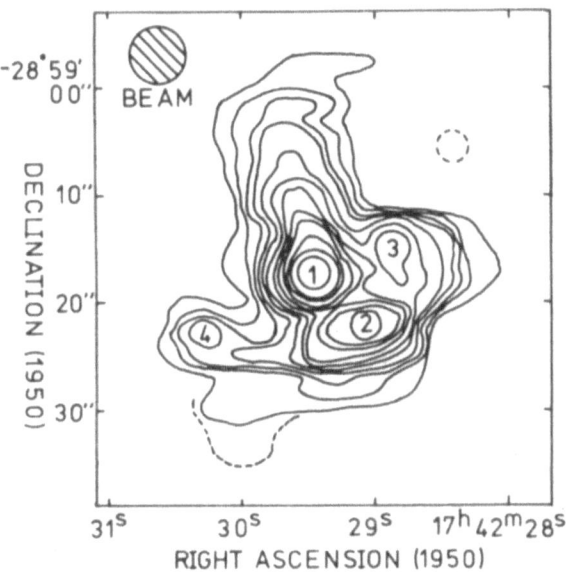

Fig. 10. Distribution of 10.5-μ radiation in the infrared core of the Galaxy (Rieke and Low (27)).

The various features discussed in this chapter are summarized in the following table.

	gaseous mass ($M_\odot$)	R (pc)	Radial motion (km/s)	Time scale (years)
3-kpc arm	10^7			
exp.arm at +135 km/s	10^7	1000-3000	100	$(5-15) \times 10^6$
high-vel.features	10^5			
outside gal.plane				
HI nuclear disk	4×10^6	800	–	--
exp.molecular clouds	10^6	100-700	200	$(1- 2) \times 10^6$
radio sources		0-700		
+50 km/s abs.feature	10^5	5-10	50	10^5
ir nuclear disk and radio fine structure	10^4:	0.5	?	10^3

Fig. 11. Radio contour map of NGC 4258 superimposed on optical
plate taken by A.R. Sandage. Contour unit 0.9°K (from van der
Kruit, Oort and Mathewson (28)).

Chapter IV

The explosive spiral NGC 4258

The following is a summary report on the phenomena observed
in this galaxy, and is largely based on an investigation by
Van der Kruit, Oort and Mathewson (28). A more complete report
is being published in the "Comptes Rendus du Colloque International
du CNRS sur la Dynamique des Galaxies Spirales", which colloquium
took place from September 16 to 20, 1974.

NGC 4258 was observed with the Synthesis Radio Telescope at
Westerbork in the course of a programme aimed at determining the
distribution of synchrotron radiation in the brighter spirals.
It showed a remarkable distribution of this radiation, such as
has not been observed in any other galaxy. As may be seen in
Fig. 11, which shows the continuum emission at 21-cm wavelength,
there is in addition to the normal synchrotron emission from the
brighter spiral arms a much stronger, as well as wider, emission
from a structure that runs approximately perpendicular to the
optical arms, and appears to cut right through these arms. The
structure consists of fairly sharp ridges of an exceptional
smoothness and quite steep on the "preceding" sides, followed by
broad plateaus of somewhat less intense radiation. Observations
made at 49 cm wavelength show exactly the same structure. The
ratio of the surface brightness at the two wavelengths is quite
similar to that observed in radio galaxies.

From optical observations made by E.M. and G. Burbidge and
Prendergast as well as by Chincarini and Walker it was known that
large deviations from normal rotation occur in this galaxy. By
far the most interesting discovery was, however, made by Courtès
and Cruvellier (29) in 1961, who showed by using a Hα interference
filter that the galaxy contains two Hα arms consisting of unusually smooth and narrow filaments starting from the central lens
and extending to about 150", or 5 kpc, from the centre (Fig. 12).
These anomalous "arms" co-incide so exactly with the anomalous
radio ridges that there can hardly be any doubt that they are
part of the same phenomenon.

The Hα filaments rotate in the same sense as the normal parts
of the galaxy, though with reduced velocity (Van der Kruit (31)).
This indicates that they, as well as the related synchrotron
ridges, lie in the equatorial plane. If this is so, the run of
these arms and ridges in this plane must be roughly as indicated
in the face-on sketch Fig. 13.

To account for these phenomena Van der Kruit, Oort and
Mathewson conceived a model in which plasma clouds were ejected
from the nucleus in directions roughly perpendicular to the
apparent major axis of the nebula and under relatively small
angles with the plane of the disk, so that the bulk of the clouds
would fall back into the disk and interact with the disk gas.

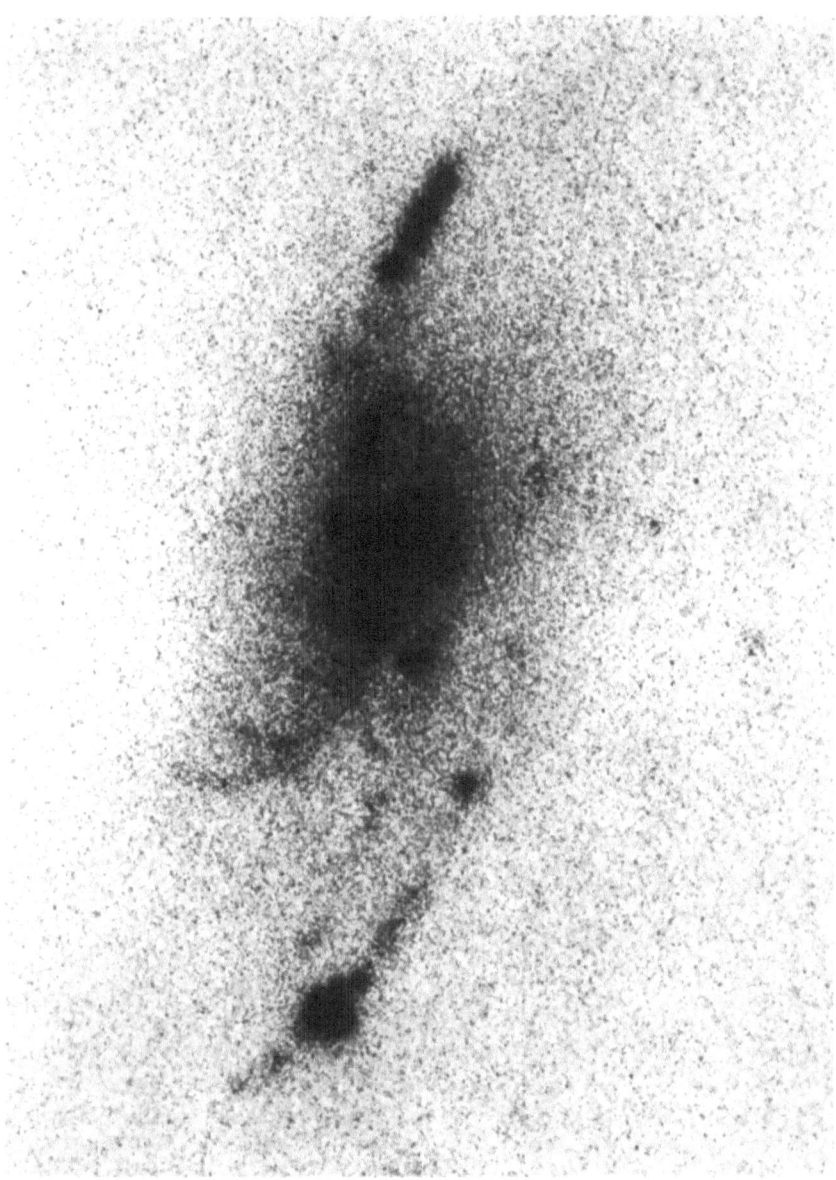

Fig. 12. Hα interference-filter plate of NGC 4258, showing the inner normal arms and the South-Eastern filamentary arms (Courtès, Viton and Véron (30)).

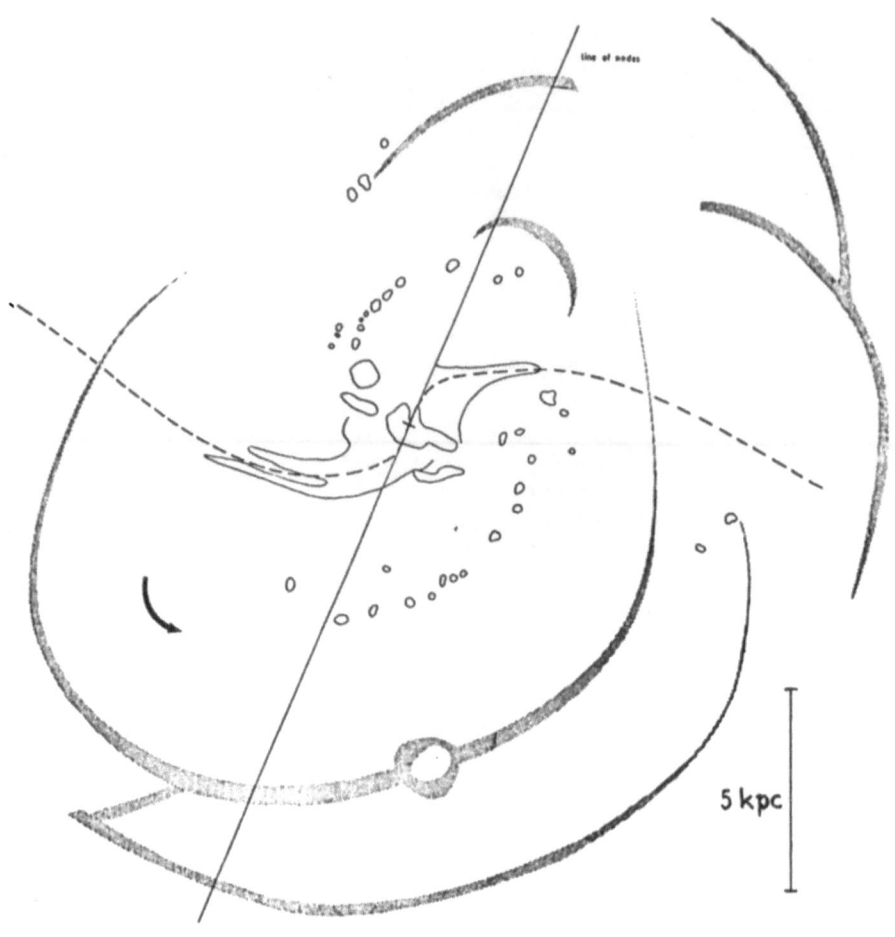

Fig. 13. Face-on sketch of radio ridges (dashed), anomalous Hα
arms (full-drawn contours), inner normal arms (separate patches)
and outer normal arms (shaded). From van der Kruit et al. (28).

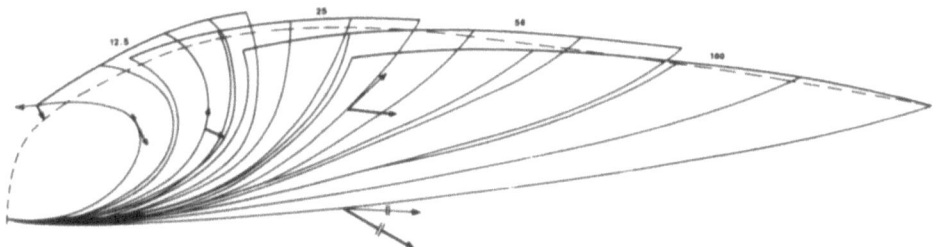

Fig. 14. Sketch of possible orbits of ejected clouds (van der Kruit et al. (28).

This would decelerate the clouds and at the same time sweep them forward in the direction of the rotation, the clouds describing orbits like those indicated in a sketchy manner in Fig. 14. The clouds ejected last (or, if the ejection took place in a wide cone, those at the side of the cone which preceded in the rotation) would now occupy the front ridges of the radio emission and the Hα arms. In the specific model proposed the ejection velocities varied between 800 and 1600 km/s, and the eruption took place about 20 million years ago. The model was extremely rough; it did not pretend to represent the actual conditions, its only purpose being to show that some sort of model can be constructed along the lines sketched.

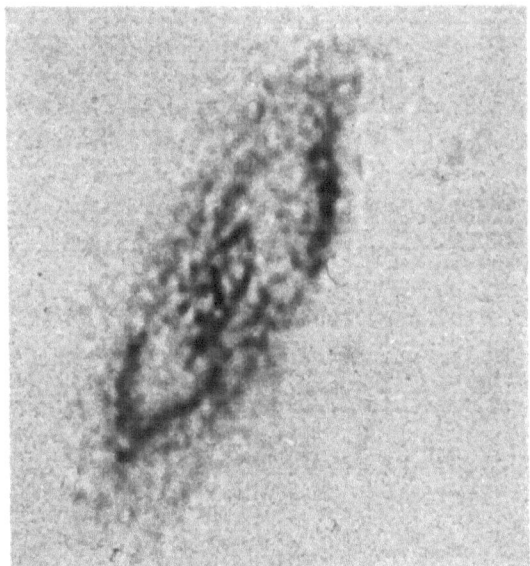

Fig. 15. Radiograph of HI column density in NGC 4258 (G.D. van Albada, unpublished).

At first it was thought that the ionization in the Hα arms might be due to the collision of the ejected clouds with the disk gas and that the increased synchrotron radiation would be caused by a compression of the interstellar gas through the interaction with the traversing or infalling clouds. The former may well be so, but the synchrotron radiation might also come from the ejected clouds themselves. Observations which have recently been made in the 21-cm line show no signs of any heaping up of the neutral hydrogen near the radio ridges and plateaus. This is clearly indicated by the HI map in Fig. 15, which is due to G.D. van Albada. It is therefore likely that the interstellar hydrogen which was ionized in the shocks has not recombined and may well have spread into a thick layer.

REFERENCES

1. J.P. Ostriker and P.J.E. Peebles, Astrophys.J. 186, 467, 1973.
2. Stellar Populations, ed. D.J.K. O'Connell, North-Holland Publ.
                                        Co., 1958.
3. Alan Sandage, The Hubble Atlas of Galaxies, Carnegie Inst. of
                  Washington 1961. Publication 618.
4. W.W. Morgan, Publ.Astr.Soc.Pacific 70, 364, 1958, and Spiral
                  Structure in External Galaxies, in: I.A.U.
                  Symposium No. 38, The Spiral Structure of our Galaxy,
                  Ed. W.Becker and G. Contopoulos, D. Reidel Publ.
                  Co., Dordrecht, Holland, 1970.
5. W.W. Morgan and N.U. Mayall, Publ.Astr.Soc.Pacific 69, 291, 1957.
6. G. and A. de Vaucouleurs, Reference Catalogue of Bright Galaxies,
                  Univ. of Texas Press, Austin,1964.
7. C.J. van Houten, Bull.Astr.Inst.Netherlands 16, 1, 1961.
8. W. Baade and L. Spitzer, Astrophys. J. 113, 413, 1951.
9. D.S. Mathewson, P.C. van der Kruit and W. Brouw, Astron.Astrophys.
                  17, 468, 1972.
10. A.H.Rots and W.W. Shane, Astron. Astrophys. 31, 245, 1974.
11. C.Balkowski, Astron. Astrophys. 29, 43, 1973.
12. L.Bottinelli, Astron.Astrophys. 10, 437, 1971.
13. A.H.Rots, Distribution and Kinematics of Neutral Hydrogen in the
                  Spiral Galaxy M81, Doctor's thesis, Groningen, 1974.
14. J.H.Oort, Astron. Astrophys. 7, 381, 1970.
15. A comprehensive survey has been given by A.M. Hulsbosch, Doctor's
                  thesis, Leiden, 1972.
16. R.D.Davies, Nature 237, 88, 1972.
17. G.L.Verschuur, Astron. Astrophys. 22, 139, 1973.
18. G.W.Rougoor and J.H. Oort, Proc.Nat.Ac.Washington, D.C. 46, 1,1960.
19. P.C.van der Kruit, Astron.Astrophys. 13, 405, 1971.
20. S.C.Simonson III and G.L. Mader, Astron. Astrophys. 27, 337,1973.
21. G.W.Rougoor, Bull.Astr.Inst.Netherlands 17, 381, 1964.
22. J.H.Oort, in Les Noyaux des Galaxies , ed.D.J.K. O'Connell,
                  Pontif. Ac.Sc. Scripta Varia 35, 1971.

23. R.H. Sanders and G.T. Wrixon, Astron. Astrophys. 26, 365, 1973.
24. R.H. Sanders and T. Lowinger, Astron. J. 77, 292, 1972.
25. Aa. Sandqvist, Astron. J. 75, 135, 1970.
26. E.E. Becklin and G. Neugebauer, Astrophys. J. 151, 145, 1968.
27. G.H. Rieke and F.J. Low, Astrophys. J. 184, 415, 1973.
28. P.C. van der Kruit, J.H. Oort and D.S. Mathewson, Astron.
                                    Astrophys. 21, 169, 1972.
29. G. Courtès and P. Cruvellier, C.R. Acad.Sci. Paris 253, 218,1961.
30. G. Courtès, M. Viton and P. Véron in Quasi-Stellar Sources and
                Gravitational Collapse, ed. Robinson et al, Univ.
                of Chicago Press, Chicago, Ill. 1965.
31. P.C. van der Kruit, Astrophys. J., 192, 1, 1974.
32. B. Balick and R.L. Brown, Astrophys.J. 194, in print, 1974.

THEORY OF SPIRAL STRUCTURE[*]

C.C. Lin

Massachusetts Institute of Technology,
Cambridge, Massachusetts, U.S.A.

1. INTRODUCTION

The late Bertil Lindblad first suggested the concept of density
waves as a basis for explaining the spiral structure in disk-
shaped galaxies. Over a number of years, he and his collaborators
attempted to establish this concept by showing that stars in epicyclic motion tend to aggregate into a spiral gravitational well,
which is then maintained in turn by the excess of stars gathered
there. Unfortunately, the mathematical method he used did not
enable him to calculate the collective behavior of the stars in a
convenient manner, and he could not produce the necessary quantitative conclusions for comparison with observations in order to
substantiate his reasoning. His ideas were therefore not widely
accepted.

Modern computing machinery provides one approach to the
quantitative treatment of stellar systems. This method was first
adopted by P.O. Lindblad (1960, 1962) for the study of spiral
structure. More recently, other investigators (Miller and
Prendergast, 1968; Hohl and Hockney, 1969; Hohl, 1970) have carried
out more extensive "numerical experiments" along similar lines.
The most extensive of these, carried out by Miller, Prendergast
and Quirk (1970) involves the consideration of both a gaseous
component and a stellar component, the latter consisting of
approximately 10^5 stars. Their general conclusions support the
concept of density waves. The numerical method has the advantage
of providing at least a qualitative description of the process

---

[*]The research work reported in these lectures was supported in
part by a grant from the National Science Foundation.

of evolution. However, it has not, as yet, yielded any quantitative results for specific comparison with observations. (Cf.
lectures by Prof. K.H. Prendergast.)

Quantitative results can be more readily obtained by analytical
methods more suitable for the study of collective modes. In order
to stay close to comparison with observations, Frank Shu and I
(Lin and Shu, 1964, 1966, 1967) approached the theoretical problem
by first formulating the hypothesis of quasi-stationary spiral
structure (QSSS hypothesis); that is, we adopt as a working
hypothesis the statement that a density wave pattern of spiral
form, however it was originated, does exist in a galaxy, simply
because the optical spiral pattern is observable. We then work
in two different directions from this central position. On the
one hand, we examine its consequences, compare them with observational data, and infer on the underlying astrophysical processes.
On the other hand, we examine the basic dynamical mechanisms
to see how such patterns can be initiated and maintained in an
almost permanent manner.

The study of basic dynamical mechanisms turns out to be-as
one would expect-close to the study of inhomogeneous electromagnetic plasmas (Lin and Shu, 1968), with magnetic field replaced by rotation. Various aspects of these problems have by
now been studied by a number of investigators in the gravitational
case, and we have almost arrived at a complete understanding of the
problem of the origin and permanence of galactic spirals. I shall
present a report on the current status of the theoretical developments in sections 3 and 4.

In view of the difficulty of the theoretical problem, it is
fortunate that, from the beginning, we have placed great emphasis
on working out the consequences of the QSSS hypothesis. In particular, after the dispersion relationship had been worked out
(Lin, 1965), we applied the theory to the calculation of the spiral
pattern of the Milky Way System. This was presented at the I.A.U.
Symposium No. 31 held at Noordwijk in 1966. It is the first
theoretical spiral pattern ever worked out on the basis of dynamical
principles. We believe that it still remains essentially correct,
although minor refinements of the model have since been introduced.
Frank Shu and his collaborators worked out the spiral pattern of
three other galaxies. More recently, in a joint paper by W.W.
Roberts, M.S. Roberts, and F.H. Shu, Shu's method has been applied
to about two dozen galaxies. These authors find that a categorization
of the galaxies according to their theoretical framework leads to
results in general agreement with the observed sequence of luminosity class and Hubble type.

Many other theoretical predictions have by now been worked
out and found to be in satisfactory agreement with observations by
myself and my collaborators: Frank Shu, William Roberts, and Chi
Yuan. Several observers, W.B. Burton (1970 a,b), W.W. Shane
(1971, 1972), S.C. Simonson, III and G.L. Mader (1973), have also
adopted the density wave theory for the interpretation of features

exhibited by 21 cm line profiles. A direct support of the density
wave theory came from observations of the distribution and motion
of neutral hydrogen in M51 and M81, as was already reported by
Prof. Oort in his lectures.

2. GENERAL SPIRAL FEATURES OBSERVED

As is well known, the first problem that faces us is the winding
dilemma; i.e., whether the spiral arms observed each contain the
same material over many revolutions of the stellar system. Super-
ficially, the answer appears to be a definite "yes", for the
brilliant young stars marking the spiral arms are definitely
material objects. However, the spiral pattern for many galaxies
must then change appreciably over a period of time of the order of
two or three revolutions of the system; for the inner parts of a
galaxy are generally rotating at a rate several times that of the
outer parts, as exemplified by the Milky Way System. It is most
unlikely that such a rapid change of appearance is actually taking
place, as the classification of spiral galaxies into Sa, Sb, and Sc
types is not only based on geometry but also on other physical
characteristics; e.g., the gas content and the mass concentration
in the nuclear region. A more subtle difficulty is the implication
that the galactic magnetic field must steadily increase in the course
of time, if indeed the material arms wind up more tightly. Clearly,
both difficulties would be avoided if the spiral structure were
associated with a wave pattern.
    Let us now consider a number of general features observed in
galactic spirals. In the study of these features, there is one
important theme to be kept in mind: coexistence[†]. The complicated
spiral structure of the galaxies indicates the coexistence of
material arms and density waves,and indeed of the possible coexistence of several wave patterns. These features influence but
do not destroy one another. When apparently conflicting conclusions
are indicated by observations, the truth might indeed lie in the
coexistence of material arms and several wave patterns. To be
sure, before taking this "easy way out", one should examine each
interpretation of observational data as critically as possible.
    Furthermore, as is well known, the different categories of
optical objects defining essentially the same spiral arm usually
appear displaced relative to one another (Morgan, 1970). The
radio features again do not necessarily coincide with the optical
features.
    Eleven prominent features in spiral galaxies, including those
discussed above are listed in Table I, together with the theoretical

---

[†]Zwicky (1957) used the term "coexistence" to describe the "blue"
and the "red" components of M51. We are using it in a broader
context.

Table I

Observed Features in Spiral Galaxies and their Theoretical Explanation
in Terms of Density Waves

Observation	Theory (QSSS hypothesis)
(1) existence of a grand design	(1) wave pattern
(2) persistence of spiral pattern	(2) quasi-stationary spiral structure
(3) spiral pattern usually two-armed	(3) $\Omega - \kappa/2$ nearly constant (Lindblad)
(4) multiple-armed in outer regions	(4) $\Omega - \kappa/m$ nearly constant in outer regions
(5) ring structure (e.g. in NGC 5364)	(5) $\lambda \approx 0$ at resonance
(6) HI distribution follows optical spiral arms[†].	(6) HI concentration at minimum of spiral gravitational potential.
(7) HII regions arranged like a string of beads.	(7) galactic shock triggering star formation
(8) dust lanes on inner side of bright spiral arms.	(8) gas compressed at galactic shock before stars form.
(9) abundance distribution of ionized hydrogen varies greatly over the disk; none inside ring, except at center.	(9) gas compression and rate of star formation vary with radius; wave pattern becomes tightly wound and terminates as resonance is approached.
(10) peak of abundance distribution of neutral hydrogen outside of HII distribution.	(10) shock mechanism needed for star formation ; not available in outer regions.
(11) magnetic field generally weak, but appears to be intensified along the dust lane in certain galaxies (e.g. M51)[†]	(11) absence of perennial stretching by differential rotation; field increased by compression of gas.

[†]Well established by WSRT observations after theoretical predictions had been made.

interpretations. The detailed discussions will be found in sections 3, 5 and 6. Other support for the density wave theory may be found from detailed observations in the Milky Way System. These data will be discussed in section 7.

3. THEORY OF DENSITY WAVES

3.1 QSSS hypothesis

To implement the QSSS hypothesis concerning density waves, we may adopt the scheme shown in the following diagram. It leads to an integral equation for normal modes, which has been formulated in the pure stellar case. (Kalnajs, 1965; Shu, 1970a). The attempted numerical solution of this equation (Kalnajs, 1970) has not yet led to results directly comparable with observations. Shu (1970a) was able, however, to use it to clarify the nature of the so-called "anti-spiral theorem", which essentially states that <u>there are no spiral modes if resonances are absent.</u> We shall not attempt to examine the issues concerning the appropriateness and the solution of this integral equation. We shall take a more physical approach to the discussion of the normal modes. We shall see that resonances between the stars and the wave pattern are important. If the angular velocity of the wave pattern is Ω_p, and the angular velocity of circular motion is $\Omega(\varpi)$ at galacto-centric distance ϖ, then resonances are defined, following Bertil Lindblad, by

$$m(\Omega_p - \Omega) = n\kappa, \quad n = o, \pm 1, \pm 2, \ldots, \tag{3.1}$$

where m is the number of arms in the spiral pattern and $\kappa(\varpi)$ is the epicyclic frequency. Mark (1971) has shown that trailing spiral waves of the fairly tight kind are absorbed at the Lindblad resonances $n = \pm 1$, and he has also (Mark, 1974a,b) worked out the detailed driving mechanism near corotation resonance $n = 0$. This is the crucial feedback process suggested in my paper at the Basel symposium in 1969. We shall discuss this driving mechanism in some detail (section 4) by using the gas-dynamical model (after this procedure is justified), since the complete treatment of the stellar system is fairly complicated. We shall see that this mechanism leads automatically to short <u>trailing spiral waves.</u>
Before we come to the discussion of these dynamical mechanisms, let us first consider the linear asymptotic theory for fairly tightly wound spiral waves-which is actually applicable to Sc galaxies-and show that the theory can be used for the interpretation of some of the observed features listed in Table I. It is actually on the basis of this asymptotic theory that one can develop the nonlinear gas dynamic theory that yields the interpretation of the other features listed in the same Table. (See section 5).

For fairly tightly wound spirals, a dispersion relationship may be obtained that connects the spacing between the spiral arms with the pattern frequency of the wave together with other known parameters in the model. Such a relationship was first obtained in a crude form (Lin and Shu, 1964) and then in the form to be presented below a little later (Lin, 1965; Lin and Shu, 1966; Lin, Yuan, and Shu, 1969).

Suppose that the spiral gravitational potential of the density wave is given by

$$\mathcal{V}_1 = -A(\tilde{\omega})\cos \chi \quad , \qquad \chi = \omega t - m\theta + \Phi(\tilde{\omega}), \tag{3.2}$$

where $(\tilde{\omega},\theta)$ are the plane polar coordinates and t is the time. Then the spiral pattern at each instant t has m arms, and the equipotentials are approximately given by the m-armed spiral curve

$$m(\theta-\theta_o) = \Phi(\tilde{\omega}) - \Phi(\tilde{\omega}_o) \tag{3.3}$$

if $\Phi(\tilde{\omega})$ is of the form $\Lambda f(\tilde{\omega})$ where Λ is a large parameter, and $f(\tilde{\omega})$ is a smooth function. The spacing between the arms is given by the "wave length"

$$\lambda = \frac{2\pi}{|k|} \quad , \tag{3.4}$$

where $k=\Phi'(\omega)$, and the pattern propagates around the galactic center at a pattern speed

$$\Omega_p = \omega/m \tag{3.5}$$

The dispersion relationship connects $|k|$ with the intrinsic frequency

$$\nu = m(\Omega_p - \Omega)/\kappa \tag{3.6}$$

where $\Omega(\tilde{\omega})$ is the angular velocity of circular motion at radial distance $\tilde{\omega}$ and κ is the corresponding epicyclic frequency. In a pure stellar disk, in which the stars obey the Schwarzschild distribution of velocities, we have the following dispersion relationship

$$(\omega-m\Omega)^2 - \kappa^2 = 2\pi G\mu|k| \quad , \tag{3.7}$$

where G is the constant of universal gravitation, and

$$\mu = \sigma_* F_\nu(x). \tag{3.8}$$

In the above formula, $\sigma_*(\tilde{\omega})$ is the projected surface density of the infinitesimally thin galactic disk, and $F_\nu(x)$ is the reduction factor defined by

$$F_\nu(x) = \frac{1-\nu^2}{x}\left\{1-\frac{\nu\pi}{\sin\nu\pi}\frac{1}{2\pi}\int_{-\pi}^{\pi}\exp\{-x(1+\cos s)\}\cos(\nu s)ds\right\} \quad (3.9)$$

where $x=k^2<c_{\bar\omega}^2>/\kappa^2$, and $<c_{\bar\omega}^2>$ is the mean square value of the peculiar velocity of the stars. Graphs for this dispersion relationship may be found as Fig. 5.2 and Fig. 5.3 in the Brandeis lectures.

If there is also a gaseous component and if it is justifiable to treat it within the scope of the linear theory, as postulated above, we need only to replace (3.8) by

$$\mu = \sigma_* F_\nu(x)+\sigma_g F_\nu^{(g)}(x_g). \quad (3.10)$$

In this formula, σ_g is the (projected) surface density of the gas, and

$$F_\nu^{(g)}(x_g) = \{1+x_g/(1-\nu^2)\}^{-1}, \quad (3.11)$$

where $x_g=k^2a^2/\kappa^2$ and a is the acoustic velocity of the gas.

In the Brandeis lectures, the reader will find the derivation of the above relationships as well as the missing details in the justification of the following remarks (unless the reference is otherwise given.)

1. Actually, it is usually not justifiable to treat the gas according to the linear theory, because $a^2 << <c_{\bar\omega}^2>$. Galactic shocks are therefore formed in the gas even when the perturbations in stellar density are small. Since the amount of gas in a spiral galaxy is usually small, we may first neglect the gas in considering the general nature of the waves, and then treat the gas according to the nonlinear theory. (See section 5.)

2. For the pure stellar disk, the dispersion relationship shows that there is a wave solution only if $|\nu|<1$, i.e.,

$$\Omega - \frac{\kappa}{m} < \Omega_p < \Omega + \frac{\kappa}{m}. \quad (3.12)$$

This is the basis for the interpretation of the observed features (3) and (4) in Table I.

3. At Lindblad resonance $|\nu|=1$, the wave length $\lambda=0$ to the approximation indicated above. This is the basis for the interpretation of item (5) in Table I.

4. To the next approximation, one can obtain the amplitude distribution of the (stationary) density waves as a function of galacto-centric distance. It has indeed been shown that this distribution conforms to a principle of conservation of density of wave action (Toomre, 1969; Shu, 1970b).

5. However, the amplitude distribution appears anomalous near the resonant points. Indeed, closer examination (Mark,1971) shows that the waves are absorbed at the Lindblad resonances $|\nu|=1$ (by a mechanism similar to Landau damping in plasma waves) and that the wave length is actually small but not zero. The condition near corotation will be discussed in section 4.

6. All the above discussions are for disks of infinitesimal thickness. The correction due to finite thickness was worked out by Shu (1968; see also Shu et al,1971), and by Vandervoort (1970).

4. THEORY OF DENSITY WAVES

4.1 Mechanism for long-term maintenance

The issue of long term maintenance of trailing wave patterns was first raised by Toomre (1969) who pointed out that short trailing waves propagate away from corotation rather quickly. A source of such waves must be provided. Lin (1969) responded to this issue by suggesting that there is a feedback mechanism from the central regions via a long trailing wave pattern (open spiral) propagating toward corotation, or via direct bar-like gravitational field associated with an oval distortion of the central regions. In line with this suggestion, Shu et al.(1971) indicated the possible existence of the open spiral by dashed curves in their photograph of M51.

As mentioned above, the detailed mechanism of this process near corotation has been worked out by Mark (1974b) for long wave driving, and earlier by Feldman and Lin (1973) for bar driving. Mark also noted the possibility of strong amplification if the corotation region is locally unstable in the Jeans sense.

In the following, we shall sketch the gas dynamical theory as worked out by Lau and Lin (1974) and presented in part by Lin (1973) at a conference on Modern Developments in Fluid Dynamics. The justification for adopting the gas dynamical treatment is two-fold. First, near corotation, the reduction factors for the gaseous and the stellar cases are within a few percent of each other if $x_g=x$. Secondly, when all the appropriate approximations are introduced in both cases, the final mathematical problem becomes identical in either case. [See Eq. (4.24)]

We adopt a coordinate system rotating with an angular velocity Ω_p of the pattern. We adopt a polytropic relationship between the pressure p and the density ρ so that

$$p=\kappa\rho^\gamma, \tag{4.1}$$

where $\gamma>1$; and we introduce the acoustic speed a and the enthalpy h by

$$a^2 = \frac{dp}{d\rho} = (\gamma-1)h. \tag{4.2}$$

We now consider an infinitesimally thin disk. We replace the density ρ by the projected density σ, but retain the same symbols for other quantities. We consider only two-dimensional motions in the plane of the disk. We study the linearized equations for small perturbations from a basic axisymmetric state (which we denote by a subscript zero.) We then consider solutions of the form

$$h = h_o(\widetilde{\omega})+h_1(\widetilde{\omega})\exp\{im(\Omega_p t-\theta)\}, \tag{4.3}$$

$$\psi = \psi_o(\widetilde{\omega})+\psi_1(\widetilde{\omega})\exp\{im(\Omega_p t-\theta)\}, \tag{4.4}$$

where ψ is the gravitational potential. Then, from the gas dynamical equations, we find that the amplitude functions $h_1(\widetilde{\omega})$ and $\psi_1(\widetilde{\omega})$ satisfy the following single ordinary differential equation of the second order:

$$L(h_1+\psi_1)+Ch_1 = 0 \tag{4.5}$$

where

$$L = \frac{d^2}{d\widetilde{\omega}^2} + A \frac{d}{d\widetilde{\omega}} + B = (\quad)''+A(\quad)'+B(\quad) \tag{4.6}$$

and the coefficients A,B,C are defined as follows:

$$\widetilde{\omega}A = - \frac{d\ln\mathscr{A}}{d\ln\widetilde{\omega}} , \quad \mathscr{A} = \frac{\kappa^2(1-\nu^2)}{\sigma_o \widetilde{\omega}} \tag{4.7}$$

$$\widetilde{\omega}^2 B = - m^2 + \frac{2\Omega}{\Omega-\Omega_p} \frac{d\ln\mathscr{B}}{d\ln\widetilde{\omega}} , \quad \mathscr{B} = \frac{\kappa^2(1-\nu^2)}{\sigma_o \Omega}; \tag{4.8}$$

$$C = - \frac{\kappa^2(1-\nu^2)}{a_o^2} . \tag{4.9}$$

Note that, if $\widetilde{\omega}_o$ is a typical scale of the galaxy,

$$|C| \gg \widetilde{\omega}_o^{-2} \tag{4.10}$$

except when $\nu^2=1$. The functions A, B and C depend only on the basic model and the pattern speed Ω_p. Note that A,B,C are regular for finite values of $\widetilde{\omega}$ except possibly for

$$\nu = 0, \pm 1. \tag{4.11}$$

The absence of resonances then precludes the existence of spiral solutions (cf. Rayleigh 1880 as quoted in Lin, 1965).
    The derivation of the above equations are given in the paper by Feldman and Lin (1973), where one may also find the determination of the proper contours to be adopted in the complex plane when a singularity in the coefficeints is encountered. In the same paper,

it is shown that if we are considering fairly tightly wound spirals either purely trailing (k<0) or purely leading (k>0), we have, according to Poisson's equation, the following relation between gravitational potential and density:

$$\frac{d\psi_1}{d\tilde{\omega}} = -i\Sigma(r)h_1 \, \text{sgn}(k) \tag{4.12}$$

where

$$\Sigma = \frac{2\pi G\sigma_o}{a_o} >> \tilde{\omega}_o^{-1}. \tag{4.13}$$

Consider now trailing spirals for which sgn(k)=-1. To get the results for leading spirals, we need only change the sign of Σ in the following discussions.

If we use (4.12) to eliminate ψ_1'' and ψ_1' in (4.5), we get

$$h_1'' + \{A+i\Sigma\}h_1' + \{i\Sigma'+iA\Sigma+B+C\}h_1 = - B\psi_1. \tag{4.14}$$

Since A is of the order of $\tilde{\omega}_o^{-1}$, $|A\Sigma| << |C|$. If we adopt similar approximations, we see that the above equation becomes

$$h_1'' + i\Sigma h_1' + Ch_1 = 0, \tag{4.15}$$

which is an equation for free waves. The associated dispersion relationship in the asymptotic theory is

$$-k^2 - \Sigma k + C = 0, \tag{4.16}$$

where C is defined by (4.9). This relationship was obtained by a slightly different derivation as Eq.(3.24) in the Brandeis lectures. The solution of (4.16) is

$$k = - \frac{\Sigma}{2} \pm (\frac{\Sigma^2}{4} + C)^{\frac{1}{2}}. \tag{4.17}$$

Since Σ>0, the upper sign corresponds to long trailing waves, and the lower sign to short trailing waves. Note that the two waves have comparable wave numbers in the cases of interest (K≈0 in the notation introduced below).

$$c_g = - \frac{d\nu}{dk} = \mp \frac{a_o^2}{\kappa^2} (\frac{\Sigma^2}{4} + C)^{\frac{1}{2}} \nu^{-1}. \tag{4.18}$$

Thus, a group of short trailing waves propagates toward the galactic center inside of the corotation circle, where ν<0; they propagate away from the galactic center, where ν>0. In either case, they propagate away from the corotation circle.

Clearly, the direction of propagation is reversed when long trailing waves are considered.

Again, there is a reversal of direction of propagation when leading waves and trailing waves are correspondingly compared.

4.2 Reduction to normal form

The equation (4.15) can be reduced to the normal form

$$\frac{du^2}{dx^2} + (\frac{\Sigma^2}{4} + C)u = 0, \tag{4.19}$$

by the standard transformation

$$h_1 = u \exp\{-i\int \frac{\Sigma}{2} dr\} \tag{4.20}$$

In (4.19) we have also introduced the variable

$$x = \tilde{\omega} - \tilde{\omega}_{co} \tag{4.21}$$

where $\tilde{\omega}_{co}$ is the corotation radius. It is further convenient to introduce the notation

$$K \equiv \frac{\Sigma^2}{4} - \frac{\kappa^2}{a_o^2} = (\mu_o^2 - 1)\alpha^2, \tag{4.22}$$

where

$$\mu_o = \frac{\pi G \sigma_o}{\kappa a_o} \quad , \quad \alpha^2 = \frac{\kappa^2}{a_o^2}. \tag{4.23}$$

It is easy to verify that Jeans instability corresponds to $\mu_o > 1$, and neutral stability to $\mu_o = 1$. In terms of K, the equation (4.19) becomes

$$\frac{d^2u}{dx^2} + (K + \alpha^2\nu^2)u = 0, \tag{4.24}$$

where $\nu^2 = 0$ at $x = 0$. Equation (4.24) has also been obtained by Mark from a study of the stellar system.

Near corotation, Eq.(4.24) is similar to that for the scattering problem in quantum mechanics with a quadratic potential hill. The solution of such problems is well known. Let us just consider the case of marginal stability $K = 0$; Eq.(4.24) can be further simplified by introducing the new variable

$$\zeta = \alpha \int_0^x \nu \, dx \tag{4.25}$$

since α is a large parameter. We get, in the lowest approximation

$$\frac{d^2u}{d\zeta^2} + \frac{1}{2\zeta} \frac{du}{d\zeta} + u = 0, \tag{4.26}$$

whose solution is a linear combination of

$$u = \zeta^{\frac{1}{4}} Z_{\frac{1}{4}}(\zeta) \tag{4.27}$$

where $Z_{\frac{1}{4}}$ is any Bessel function of order $\frac{1}{4}$.

The solution

$$u = \zeta^{\frac{1}{4}} H_{\frac{1}{4}}^{(2)}(\zeta) \qquad \qquad (4.28)$$

corresponds to a pure short trailing wave outside of the corotation circle (arg z=0), since it has the asymptotic behavior

$$u(x) \sim \left(\frac{2}{\pi}\right)^{\frac{1}{2}} \zeta^{-\frac{1}{4}} \exp\{-i(\zeta - \frac{3\pi}{8})\}. \qquad \qquad (4.29)$$

Inside of the corotation circle (arg z=2π), it has the behavior

$$u(x) \sim \left(\frac{2}{\pi}\right)^{\frac{1}{2}} |\zeta|^{-\frac{1}{4}} \{2^{\frac{1}{2}} \exp\{i(\zeta + \frac{3\pi}{8})\} + \exp\{-i(\zeta + \frac{3\pi}{8})\}\}. \quad (4.30)$$

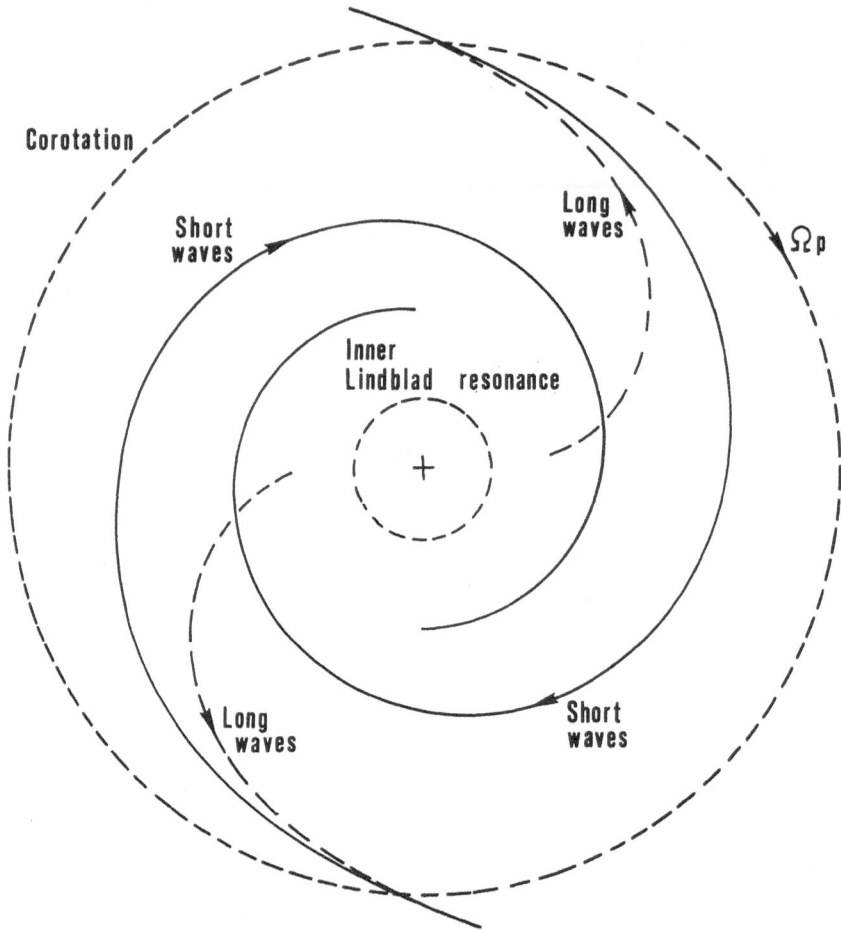

Fig. 1. Interaction of long spiral waves and short spiral waves near corotation. (After Shu et al, 1971, slightly modified; original figure reproduced from their preprint in article by R. Wielen, 1974.)

This represents a combination of two waves: a short trailing wave with amplitude $2^{1/2}$ times as large as that in (4.29) and a long trailing wave with the same amplitude as that in (4.29). Physically, it means that a long trailing wave approaching corotation excites two short trailing waves departing corotation. In particular, the one going towards the galactic center has twice as much energy as the original long wave. (See Fig. 1, which is a slight modification of a figure prepared by Shu et al. The figure was omitted in their final published version, but published by R. Wielen in his review articles.)

In the above discussion, we just exhibited one specific example. Mark has shown that there is a strong amplifying effect, if K>0, or if there is gas present to cause a similar net effect. In the latter case, the parameter involved is $1-Q+1.82(\sigma_g/\sigma_*)$, where Q is proportional to the stellar dispersive velocity. If Q=1, and $\sigma_g/\sigma_*=0.05$, he finds that the short wave travelling from the corotation circle towards the galactic center has four times the energy density as the approaching long wave.

4.3 Discussion

Resonant stars at corotation. The mechanism discussed here is different from that considered by Lynden-Bell and Kalnajs (1972), who studied the effect of resonant stars near corotation. There is some indication that their effect is related to the singular term in the coefficient B defined by (4.8), (see Feldman, 1973) since it vanishes (to a good approximation) when

$$\frac{d\ln \mathcal{B}}{d\ln \varpi} = 0 \text{ at } \nu=0. \tag{4.31}$$

The continuous supply of resonant stars is a problem in maintaining this mechanism. Also, in the present investigation, it appears that this resonance mechanism is not as effective as the feedback mechanism considered above.

The issue of infinite amplitude. A remark should be made concerning the infinite amplitude found by Shu (1970b) near corotation. This is a familiar situation near turning points when one insists on having only one wave passing through. (In the case of Schrödinger equation, the amplitude factor $[E-V(x)]^{-1/4}$ becomes infinite at the turning point.) The removal of this difficulty depends on the recognition of a system of three waves.

Normal Spirals and Barred Spirals. In the original Hubble classification, a distinction is made between normal spirals and barred spirals, but it has long been recognized (perhaps by Hubble himself) that there is a gradual transition between these two categories. The analytical theory has been well developed only for the discussion of normal spirals. Much work remains to be done for our understanding of barred spirals. At this point, we would like only to offer the following conjectures.

A normal spiral galaxy possesses a substantial nuclear bulge,

or a halo, or both. It is consequently stabilized against bar-
like distortion in the central regions. A prominent "short-wave"
pattern may therefore exist according to the theory presented
above.

A barred spiral galaxy is not so stabilized. The angular
velocity (averaged with respect to galacto-centric longitude) is
presumably more uniform in such a galaxy near its central region.
The bar is presumably rotating at an angular velocity slower than
that of the material there. Corotation occurs somewhere in the
outer region. The increased non-circular velocity of the stars-
due to the influence of the non-axisymmetrical gravitational field
of the bar-prevents the "short-wave" spirals from becoming prominent.

Clearly, the transition is gradual between the two categories
of galaxies, if we adopt this dynamical interpretation.

These conjectures are based on the following arguments.

1. Normal spirals. The following facts are known and the
inferences can be made about a typical normal spiral.
(a) There is a prominent "short-wave" pattern, but there does not
appear to be a prominent "long-wave" structure (if it exists at
all).
(b) There does not appear to be any substantial deviation from
axial symmetry in its mass distribution.
(c) Non-circular peculiar velocities of the stars appear to be
limited to the general level of that required to counteract local
Jeans instability.

There is direct observational support of the general correct-
ness of the last statement in the solar vicinity. In external
galaxies, this has to be inferred from the existence of "short-
wave" spiral structure not too far from corotation region. If
there had been a substantial bar-like distribution of mass in the
galaxy, the increased non-circular velocity of the stars would
tend to wipe out these "short-waves" near corotation.

We infer that the mass distribution in normal spiral galaxies
must be such that a bar-like distortion does not occur easily,
as stated above. The relative magnitude of the nuclear mass
versus the disk mass is important in determining the morphology and
the character of the spiral structure (cf. Roberts, Roberts, and
Shu, 1974).

2. Barred spirals. Now consider the opposite case. Suppose
that the system does not possess enough spherical or spheroidal
component in its mass distribution, and is consequently not stabilized with respect to distortion into a bar-like shape in the
central regions. Because of the non-axisymmetrical field, the stars
would obtain large non-circular peculiar velocities. The propagation of short waves becomes impossible for a rather wide region
around corotation; even driving by the bar and its associated
"long wave" becomes ineffective. The resultant spiral structure
would then become similar to that observed in barred galaxies.

This is the type of results obtained in the numerical experiments. There is a strong tendency for non-circular velocities

of the stars to increase when the bar-like distribution of mass is
established. Generally only quite open spirals are found. Artificial "cooling" of a "gaseous component" would however permit
relatively tightly wound spiral features to appear.

Analytical efforts in search of normal modes have so far
also produced only relatively open spirals whenever non-circular
stellar velocity or pressure is included in the system.

5. GALACTIC SHOCKS AND STAR FORMATION

We now return to the study of physical implications of the density
wave theory and begin with an examination of the behavior of the
gaseous component in the spiral gravitational field. We shall
find that this leads to our understanding of the six observational
features (6)-(11) in Table I.

The first study of this kind was made by Fujimoto (1966).
At that time, there was no guide on the choice of some of the
parameters such as the pattern speed. Because of this and certain
other limitations (see Roberts 1969, p. 140), Fujimoto was not
able to obtain a two-armed spiral shock pattern, although he did
find a strong concentration of the gas in the neighborhood of the
minimum of the gravitational potential.

With the parameters of the problems suggested by those obtained from a study of our own galaxy (Lin, Yuan, and Shu, 1969;
see also section 7), successful calculations of the gaseous behavior were made that show the two-armed spiral shock pattern.
Roberts (1969) studied the case of the isothermal shock (a hypothesis already used by Fujimoto.) Roberts and Yuan (1970) examined
the effect of the magnetic field. They found, as expected, that
the magnetic field is strengthened by the compression processes.
(cf. item (11)) in Table I . During such early work, the physical
conditions in the gas was unclear. We now know that the gas has a
multi-phase structure. A two-phase model was used by Shu et al
(1972). They found that the "hot phase" behaves very much as
before, apart from a small modification due to the more easily
compressed "cold phase". The cold phase essentially behaves in
response to the change of external pressure in the hot phase.

Shu and Roberts also found that the solutions with shock are
a continuation of the solutions without shock, thus laying the
foundation for dynamical interpretation of the luminosity classification of the galaxies. Time-dependent calculations leading
to the formation of shocks was made by Paul Woodward (1974).

We shall give only a brief description of the results here.
The reader should consult the two review papers by Shu (1973)
and by Roberts(1973).

Consider an observer in a coordinate system corotating with
the wave pattern. To this observer, the pattern is fixed, and
material objects are moving through its gravitational field. We
may visualize them as moving along imaginary stream tubes.

Calculations show that the gas is suddenly compressed near the minimum of the gravitational potential. We shall call this compression along a wide front of many kiloparsecs a <u>galactic shock</u>. At a shock front, a streamline suddenly changes its <u>direction</u>, and eventually closes on itself approximately after another similar turn at the other shock. The sudden compression of the gas collects the existing dust particles into a prominent <u>dust lane</u>, whose strength may be further enhanced by the formation of more dust particles, induced by increased gas density. This dust lane is very <u>narrow</u>, since the calculated results show that the sudden compression of the gas at the galactic shock is followed by a rather rapid decompression.

This same compression process also provides a triggering mechanism for star formation, by bringing the individual gaseous clouds into a state of continuing gravitational collapse. Once begun, this process will continue for each cloud complex and for the individual clouds of the complex, even after decompression sets in on the scale of the spiral arm. Brilliant stars are therefore formed almost simultaneously <u>over a wide front</u> of many kiloparsecs. These stars and the associated HII regions are aligned as a narrow arm, like beads on a string, for their lifetime is on the order of 10^7 years, and there is very little radial displacement over such a short period of time. <u>The observed spiral arm is therefore the brilliant manifestation of very young objects whose location and arrangement are controlled by the invisible gravitational field determined by the older stars</u>,—an establishment behind the scenes.

The formation of these bright young objects is triggered off by compression on a large scale. However, this compression does not guarantee the formation of stars, since the initial conditions of the gas also play an important role. Thus, the distribution of HII regions is often found to be <u>patchy</u>, and the spiral arms are often better delineated by the dust lanes (Lynds, 1970). In the case of less massive gaseous globules and proto-stars, the contraction time may be quite long because of internal pressure. This would account for the <u>possible existence of young M-dwarfs</u> recently discovered and studied in the solar vicinity (Murray and Sandulak, 1972; Weistrop, 1972). Since the stars have peculiar velocities of the order of 10 km/sec. the clustering of the stars would not be apparent after 10^8 years. There <u>may</u> therefore be more field stars of low masses than indicated by the study of open clusters containing massive stars. At the same time, there is no reason as yet to suspect a predominance of such young M-dwarfs in the solar vicinity.

The bright stellar arm is expected to be somewhat separated from the dust lane where the compression occurs, for there is a time of the order of 30-50 million years required for the collapse of the gaseous clouds into stars. It may be shown that this separation is generally of the order of several hundred parsecs. The exact extent of the separation depends, among other things,

upon the component of the velocity at which the gas rushes through
the dusty region in direction normal to the shock. As it turns
out, this relative velocity decreases with increasing distance
from the galactic center, and hence the separation is minimal towards the end of the spiral pattern. As we move inwards, greater
separation is expected. Such a change can indeed be seen in M51.
At the inner resonance ring, where the spiral features are tightly
wrapped together, there may again be very little separation.
However, at least in M51, up to the point where a dust lane may
be traced, the winding is not yet tight so that a substantial
separation is still noticeable.

The process of star formation described above also explains
why the abundance distribution of neutral hydrogen does not always
match that of ionized hydrogen, for the latter can be readily
produced only by density waves, and the strength of such waves
varies greatly over the galactic disk. In this picture, the tip
of the spiral arm is roughly the location where the material objects co-rotate with the wave pattern. Shu suggested (see Fig.
13 in Shu, 1973) that the ratio of HII/HI may be roughly proportional to $(\Omega-\Omega_p)$ x (compression ratio)^2. This is reasonably borne
out by data in our own galaxy.

6. EXTERNAL GALAXIES

Shu et al (1971) applied the density wave theory to the calculation
of the spiral patterns of M33, M51, and M81 and concluded that the
pattern speed is close to that of the material speed in the outer
parts of these galaxies, in agreement with the statement made above.
It should be noted that the dust lane should be outside of the
optical spiral arm whenever $\Omega_p>\Omega$, the local angular velocity for
circular motion. There is no such indication in NGC 5194(M51).

To be sure, the outer parts of NGC 5194 are presumably very
much disturbed by the passage of the companion NGC 5195, and
Toomre and Toomre (1972) suggested that the "tidally provoked
material arms near the periphery of a disk" might have caused a
transient rather than a long-lived spiral wave. However, there is
no indication that the present inner spiral structure of M51 is
induced by the last passage of the companion, for the time required
is too long for the propagation of such an influence into the central
regions. Even in the outer parts, there is the distinct possibility
that the companion induced a rather broad disturbance of the
material (cf. photographs by Van den Bergh), and that the observed
blue stars and dust lane are produced by a density wave over this
disturbed disk, in a manner similar to that in the interior regions.

About two dozen external galaxies have now been studied by
Roberts, Roberts, and Shu (1974), as mentioned at the beginning of
these lectures. Their results are compatible with the thoughts
mentioned above; i.e., the spiral patterns are quasi-stationary
patterns sustained by the feedback mechanism such as that described

in section 4, and the density wave pattern generally corotates with the material near the tip of the observed optical spiral arms. From their studies, Roberts, Roberts, and Shu were able to correlate the luminosity classification and the character of these galaxies with dynamical parameters in a manner <u>insensitive to the exact determination of the distances</u> of these galaxies.

Radio mapping of the continuum emission of M51 (Mathewson et al, 1972) provides one of the most spectacular verifications of the general concepts of the density wave theory. Quantitative theoretical studies of the strength of the synchrotron emission remain to be carried out. Concentration of neutral hydrogen in the spiral arms was already indicated in the study of M31 by M.S. Roberts (1966a), but the contrast is not as clear-cut, because of insufficient resolution. There are also two other physical reasons for reduction of the contrast: (1) The gas in the most dense clouds has largely turned molecular. (2) There is self-absorption in the dense clouds reducing the amount detectable, possibly by a factor of two, according to preliminary estimates made by W.W. Shane and the present writer during the summer of 1972 (unpublished). But new WSRT results obtained by Shane and Rots in M81 show clearly the fact that HI distribution follows optical spiral arms. (cf. lectures by Prof. J. Oort for other observational results.)

7. THE MILKY WAY

The theory discussed above can be tested against more detailed observations in the Milky Way System. Indeed, many such data clearly suggest the existence of density waves. However, it should be made clear at the very outset that the determination of the spiral structure of the Milky Way is intrinsically a very difficult task. The observer, being inside of the system, can be easily distracted by local features. Under such circumstances, it is difficult to distinguish a double two-arm spiral pattern from a single two-arm spiral pattern which is doubly as tight. What I shall present is therefore based on a self-consistent dynamical model which is in reasonable agreement with a number of observations of various kinds. Any other suggestion should also be checked against the same set of observational data.

I shall not record all my discussions of these data presented at Erice, since an earlier discussion (Lin, 1970) was published. In that paper presented at Brighton, I prepared the following table (Table II), which is to be used together with the spiral structure as first presented at Noordwijk (1966). There have been more studies of the spiral structure of the Milky Way, carried out or projected. The following is just a brief survey of the current situation. Readers wanting to know more details should consult the Brighton paper (where earlier references may be found) and the more recent publications.

7.1 Neutral Hydrogen

On the basis of the density wave theory, it is expected that,
together with associated radial motion, there exists a higher
circular speed of gaseous motion on the outer side of a concentration of hydrogen gas and a lower speed on the inner side of
such a concentration, than would have been observed in its absence. This theoretical result can be used (Lin, Yuan, and Shu,
1969; Yuan, 1969a) to correlate the oscillations in the rotation
curve with the location of the Sagittarius arm and the Norma-
Scutum arm, and thereby to derive the magnitude of the spiral
gravitational field in the solar vicinity (as shown in Table II).
This value is reconfirmed by the studies of star migration. (See
below.) Earlier, Shane and Bieger-Smith (1966) represented apparent
irregularities in the measured rotation curve by attributing them
to non-circular motions. A high stream of gas outside of the
Sagittarius arm was noted by Burton (1966).

    (1) Once non-circular motions are admitted, the analysis of
the observational data becomes very complicated, and a process of
successive approximation (or trial-and-error) has to be adopted.
From a theoretical standpoint, one would then begin by calculating
a spiral pattern and then attempt to reproduce the observational
data with theoretical profiles. Such calculations were made by
Yuan (1970) over a wide range of longitudes, and by Burton (1970,
1971, 1972) and by Shane (1970, 1971, 1972) for a limited range
of longitudes essentially covered by their own observations.
The latter authors also constructed detailed theoretical contour
maps in the longitude-velocity plane and compared these with
their observational data, (See, for example, Burton, 1972, Fig.4
versus Fig. 1).

    (2) "Velocity crowding" has been established in the above-
mentioned work as a very important effect in producing peak intensities in these velocity profiles. To interpret the classical
mystery in the Perseus arm (see G. Münch, 1965), Roberts (1972)
constructed a flow field including the shock and showed that the
two intensity peaks observed can be interpreted in terms of velocity crowding. (See also recent observations of interstellar
absorption lines in the Southern Hemisphere by J. Rickard.)

    Another most dramatic consequence of the flow field was noted
by Yuan and Liebovitch (1971) in the longitude directions around
ℓ=315°-330°. On the basis of observational data, the circular
model yields apparently broken spiral arms at a distance of about
8-10 kpc from the galactic center. The same data were shown,
however, to be compatible with a smooth continuation of the
Sagittarius arm including streaming motions.

    (3) Shane (1971) and later Simonson and Mader (1973) have
studied the 3 kpc arm in terms of Lindblad resonance, consistent
with the Noordwijk model. They showed that the observed features
can be accounted for quite naturally in this picture. (Simonson
and Mader have also attempted to describe other features in the

Table II

Certain Physical Parameters in the Milky Way

(mostly near the Sun)

1. rms radial velocity predicted for stars in the solar neighborhood	$\leqslant 37$ km sec^{-1}
2. Spiral pattern (primary component), pattern speed	13.5 km sec^{-1}kpc^{-1}
3. In the solar vicinity	
(a) arm spacing (between Perseus and Sagittarius arms)	3.5 kpc
(b) amplitude of spiral gravitational field	5% of mean field
(c) amplitude of variation of projected mass density	10% of mean
(i) in stars	5%
(ii) in gas	5%
(d) rms turbulent velocity of the gas (adopted)	7 km sec^{-1}
(e) magnetic field (adopted for dynamical consistency)	5 μG

central regions, but these are not related to the density wave theory.)

(4) Yuan and Wallace (1973) have explained the rolling motion in spiral arms as only apparent motions caused by the combined effect of the bending of the galactic plane and the differential rotation.

7.2 Ionized hydrogen and molecular lines

The continuum survey by Westerhout (1958) shows that the bulk of the galactic continuum radiation comes from a narrow range of latitudes centered about the galactic plane, and that most of the ionized hydrogen must be concentrated in a ring somewhat outside of the 4 kpc radius ($\ell=\pm26°$). From this distance outwards, there is a decline of the density of ionized hydrogen to practically nothing at 13 kpc., while the amount of neutral hydrogen increases to a peak value. As mentioned above, this feature is consistent with that discovered by M.S. Roberts (1966a, 1968) in a number of external galaxies, and can be accounted for in terms of the process of star formation induced by galactic shocks. (See data presented by Mezger and given a slightly modified interpretation by Lin, 1970).

In the gravitation theory of spiral structure, one may expect that, on a large scale, ionized hydrogen and dark clouds (which contain very little atomic hydrogen but can be observed through

the molecular lines) should have essentially the same kinematics
as the peak concentration of neutral hydrogen. Kerr (1970) surveyed the data obtained and analyzed the data presented by a
number of authors on ionized hydrogen. Minn and Greenburg (1973)
studied the lines of H_2CO which are expected to be present in
dark clouds. In both cases, the expectations are confirmed.
Carbon monoxide lines present another potential tool for the
study of spiral structure.

7.3 Stars

Roberta Humphreys (1972) observed streaming motions of supergiants
along the Carina arm in general agreement with the expectations
from the density wave theory.

Yuan (1969b), continuing the earlier work of Strömgren and
Contopoulos on the migration of stars, has shown that if there
exists a spiral gravitational field on the order of 4-7% of the
axisymmetric field in the solar vicinity, many of the stars now
found in the solar vicinity would have their birth places in the
spiral arms. This is based on a limited sample of 25 stars, but
the results are fairly conclusive. Below 4%, the field is not
strong enough to accomplish the purpose; above 7%, the peculiar
velocities of the stars would be too large at birth. Naturally,
it has been assumed that the spiral structure is of the form
discussed above. The effects of secondary spiral arms have been
omitted. Note that the value 5%, as recorded in Table II, is also
consistent with that obtained from the study of streaming motions.

Yuan noted that the calculation of the birth places becomes
unreliable when the age of the stars exceeds 3×10^8 yrs. It is
then natural to try to use these calculations to account for the
"vertex deviation" of the velocity distribution of local stars,
which is especially noticeable in the A-stars. Yuan found indeed
very satisfactory results. (See Oort, 1965, where A.D. Young's
earlier suggestion was quoted.)

7.4 Oort constants

The existence of the spiral gravitational field would naturally
raise the question: Are the Oort constants properly determined
by the conventional methods? The answer is not obvious, but Lin,
Yuan, and Roberts have shown that the modifications are indeed
small. During these studies, they also found good theoretical
basis for known observed deviations from results expected from
a smooth rotation curve, if the data were interpreted on the
basis of circular motion models. For example, it is known that
the stars in the Perseus arm tend to show equivalent circular
velocities lower than those expected from a smooth rotation
curve. This is easily accounted for in the present study. The
reader is referred to the original paper for further details.

REFERENCES

General references

1. Galactic Structure, Volume V of Stars and Stellar Systems;
    edited by A. Blaauw and M. Schmidt (University of Chicago
    Press, Chicago, 1965)
2. Galactic Structure (Ithaca, 1965), Volume IX of Lectures in
    Applied Mathematics; edited by Jürgen Ehlers (American
    Mathematical Society, Providence, R.I., 1967)
3. Brandeis University 1968 Summer Institute in Theoretical
    Physics: Astrophysics and General Relativity, Volume 2;
    edited by M. Chrétien, S. Deser, and J. Goldstein (Gordon
    and Breach); pp. 239-329.
4. Proceedings of IAU Symposium No. 38, Basel, Switzerland,
    August 29-September 4, 1969; edited by W. Becker and G.
    Contopoulos (Reidel, The Netherlands)

Specific references to individual articles

    References to articles in the above volumes will be made
with the following abbreviations:
1. Galactic Structure (1965)
2. Galactic Structure (Ithaca, 1965)
3. Brandeis Lectures (1968)
4. Basel Symposium (1969)

Burton, W.B., 1966, B.A.N. 18, 247
Burton, W.B., 1970a, Astr. Astrophys. Suppl., 2, 261
Burton, W.B., 1970b, Astr. Astrophys. Suppl., 2, 291
Burton, W.B., 1971, Astronomy and Astrophysics, 10, p. 76
Burton, W.B., 1972, Ibid. 19, 51
Burton, W.B. and W.W. Shane, 1970, Basel Symposium, p. 397
Contopoulos, G., 1970, Basel Symposium, p. 303
Feldman, Stuart, 1973, Ph.D. Dissertation, M.I.T., esp., pp. 105-107
Feldman, Stuart and C.C. Lin, 1973, Studies in Applied Mathematics,
52, p. 1
Fujimoto, M., 1966, in IAU Symposium No. 29, p. 453
Hohl, F., 1970, NASA TR R-343
Hohl, F., 1973, Ap. J. 184, 353 and earlier references
Hohl F. and R.W. Hockney, 1969, J. Comput. Phys. 4, 306
Humphreys, Roberta M., 1972, Astron. Astrophys. 20, 29
Kalnajs, A.J., 1965, Ph.D. Thesis, Harvard University
Kalnajs, A.J., 1970, Basel Symposium, p. 318
Kato, S., 1973, Publ. Astron. Soc. Japan 25, 231
Kerr, F.J., 1969, Annual Reviews Astron. Astrophys. 7, 39
Kerr, F.J., 1970, Basel Symposium, p. 95
Lau, Y.Y. and C.C. Lin, 1974, Preprint, to be published
Lin, C.C., 1965, Galactic Structure (Ithaca), p. 66
Lin, C.C., 1966, SIAM J. Appl. Math. 14, 876

Lin, C.C., 1967, Ann. Rev. Astron. and Astrophys. 5, 453
Lin, C.C., 1968, in Galaxies and the Universe (Columbia), p. 33
Lin, C.C., 1969, Basel Symposium, p. 377
Lin, C.C., 1970, in Highlights of Astronomy (IAU, Reidel 1971)
pp. 88-121
Lin, C.C., 1973, Haifa Symposium, in press.
Lin, C.C. and F.H. Shu, 1964, Ap. J. 140, 646
Lin, C.C. and F.H. Shu, 1966, Proc. Nat. Acad. Sci. 55, 229
Lin, C.C. and F.H. Shu, 1967, IAU Symposium No. 31 (Noordwijk),
p. 313
Lin, C.C. and F.H. Shu, 1968, Brandeis Lectures
Lin, C.C., C. Yuan and F.H. Shu, 1969, Ap. J. 155, 721
Lin, C.C., C. Yuan and W.W. Roberts, 1974, Highlights of Astronomy,
in press.
Lindblad, B., 1963, Stockholm Obs. Ann. 22, No. 5
Lindblad, P.O., 1960, Stockholm Obs. Ann. 21, No. 4
Lindblad, P.O., 1962, in Interstellar Matter in Galaxies (Benjamin,
New York), p. 222
Lynden-Bell, D. and A.J. Kalnajs, 1972, MNRAS 157, p. 1
Lynds, B.T., 1970, Basel Symposium, p. 26
Mark, James W-K., 1971, Proc. Nat. Acad. Sci. USA, 68, 2095
Mark, James W-K., 1974, Preprints I and II
Mathewson, D.S., P.C. van der Kruit and W.N. Brouw, 1972, Astron.
Astrophys. 17, 468
Mezger, P.G., 1970, Basel Symposium, p. 106
Miller, R.H. and K.H. Prendergast, 1968, Ap. J. 151, 699
Miller, R.H., K.H. Prendergast and W.J. Quirk, 1970, Basel
Symposium, p. 365
Minn, Y.K. and J.M. Greenberg, 1973, Astron. and Astrophys. 22,
13; 24, 393
Morgan, W.W., 1970, IAU Symposium No. 38, p. 9
Münch, G., 1965, Galactic Structure, p. 203
Murray, C.A. and N. Sandulak, 1972, MNRAS 157, 273
Roberts, M.S., 1966a, Ap. J. 144, 639
Roberts, M.S., 1966b, IAU Symposium No. 31, p. 189
Roberts, W.W., 1969, Ap. J. 158, 123
Roberts, W.W., 1974, Highlights of Astronomy, in press.
Roberts, W.W. and C. Yuan, 1970, Ap. J. 161, 877
Roberts, W.W., M.S. Roberts and F.H. Shu, 1974, IAU Symposium
No. 58; also preprint.
Shane, W.W., 1971, Astron. Astrophys. Suppl. 4, p. 1, 315
Shane, W.W., 1972, Ibid. 16, p. 118
Shane, W.W. and G.P. Bieger-Smith, 1966, B.A.N. 18, 263
Shu, F.H., 1968, Ph.D. Thesis, Harvard University. See also
Brandeis Lectures, p. 314.
Shu, F.H., 1969, Ap. J. 158, 505
Shu, F.H., 1970a, Ap. J. 160, 89
Shu, F.H., 1970b, Ap. J. 160, 99
Shu, F.H., 1970c, Basel Symposium, p. 323
Shu, F.H., 1973, American Scientist 61, 524

Shu, F.H., V. Milione, W. Gebel, C. Yuan, D.W. Goldsmith and
W.W. Roberts, 1972, Ap. J. 173, 557
Shu, F.H., V. Milione and W.W. Roberts, 1973, Ap. J. 183, 819
Shu, Frank H., Robert V. Stachnik and Jonathan C. Yost, 1971,
Ap. J. 166, 465.
Simonson, S.C. and G.L. Mader, 1973, Astron. Astrophys. 27, p. 337
Strömgren, Bengt, 1967, IAU Symposium No. 31, p. 323
Toomre, A., 1969, Ap. J. 158, 899
Toomre, A., 1970, Basel Symposium, p. 334
Toomre, A. and J. Toomre, 1972, Ap. J. 178, 623
Vandervoort, P.O., 1970, Ap. J. 161, p. 87
Weistrop, Donna, 1972, A.J. 77, p. 849
Westerhout, G., 1958, See Galactic Structure, 1965, p. 196
Wielen, Roland, 1974, Publications of the A.S.P., in press.
Woodward, Paul R., 1974, Center for Astrophysics Preprint No. 100
Yuan, C., 1969a, Ap. J. 158, 871
Yuan, C., 1969b, Ap. J. 158, 889
Yuan, C., 1970, Basel Symposium, p. 391
Yuan, C. and L.S. Liebovitch, 1971, B.A.A.S. 4, 168
Yuan, C. and L. Wallace, 1973, Ap. J. 185, p. 453

INTERSTELLAR MATTER

P. G. Mezger

Max-Planck-Institut für Radioastronomie,
Bonn, FRG

The best available textbooks on interstellar matter are
those by Spitzer (1968) and Kaplan and Pikelner (1970). Recent
reviews on observations pertaining to the interstellar medium
are given in two proceedings of Advanced Study Institutes by
Wickramasinghe, Kahn and Mezger (1972) and by Pinkau (editor,
1974). In Section 1 of this paper, I give a short summary of
(primarily observational) results pertaining to interstellar
matter and radiation fields. For a more detailed discussion, I
refer to my review in Wickramasinghe et al. (1972) and to the
reviews by Lequeux and Pottasch in Pinkau (editor, 1974). In
Section 2, I deal in more detail with some selected topics in
which important progress has been made during the past years.
Section 3 deals with HII regions and with the rate and efficiency
of star formation in our Galaxy.

1. INTERSTELLAR MATTER: AN OVERVIEW

Most of our knowledge about the physical state of interstellar matter is obtained from observations in our own Galaxy. There
is some controversy concerning the Hubble classification of the
Galaxy. While Schmidt-Kaler and Schlosser (1973) and van den
Bergh (1968) classify it as Sb, Arp (1965) and Georgelin (1971)
classify it as Sc. The total mass of our Galaxy is $2 \cdot 10^{11} \ M_\odot$,
and interstellar matter accounts for approximately 10 %. The degree of activity in the nuclear region of the Galaxy may be
classified between quiet and mild.

G. Setti (ed.), Structure and Evolution of Galaxies, 143–177. All Rights Reserved.
Copyright © 1975 by D. Reidel Publishing Company, Dordrecht-Holland.

1.1 Element Abundances

H and ^{4}He are the main constituents of the interstellar matter; these two elements together account for more than 98 % of the total mass.

Table 1.1

Element Abundances in the Solar Vicinity

Element	N(Element)/N(H)	Mass Fraction
H ^{4}He N, O, C, Ne Mg, Si, S, Fe A > 4	1.00 0.10 $\sim 10^{-4}$ $\sim 10^{-5}$	X = 0.701 Y = 0.278 0.016 0.0036 Z = 0.020

The mass fractions given in Table 1 are precise; for precise number abundances for elements with A > 4 (i.e. elements heavier than ^{4}He) see, e.g., Cameron (1973).

In discussing the origin of elements, three facts have to be borne in mind:

i) Of all cosmological models, the "big bang" model fits most observations best. In this model, the universe starts from a state of high density and high temperature (T > 10^{10} K) but cools rapidly by expansion.

ii) Stars and interstellar matter (IM) in a galaxy are not separate entities. Young stars form out of the IM and old stars in their post main sequence (MS) evolution return "processed" gas to the IM.

iii) The interstellar space is permeated by cosmic rays, which interact with the IM.

In the following few paragraphs, I summarize some generally accepted theories concerning the origin of elements. Light elements (H through ^{4}He) and possibly ^{7}Li have been formed primarily through neutron capture about 300 s after the big bang. Subsequent processing of the primordial gas through stars has slightly increased the abundance of ^{4}He but has decreased the abundance

of ^{2}H. For a comprehensive review of "big bang" element production, I refer to the book by Peebles (1971).

^{6}Li, part of ^{7}Li, ^{9}Be, ^{10}Be and ^{11}Be are formed primarily
through spallation, i. e. through collisions of C, N, O nuclei
in the cosmic rays with H and ^{4}He atoms in the IM. For a review
of these spallation processes, I refer to the lectures by Reeves
published in Pinkau (1974) p. 267.

C, N, O and heavier elements are formed by nucleosynthesis
in stars. The abundances of these elements are therefore directly related to the fraction of IM processed through stellar in-
teriors (see Sect. 1.3).

It is not quite clear to which extent solar system abundances (e.g. as compiled by Cameron, 1973) represent galactic or
cosmic abundances. Young galaxies may be underabundant in heavy
elements. A number of observations indicate an increase in the
abundance of N and possibly of O with decreasing distance to the
center of external galaxies.

1.2 Dust

The interstellar gas is mixed with dust particles which
cause extinction of transmitted light by scattering and absorption. Figure I.1 shows a typical interstellar extinction curve.

It is customery to relate the color excess $E = A_B$ (λ =
4400) $- A_v$ (λ = 5500) with the visual extinction A_v by: $A_v = RE$.
In the general IM, R is generally found to be near 3; however,
in some regions R may attain higher values.

At a given wavelength, the extinction cross section σ_e is
the sum of absorption and scattering cross sections

$$\sigma_e = \sigma_a + \sigma_s$$

The albedo is defined as $\Gamma = \sigma_s/\sigma_e$; consequently $\sigma_a = \sigma_e(1 - \Gamma)$.
While the extinction curve can be explained by a mixture of di-
electric grains (radius a = 0.15μ), graphite grains (a = 0.05μ)
and iron grains (a = 0.02μ), the very high albedo of $\Gamma \simeq 0.9$ at
λ = 1500 A cannot be explained by this or any other grain mix-
tures suggested to date.

The relation between extinction in the visible and column
density of hydrogen is, for R = 3.0 (Jenkins and Savage 1974),

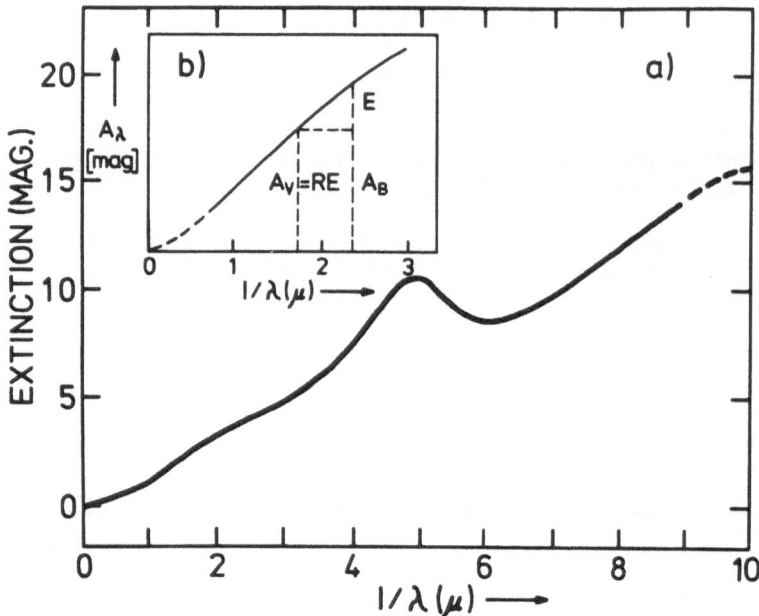

Figure I.1   a) Typical interstellar extinction curve

          b) Schematic representation of the optical
             section of the extinction curve

$$N_H/A_v = 2.5 \ 10^{21} \ \text{H-atoms cm}^{-2} \ (\text{vis. mag})^{-1}$$

The general relation between magnitude A and optical depth, τ, is

$$\tau = \frac{1}{1.086} \left[\frac{A}{\text{mag}}\right]$$

1.3 Star Formation and Mass Exchange

About $1.1 \ 10^{10}$ years ago, the Galaxy condensed out of the expanding primordial gas. It appears that the first stars to form were those in globular clusters, which form a spherical sub-system (see Fig. 1.2). The gas then collapsed and formed a flat disk. The metal content of stars in globular clusters differs from that of stars in the galactic disk by as much as a factor of 100, while their He-abundance appears to be roughly the same.

Stellar evolution starts at gas densities of $\sim 10^6$ H-atoms cm^{-3}. As long as the star is primarily a contracting neutral cloud, we call it a protostar. Its pre-MS contraction phase ends when nuclear reactions start in the core. The star spends most of its lifetime on the MS, burning H into He in its core.

At present, stars are observed in the mass range 0.08 to about 50 $M_\odot$. Less massive stars never attain a high enough central temperature for nuclear reactions to start. The upper end of the mass spectrum appears to be determined by the initial conditions of star formation such as dust content, fragmentation and subsequent initiation of free-fall contraction etc., rather than by pulsational instabilities of the MS star. For stars $M \gtrsim 0.5 \ M_\odot$, the mass spectrum (i.e. number of stars with mass between M and M + dM) can be approximated by Salpeter's law (Salpeter, 1955)

$$\xi(M) = \text{Const. } M^{-1.35}$$

For stars $M < 0.5 \ M_\odot$, the observed numbers appear to decrease again. Salpeter's law was originally determined for stars in the solar vicinity. It is often used as a law governing star forma-tion at any time and in any galaxy, although there is, to my knowledge, no justification for this assumption. In the following, we refer to the corresponding luminosity function as Salpeter's original luminosity function (SLF).

The MS lifetime of a star depends strongly on its mass; the MS lifetime of an O-star (50 $M_\odot$) is only some 10^6 years, while

the MS lifetime of a star slightly less massive than the sun is
$\sim 10^{10}$ years. Conventional theories of stellar evolution yield
a pre-MS evolutionary time of about 10^{-2} times the MS lifetime
of the star.

The post MS evolution of stars, especially of massive stars,
is not yet fully understood. We know that, to be stable, dead
stars, like white dwarfs or neutron stars, cannot exceed masses
of slightly more than one solar mass. Therefore, the assumption
is usually made that, in the course of post MS evolution, a
massive star returns $(M_* - 1M_\odot)$ of its original mass M_* to the
interstellar gas. Mass loss phenomena are observed in super
novae (SN), novae, planetary nebulae, blue super giants and red
giants. In our Galaxy, a mass return of about 1 $M_\odot$/year is
estimated. It is through this mass return that the IM is enriched
in heavy elements. Detailed theoretical investigations of this
process have been made by Talbot and Arnett (1973; 1974).

The most massive stars have the shortest pre MS evolution
time. Once O-stars reach the MS, their effective temperatures
attain values in the range from $(3.7 - 5.2) \; 10^4$ K. Therefore
these stars emit about 50 % of their total luminosity in the
Lyman continuum (Lyc) range. Lyc-photons can ionize hydrogen, and
young O-stars are usually surrounded by an HII region, i. e. a
volume of interstellar space in which H (and most other elements)
are nearly fully ionized. HII regions can be easily detected in
the optical, IR and radio range. They are the best indicators
of star formation. Based on the number of O-stars, one estimates
that about 1 $M_\odot$/year of IM is transformed into stars (see Sect.
3.4). This suggests that mass loss from stars and formation of
stars out of the IM in the Galaxy are in equilibrium. However,
the exact coincidence of these two figures appears to be
fortuitous.

Fig. 1.2 shows a schematic cross section through the Galaxy.
Stars and gas form a very flat disk, while the oldest objects,
globular clusters, still outline the spherical cloud out of
which our Galaxy condensed. The spherical concentration of stars
around the center of the Galaxy is known as the nuclear bulge.

The upper left part of Fig. 1.2 shows, quantitatively, the
distribution of stars, neutral gas and HII regions. While the
stellar density decreases rapidly with increasing distance from
the center, the neutral atomic hydrogen attains its maximum surface density between 5 and 15 kpc. HII regions, which are indicators of star formation, attain a maximum between 4 and 8 kpc.
Pertaining to star formation, we draw the following conclusions:

i) The gas-to-star ratio in the Galaxy varies from about
10^{-3} in the center region to as much as 1 in the solar vicinity.

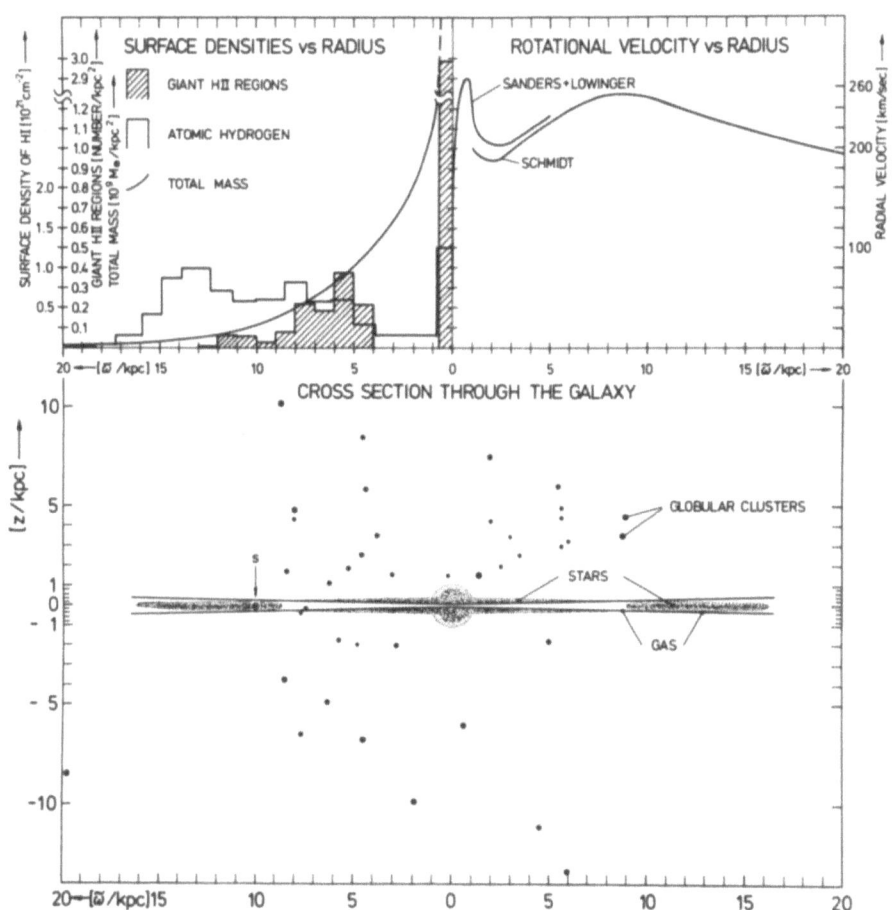

Figure I.2: Distribution of stars and interstellar matter in
the galaxy. The lower part of the diagram is a
cross section perpendicular to the galactic plane.
The upper left part of the diagram shows the column
density of stars, hydrogen gas and HII regions as
a function of the distance $\tilde{\omega}$ from the galactic
center. The rotational (orbital) velocity is given
as a function of $\tilde{\omega}$ in the right part of the upper
diagram.

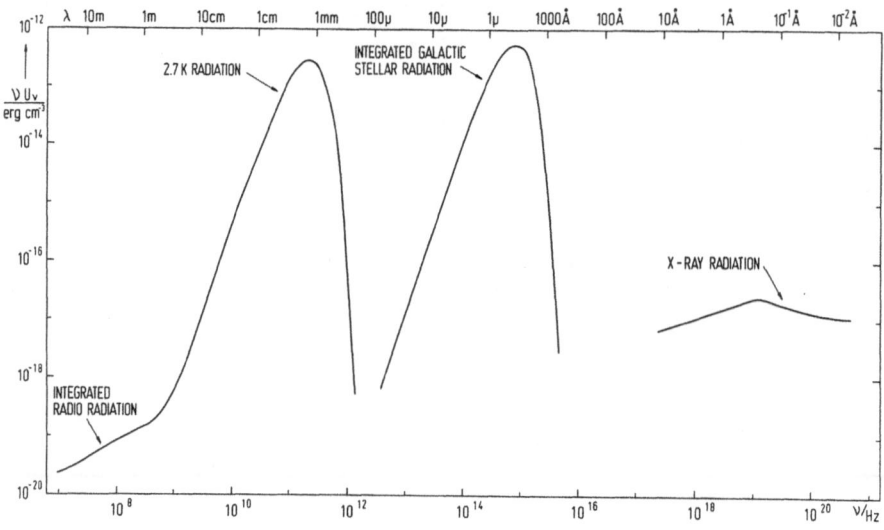

Figure I.3   Radiation density per frequency interval νu_ν, as
             estimated or measured for the vicinity of the sun.
             Most of the radio radiation is of galactic origin.
             The 2.7 K background radiation and probably part of
             the X-ray radiation is of cosmological or extragalac-
             tic origin, while the integrated starlight strictly
             pertains to the vicinity of the sun.

   ii) A high gas density is a necessary, but not a sufficient
condition for star formation.

   iii) If there is an equilibrium in our Galaxy between star
formation and mass return, the observed distribution of gas can
only be maintained by some radial gas flow.

   Some new observational results pertaining to the formation
of massive stars are discussed in Sect. 3.4.

1.4 Radiation Density and Equilibrium Temperatures

   Densities of the isotropic radiation in the solar vicinity
are shown in Fig. I.3. I have plotted energy density u_ν
$[erg/cm^3\ Hz]$ times frequency $\nu[Hz]$ vs log ν. Since $\int u_\nu d\nu =$
$\int \nu u_\nu d\ln\nu$, νu_ν represents radiation density per logarithmic frequency interval. At low frequencies, the diffuse non-thermal
galactic radiation dominates. In the mm-wavelength range, the
3 K background radiation dominates. Its radiation density is
roughly the same as that of the integrated stellar light, which
is approximated by a Planck curve for $T = 10^4$ K, multiplied by
a dilution factor of $W = 10^{-14}$. At wavelengths below 100 A, one
observes a diffuse X-ray emission, which, at least in part, appears to be of galactic origin. Between 912 A, the Lyc limit,
and about 100 A, the absorption cross sections of H and He are
so high that, until recently, it was thought that all photons
would be absorbed in the immediate vicinity of their sources
of emission.

   Cosmic radiation consists of highly accelerated nuclei and
electrons. The nuclei maintain a certain amount of ionization
of the interstellar gas; the cosmic electrons are responsible
for the diffuse galactic synchrotron radiation.

   Temperatures of the various constituents of the IM are determined by the fact that, under equilibrium conditions, Gain
= Loss. In the case of interstellar dust grains, the gain is due
to absorption of stellar UV photons, the loss is due to quasi-
blackbody radiation in the far IR. For the stellar radiation
density shown in Fig. I.3, the equilibrium temperature of dielectric grains falls in the range between 10 and 20 K. Close
to O-stars, however, grain temperatures can become one to two
orders of magnitude higher.

   The neutral gas gains   energy through ionization by cosmic
rays, X-rays and Lyc-photons. For X-rays and cosmic rays, this
gain is proportional to the gas density and independent of the
gas temperature. Loss is provided by collisional excitation of,

primarily, Lyman alpha, OI $\lambda 63\mu$ and CII $\lambda 156\mu$ lines; this loss is approximately proportional to the square of the gas density. Therefore, cooling is more effective for higher gas density, and the kinetic temperature of the neutral interstellar gas is expected to decrease from 10^4 K for $n_H < 2 \; 10^{-1}$ cm^{-3}, to values $< 10^2$ K for $n_H > 1$. This explains how cool dense clouds can be in pressure equilibrium with hot, tenuous intercloud gas (two-component model).

In an HII region, the photo-electrons carry away the excess energy of the ionizing photon $<E_{Lyc}> - h\nu_1$. Here, $<E_{Lyc}>$ is the average energy of the ionizing stellar Lyc-photons, and $h\nu_1$ is the ionization energy. Electrons lose energy primarily through collisional excitation of forbidden lines of ions of C, N, O and Ne. Theory and observations now agree that the electron temperature of HII regions is very close to 10^4 K with only small variations (Seaton, 1974).

For the interpretation of radio spectral lines, the excitation temperature T_{ex} is an important quantity. Consider a two-level atom (or molecule) with n_u and n_ℓ denoting the number of atoms with an electron in the upper and lower level, respectively and g_u and g_ℓ the statistical weights of these levels; the excitation temperature for these two levels is defined by

$$(n_u/n_\ell) = (g_u/g_\ell) \; \exp(-h\nu/kT_{ex})$$

Statistical equilibrium requires that the number of transitions $u \rightarrow \ell$ equals the number of inverse transitions.

$$n_u\{A_{u\ell} + B_{u\ell} \; u_\nu + C_{u\ell}\} = n_\ell\{B_{\ell u} \; u_\nu + C_{\ell u}\}$$

Note that, in thermodynamic equilibrium (TE), the same equation applies to each pair of levels in a multilevel atom.

Here, The A's and B's are the Einstein coefficients for spontaneous and induced emission and for absorption; the C's are the coefficients describing collisionally induced transitions:

$$C_{u\ell} = n \; <\sigma_{u\ell}v>$$

i.e. the C-coefficients depend on the density n of the colliding particles (mainly H-atoms or H_2-molecules) and on their kinetic temperature T_K, since $<\;>$ means a weighting of the collision cross sections over a Maxwellian velocity distribution of the colliding particles. u_ν is the radiation density, for which we substitute the value $(4\pi/c)B_\nu(T_r)$ corresponding to Planck radiation; in the general IM, $T_r = 2.7$ K (Fig. I.3). It can readily be seen that

$$T_{ex} = \begin{cases} T_r = 2.7 \text{ K for } C_{u\ell} \ll A_{u\ell} \\ T_K \text{ for } C_{u\ell} \gg A_{u\ell} \text{ or } n \gg A_{u\ell}/<\sigma_{u\ell}v> \end{cases}$$

Thus, so long as TE is a reasonable approximation, one expects the excitation temperature to lie somewhere between the kinetic gas temperature T_K and the radiation temperature $T_r = 2.7$ K. However, in the radio range, deviations from TE populations are the rule. Observed excitation temperatures of molecules range from 10^{12} K for H_2O masers to 1.7 K for H_2CO molecules. In general, however, deviations from TE populations are mild. Line enhancements of order of a factor two typically result.

2. INTERSTELLAR MATTER: SELECTED TOPICS

2.1 The Abundances of Light Elements and their Cosmological Implications

As outlined in Section 1.1, the light elements, H, ^{2}H, ^{3}He, and ^{4}He are predominantly formed between 1 and 600 s after the big bang. Their abundance depends on the barionic density at that time, which can be expressed in terms of the present barionic density of the universe (see Lectures by H. Reeves, this volume). Figure II.1 shows abundance computations by Wagoner (1973).

An observational determination of the primordial abundance[*] of light elements thus yields immediately the mean barionic density of the universe. If this density is larger than a critical density ρ_c, the universe is closed; if it is equal to or smaller than ρ_c, the universe is open and expands forever. This critical density, for a value of the Hubble constant $H_o = 55$ km s^{-1} Mpc^{-1},

---

[*] Note, the relation between mass fraction and number abundance:

mass fraction of element $E = N(E)m_E / \sum_i N_i m_i$

$$\simeq (1 - Z) \frac{N(E)m_E}{N(H)m_H + N(^4He)m_{He}} = 0.70 \, N(E) \frac{m_E}{m_H}$$

Here, $N(E)$ is the number abundance relative to H, m_E and m_H are the atomic weights of elements E and H and the numerical value holds for $Z = 0.020$ and $N(^4He) = 0.10$.

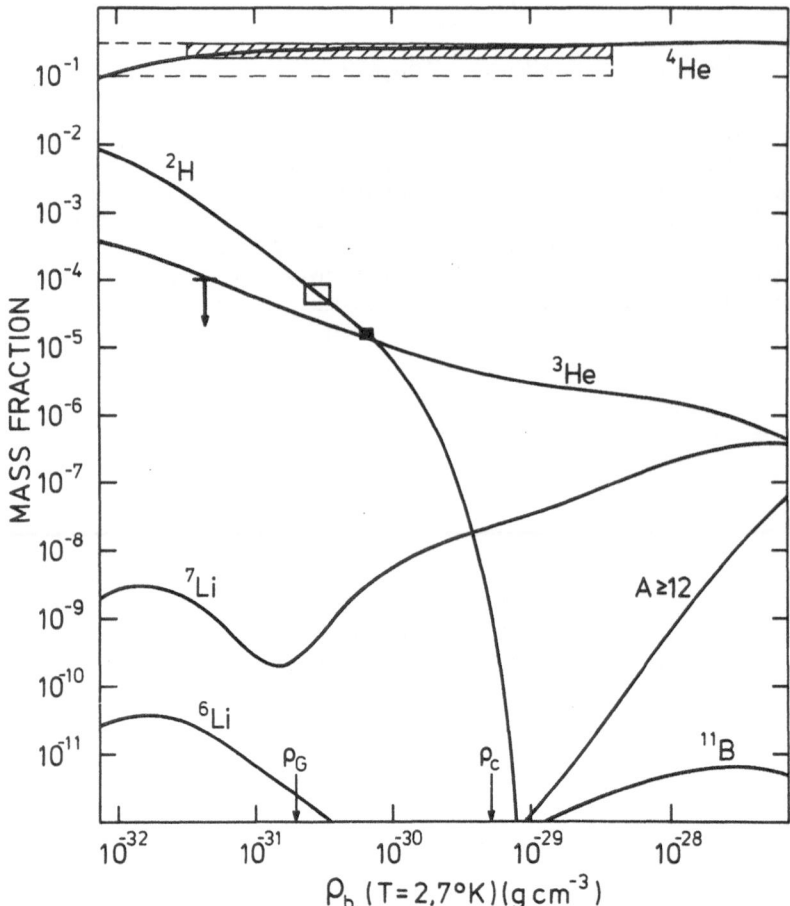

Figure II.1 Abundances of light elements (in mass fractions)
             as a function of the present barionic density of
             the universe (Wagoner, 1973).

is shown in Fig. II.1; also shown is the smoothed-out density ≃
0.035 ρ_c due to "visible" mass contained in galaxies.

    Recently, we have determined the abundance of ^{4}He in fourty
galactic HII regions (Churchwell et al. 1974) and derived, for
the Galaxy, $N(^4He) = y_{gal} = 0.10$ by number or $Y = 0.28$ by mass.
If this abundance were primordial, the barionic density would be
7.1 ρ_c, a value, which is incompatible with independent estimates
of the age of the universe. We therefore concluded that the present galactic He-abundance is the sum of primordial He plus He-
enrichment by galactic evolution. We separated the two terms by

comparison with abundance determinations of ^{2}H from Lyman lines
by Rogerson and York (1973). Their observed abundance value together with its error limits are shown as a filled square on the
^{2}H curve in Fig. II.1. To obtain the primordial ^{2}H-abundance this
value has to be corrected for depletion by processing of the
primordial gas through stars. Allowing for the uncertainties in
this correction factor, we obtain the open square for the
primordial ^{2}H-abundance. The corresponding primordial ^{4}He-abun-
dance is $y_{prim} = 0.08$ by number or $Y_{prim} = 0.24$ by mass. The
corresponding barionic density $\rho_b = (3.14 \pm 0.73) \ 10^{-31}$ g cm^{-3}
is only about twice the smoothed-out mass contained in galaxies.
This implies an open universe with an age of $t_o \simeq H^{-1} \simeq 1.8 \ 10^{10}$
years; it further implies that the amount of "unseen mass" in
galaxies or intergalactic matter cannot greatly exceed the
currently estimated mass in observed galaxies.

2.2 The Nature of Dust Grains

    Until about ten years ago, only the optical part of the
interstellar extinction curve (Fig, I.1) was known. This part of
the curve could be reproduced by dielectric grains of radii
$a \simeq 0.15\mu$. Average extinction in the visual is found to be A_V
$\simeq 1$ mag kpc^{-1}. Using the Kramers-Kronig relation, Purcell (1969)
could show that, for dielectric grains of density 1 g cm^{-3}, an
extinction of $A_V = 1$ mag requires a minimum of $7.5 \ 10^{-6}$ g cm^{-2}
in the form of dust particles along the line of sight. On the
other hand, $A_V = 1$ mag is observed to be associated with a
column density of $2.5 \ 10^{21}$ H-atoms cm^{-2} (Jenkins and Savage,
1974); multiplication of this value by 1.43 m_H, the mean atomic
mass of the IM, yields a dust-to-gas mass ratio $> 10^{-3}$. Estimates
based on more specific models (Greenberg, 1974) yield a dust-to-
gas ratio of $\sim Z/5 = 4 \ 10^{-3}$ (see below).

    Measurements from balloons and satellites have in the mean-
time extended the excitation curve nearly to the Lyc-limit (Fig.
I.1), The behaviour of the observed extinction curve can be explained by a mixture of grains of different sizes and chemical
compositions. However, these composite grain models do not yet
satisfactorily explain the high albedo of dust between 2000 and
1500 A observed by Witt and Lillie (1973).

    UV-observations from the Copernicus satellite of interstel-
lar absorption lines show a general under-abundance of C, N, O
and of some heavier elements by about a factor of ten compared to
solar system abundances (Morton et al., 1973). This agrees qualitatively with the idea that the heavier elements may be largely
tied up in particles. Quantitative comparisons were made by Field
(1974) and by Greenberg (1974). Field finds that the

extinction and line spectrum of ζ Oph can be understood if dust
grains containing silicate and graphite cores and mantles of H,
C, N and O compounds use up substantial amounts of heavy elements. The cores may condense in stellar atmospheres under equilibrium conditions, while the mantles are added in interstellar
space.

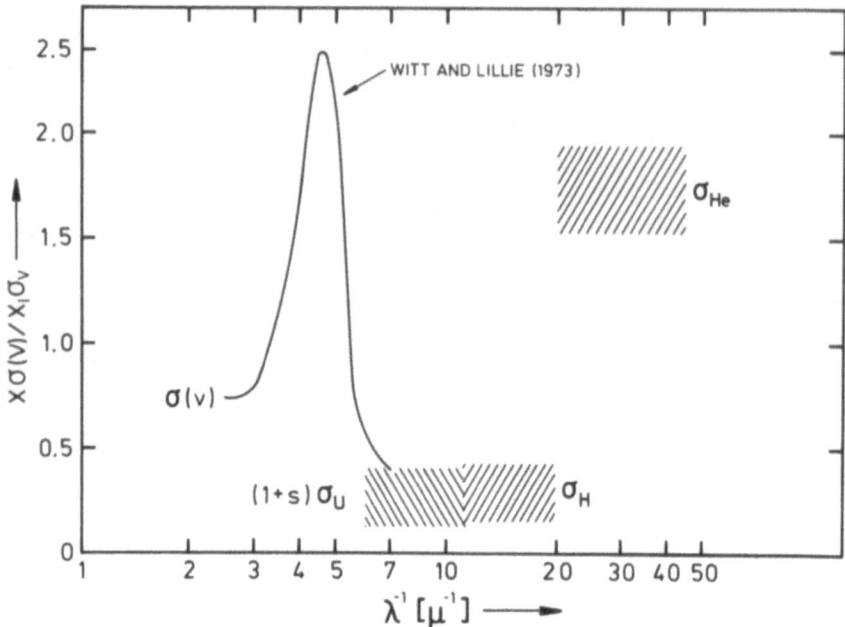

Figure II.2   The absorption cross section of interstellar dust
              normalized to the cross section for visual ex-
              tinction. Hatched areas represent the range of
              possible values for the effective absorption cross
              section in the wavelength regions $\lambda > 912$ Å;
              $912 > \lambda > 504$ Å; $504 > \lambda > 228$ Å.

Greenberg, on the other hand, comes to the conclusion that
underabundances of Fe, Mg and Si provide restrictive bounds on
possible grain models; however, the underabundances of the common
intermediate weight elements O, C, N are too great to be explained by his grain model. He suggests the existence in interstellar space of "snowballs", intermediate in size between dust
grains and comets, or alternatively, the existence of very complex molecules not yet detected by their radio or UV lines as a
sink for the missing atoms of C, N, O.

As mentioned above, the observed high albedo (or low absorption cross sections) of dust grains in the UV may be another
clue to the nature of interstellar dust grains. Direct observations by Witt and Lillie (1973) stop at 1500 Å. Mezger et al.
(1974) used an observed correlation between the He^+-abundance
and the IR-luminosity of HII regions (see Sect. 3) to determine
the absorption cross sections of the dust shortwards of the
Lyman limit. Their results are shown in Fig. II.2.

The low absorption cross sections of interstellar dust below 2000 Å derived by Witt and Lillie appear to continue right
into the Lyc region to about 500 Å; below this wavelength, the
absorption cross section increases rapidly again.

Some of the consequences of this behaviour of the absorption cross section of dust will be discussed in Sect. 3. Pertaining to the nature of dust grains, Greenberg argues that dielectric grains made of mantles of H, C, N, O compounds (a_m =
0.12 ÷ 0.15µ) on a silicate core (a_c = 0.06 ÷ 0.08µ) are required to explain the visual extinction and polarization, while
"bare" grains of either silicate or graphite (a_b = 0.005µ) can
explain the UV extinction. The ratio of the numbers of "bare"
and core-mantle grains has to be of order 10^3. Andriesse (in
Pinkau, 1974, page 187), on the other hand, feels that the optical behaviour of dust grains cannot be understood on the basis
of laboratory measurements on macroscopic species but that surface physics may play the dominant role. In another paper, Andriesse and de Vries (1974) suggest that radicals of mean size
10 Å can explain both the high extinction and high albedo observed in the far UV.

It should be pointed out that Churchwell et al. (1974) find
that either the dust-to-gas ratio must increase or the chemical
composition of dust grains must change with decreasing distance
from the galactic center. This supports the existence of an
abundance gradient of heavy elements in our Galaxy as it has
been observed in external galaxies (Searle, 1971).

2.3 Clouds and Intercloud Gas

a) Interpretation of Line Observations. For the derivation
of the following equations, I refer to a review paper by Winnewisser et al. (1974) on "Interstellar Molecules". Practically
all information about the interstellar gas has been obtained by
means of spectroscopy, i. e. observations mainly in the radio or
in the UV range. In the interpretation of line observations, the
important quantities are the optical depth τ and the excitation
temperature T_{ex}. The physical meaning of the latter quantity has

already been discussed in Sect. 1.4. It is often overlooked, how-
ever, that for gas densities $n_H \simeq 0.02$ cm^{-3} and $T_K \simeq 10^4$ K, which
are expected for the intercloud gas, T_{ex} is considerably lower
than the kinetic gas temperature ($T_{ex} \sim T_K/2$). The optical depth,
integrated over the line profile, is

$$\int_0^\infty \tau_\nu d\nu = \frac{c^2 h}{8\pi k\nu} \frac{g_n}{g_\ell} A_{u\ell} N_\ell \{1 - \frac{g_\ell}{g_u} \frac{n_u}{n_\ell}\}$$

The atomic constants have their usual meaning (see also Sect.
1.4).

$$\{1 - \frac{g_\ell}{g_u} \frac{n_u}{n_\ell}\} = \{1 - \exp(-h\nu/k\, T_{ex})\} = \begin{cases} h\nu/kT_{ex} & ; \; kT_{ex} \gg h\nu \\ 1 & ; \; kT_{ex} \ll h\nu \end{cases}$$

is the correction for stimulated emission (or negative absorption). This correction plays a dominant role at radio frequencies; very small deviations from TE of the differential population of energy levels can lead to a negative optical depth and
hence to a "masering" of the radio lines.

$$N_\ell = \int_0^L n_\ell \, ds$$

is the column density of atoms or molecules with an electron in
state ℓ. Usually one is interested in the total column density of
molecules N_o. In the case of a Boltzmann distribution

$$N_\ell = N_o \, g_\ell \, \exp\{-E_\ell/kT_{ex}\}/Q$$

with

$$Q = \sum_i g_i \, \exp\{-E_i/kT_{ex}\}$$

the partition function. For UV lines, all electrons are usually
in the ground state ($Q = g_\ell$); for hyperfine structure lines in
the radio range, the electrons are approximately equally distributed between the upper and lower levels ($Q = g_\ell + g_u$). However, for rotational transitions, one has to sum over all rotational states and obtains $Q_r \simeq kT_{ex}/hB$, with B the rotational
constant.

Absorption lines yield τ, emission lines usually yield the
product of excitation temperature and optical depth, since, for
$\tau \ll 1$, $T_{ex}(1 - e^{-\tau}) \simeq T_{ex}\, \tau$.

The relationships of these observable quantities to N_o and T_{ex} are given in Table II.1.

Table II.1

Relation between observable quantities, excitation temperature
and total column density

Type of transition	$\tau \propto$	$T_{ex}\tau \propto$
Hyperfine structure lines (radio, cm range)	$N_o T_{ex}^{-1}$	N_o
Rotational lines (radio, mm range)	$N_o T_{ex}^{-2}$	$N_o T_{ex}^{-1}$
Electronic transitions (optical, UV range)	N_o	--

It may be seen from Table II.1 that only hyperfine structure lines (such as the H λ21cm line) observed in emission or UV lines (such as the Lyman alpha line) observed in absorption yield directly the total column density. In all other cases, the un- certainty of the excitation temperature introduces a large un- certainty in derived total column densities. However, if $\tau \gg 1$ the observed line brightness temperature is identical with the excitation temperature. This yields, in special cases, an inde- pendent determination of T_{ex}. Detailed models for the excitation of the CO and CS molecules have been worked out by Turner et al. (1973), Scoville and Solomon (1974) and Goldreich and Kwan (1974) and for the excitation of NH_3 by Morris et al. (1973).

b) Physical Characteristics of Clouds and Intercloud Gas.
The average gas density in the Galaxy varies from 0.3 to 0.7 H- atoms cm^{-3}; densities as high as 10^7 H_2 cm^{-3} have been observed in some massive clouds. Observed kinetic temperatures of the neutral gas range from 5 K to 6 10^3 K. There is nothing like a "standard cloud" in the interstellar gas.

In the interarm region, the gas observed in the H λ21cm line shows a smooth distribution. It appears that the interarm region consists predominantly of a hot and tenuous gas, in which small "cloudlets" of higher density and considerably lower temperatures

are embedded. This "two-component" structure may be explained by
a thermal instability of the interstellar gas, which is a consequence of the quadratic density-dependence of radiation
cooling, i. e. a decrease of the kinetic temperature with increasing gas density (see also Sect. 1.4).

Clouds (in contrast to the above-mentioned cloudlets) of
interstellar gas appear to be related to the large scale structure of the IM, such as density wave and material spiral arms.
In the vicinity of the sun and probably connected with the Orion
arm (which is supposed to be a material arm) and with Goulds Belt,
one optically observes large dust clouds, which contain cool
($T_k \gtrsim 5$ K) and dense ($n_H \sim 10^4$ cm^{-3}) condensations which are
usually referred to as "dark clouds". The very dense and massive
molecular clouds are usually connected with density wave spiral
arms or with the inner part of the nuclear disk in the galactic
center (see Sect. 4). Observations, at present, yield the
following characteristics (the numbers should be considered with
utmost caution); < > implies an average value.

i)   Intercloud gas

$$T_k = 500 \text{ to } 10000 \text{ K}$$
$$<n_H> = 0.2 \text{ cm}^{-3}$$
$$<n_e> = 0.03 \text{ cm}^{-3}$$

ii)   Cloudlets

$$T_k = 30 \text{ to } 400 \text{ K (and possibly higher)}$$
$$n_H = 10 \text{ to } 100 \text{ cm}^{-3}$$
$$<M_H> = 20 \text{ M}_\odot$$
$$<r> = 1.5 \text{ pc}$$

iii)   Dust clouds and dark clouds

$$T_k = 5 \div 100 \text{ K}$$
$$n_H = 10^2 \text{ to } 10^4 \text{ cm}^{-3}$$
$$M_H = 10^2 \text{ to } 10^4 \text{ M}_\odot$$
$$R = 1 \text{ to } 6 \text{ pc}$$

iv)   Molecular clouds

$$<T_k> = 30 \text{ K}$$
$$<n_H> = 10^3 \text{ cm}^{-3} \text{ but } n_H \lesssim 10^7 \text{ cm}^{-3}$$
$$M = 10^2 \text{ to } 10^6 \text{ M}_\odot$$
$$R = 10 \text{ to } 30 \text{ pc}$$

We see that the range of parameters overlap.

c) Recent Observations of the Intercloud Gas. The "two-component" model of the interstellar gas has been worked out in considerable detail (see, e. g. Dalgarno and McCray, 1972; Radhakrishnan et al., 1972; Mebold, 1972). It suffers from the fact that the prime cooling lines at medium and high density (OI $\lambda 63\mu$; CII $\lambda 156\mu$) have not yet been observed; and that, of the prime sources of heating (Cosmic rays (CR) with E $\sim$ 2 MeV; X-rays (XR) with E $\sim$ 100 eV), there exists only one very uncertain X-ray observation. This, and the fact that the observed distribution of interstellar clouds does not show the expected maxima in the two stable (hot and cold) states as predicted by the two-component model, brought this model into disfavour, at least as a quantitative description of the physical state of IM.

During the past few years, however, a number of highly relevant observations were made which induced increased theoretical work on the physics of the IM. I will try to briefly review these observations.

i) In their analysis of observations pertaining to the Gum Nebulae, Brandt et al. (1971) were led to the hypothesis that this is a "fossile HII region", generated by a SN-flash some 10^4 years ago when the Vela pulsar was formed. This is strong observational support for theoretical investigations of time dependent sources of ionization and heating of the interstellar gas. For a discussion of recent theoretical work pertaining to the heating, ionization and cooling of the IM, including time-dependent models, I refer to a review paper by Grewing (1974).

ii) Results of the Princeton experiment aboard the Copernicus satellite revealed an underabundance of heavy elements in IM (Morton et al. 1974) and showed that there are apparently no ionization processes effective which would produce large amounts of highly ionized species in the IM. Especially the observed abundance of nitrogen ions disagrees with predictions of models in which the observed X-ray background for E > 150 eV plus hypothetical spikes of 100 eV X-rays or 2 MeV subcosmic particles are the source of ionization (and heating) of the IM (Rogerson et al., 1973).

Very recently it has been shown by Grewing and Jenkins (1974) that these results are compatible with B1 stars being the principal source of heating and ionization. It should be pointed out, however, that the Copernicus observations represent only the solar neighbourhood out to about 500 pc.

iii) Observations of soft X-rays in the energy range 100 $\div$ 300 eV (Yentis et al., 1972) suggest a sharp rise in the spectrum towards the lower edge of this energy band. This X-ray

spectrum, combined with the depletion of heavy elements observed
in the IM (Morton et al., 1973) can explain some of the observed
characteristics of the IM, i.e., a kinetic temperature of
$\sim$ 6000 K and a gas density of $n_H \sim 0.2$ cm^{-3} (Grewing and Walmsley, 1974). Recently a turn-up of the sub-cosmic particle flux
in the energy range 0.5 $\div$ 1.8 MeV has been found (Simpson and
Tuzzolino, 1973). It is not yet clear whether these particles
are of solar or of interstellar origin, and if they are of importance for the heating of the interstellar gas.

iv) In my review paper on IM (Wickramasinghe et al., 1972),
I have compiled and discussed observational evidence for the
existence of a diffuse, extended HII region between galactic
radii 3 and 9 kpc in our Galaxy. The correlation of this extended HII region with the distribution of giant HII regions (Fig.
I.2) suggests that the latter are density-bounded and that Lyc-
photons leaking out into the general interstellar space are
responsible for the ionization of the diffuse ionized gas.
Recently, Reynolds et al. (1973) observed diffuse galactic Hα
and [NII] emission. They conclude that this emission comes from
an extended and nearly fully ionized intercloud component of
the interstellar gas. Electron temperatures in the range 3000 K
to 8000 K and electron densities of about 1 cm^{-3} (Perseus and
Sagittarius arm) 0.05 (Orion arm) and 0.007 (interarm region)
are estimated. The observations tend to confirm the existence of
extended low-density HII regions both in spiral arms and in
interarm regions. On the other hand, it appears that these thermal electrons cannot account for the total rms density of free
electrons implied from pulsar dispersion measurements (Sect.
2.3.b).

Monnet (1971) finds similar extended disks of ionized
hydrogen in a number of Sc galaxies and in an Sb galaxy (M31).
He concludes that this low-density HII region accounts for
practically the whole interarm region out to a distance of about
3 kpc from the center. More recently Comte and Monnet (1974) investigated HII emission from M33. The observations suggest that
the extended interarm HII region is ionized by Lyc-photons
912 > λ/Å > 504 which are supplied by density-bounded HII regions located in the spiral arms. Selective absorption of Lyc-
photons by dust (see Sect. 2.2 and 3.3) can explain the absence
of photons λ < 504 Å.

v) Recent OAO-C observations reveal the wide-spread presence
of broad shallow absorption lines of the O VI ion (Jenkins and
Meloy, 1974). Temperatures of $\simeq 10^6$ K and ion densities of 1.7
10^{-8} cm^{-3} are deduced. The data suggest the presence of a low-
density, high-temperature phase of the IM, possibly produced by
SN explosions, which could be in pressure equilibrium with the
normal interstellar gas. This hot gas may also be responsible

for the soft X-ray background radiation.

In summary, the observations yield a rather complex picture of the physical state of the intercloud gas. There is obviously not a single dominant source of ionization and heating, but different ionization sources dominate in different regions of the Galaxy. As a consequence, there appear to exist several ionized components of the IM with quite different physical characteristics. This complicates the interpretation of observations, since different observations are heavily weighted towards different components of the IM (see, e.g. Smith, 1974).

d) Molecular Clouds. While UV and optical observations of molecular absorption lines pertain primarily to cloudlets and to the more tenuous dust clouds, radio molecular lines, especially emission lines in the mm-wavelength range, come from very dense and often very massive clouds, which are of special interest in the context of star formation. More than a short review of the observations is beyond the scope of this lecture. For a more detailed discussion I refer to two recent review papers by Winnewisser et al. (1974) and by Zuckerman and Palmer (1974).

Originally, surface reactions were considered to dominate the formation of interstellar molecules. This assumption appears still to be correct for the formation of H_2. But for other molecules, such as OH and CO, gas phase reactions of the type

$$O + H^+ \leftrightarrow O^+ + H$$
$$O^+ + H_2 \rightarrow OH^+ + H$$

and

$$C^+ + OH \rightarrow CO + H^+$$

can account for the observed abundances (O'Donnel and Watson, 1974; Barsuhn and Walmsley, 1974). This eliminates the difficulty of explaining how these molecules - if formed on the surface of grains - can be desorbed.

The abundance of H_2 molecules is determined by an equilibrium between their formation and destruction by radiative dissociation. The destruction rate is primarily determined by self-shielding and therefore depends on the gas density. Barsuhn (1974), by fitting OAO-C observations of $n(H_2)/n_{total}$ and τ_{Dust} to model computations, finds that densities of the cloudlets containing H_2 must lie in the range $50 < n(H_2) < 1000$. Similar model-fitting, applied to the observed ratio of ortho-to-para H_2 yields gas temperatures of ~ 80 K.

Charge exchange reactions are so effective that the observed abundance of HD can be used to determine the ionization
rate in the cloudlets. Thus, it has been concluded that the
ionizing agent for the intercloud gas does not generally pene-
trate the cloudlets (O'Donnell and Watson, 1974). Brown (1973),
from an investigation of recombination line emission from dark
clouds, arrives at a similar conclusion.

CO is of special importance in the context of the large-
scale distribution of gas at medium densities($n(H_2) \sim 10^3$ cm^{-3}),
both due to the high abundance of this molecule and due to the
fact that it is thermalized at relative low gas densities. Preliminary results of surveys of the $\lambda 2.6$mm line of CO are somewhat conflicting. Schwartz et al. (1973) find that the distribution of CO is similar to that of atomic hydrogen (as observed
in the $\lambda 21$cm line) Scoville and Solomon (as quoted by Solomon
and Stecker, 1974), on the other hand, find that the CO-distri-
bution peaks at a galacto-centric distance of about 5 kpc and
thus is similar to the distribution of giant HII regions (see
Fig. I.2). Into the interpretation of these observations enter
the problems of opacity of the CO-line and of the clumpiness of
the dense component of the interstellar gas as, for example,
found in the nuclear disk (see Sect. 4).

In very dense and massive molecular clouds, one would expect that molecules are in equilibrium with the 3 K background.
This appears to be the case in some dark dust clouds. For the
massive clouds associated with giant HII regions, however,
kinetic gas temperatures of $\sim$ 30 K appear to be more typical.
Goldreich and Kwan (1974) solved the coupled equations of statistical equilibrium and radiation transfer for a collapsing
cloud, and found that the rate at which energy is radiated in
the rotational lines of CO exceeds the rate at which work is
done by adiabatic compression of the collapsing gas. This result
implies the existence of an energy source which maintains the
temperatures of the gas against the cooling due to radiative
energy losses. Protostars may provide these sources of energy.
In fact, IR observations by Grasdalen et al. (1973) and Simon
et al. (1973) reveal the existence of protostars in the Ophiuchus
dark cloud region.

3. HII REGIONS AND STAR FORMATION

3.1 General Considerations

HII regions, in the usual meaning, are relatively dense
($n_H \gtrsim 10$ cm^{-3}) regions of the IM which are nearly completely
ionized by Lyc-radiation from stars with high effective tempera-

tures ($T_{eff} \gtrsim 10^4$K). Only very massive (and therefore young) MS stars of spectral type BO - O4 reach these temperatures. As outlined in Sect. 1.3, HII regions are therefore the best indicators for star formation. In the following, I refer to an "HII region" in the historical meaning as an ionized region of IM. I use the notation He^{++}, He^+, H^+ and C^+-regions specifically for the volumes in which these ions are the predominant form of the element.

Table III.1

First and second ionization potential of the most abundant elements

Element	1st.		2nd ioniz. pot.	
	eV	Å	eV	Å
H	13.6	912.0		
He	24.6	504.4	54.4	227.9
C	11.3	1101.2	24.4	508.6
N	14.5	853.3	29.6	419.6
O	13.6	910.7	35.1	353.1
Ne	21.6	574.1	41.1	301.9
S	10.4	1197.1	23.4	529.8

Since the Lyc-spectra of O-stars can be approximated by black-body radiation, one expects to find in HII regions that each element listed in Table II.1 is predominantly in one of its ionization states only (including the neutral state). Fig. III.1 is a schematic representation of a typical ionization-bounded HII region. (Ionization-bounded means that all Lyc-photons emitted by the ionizing star(s) are absorbed in the surrounding HII region. The alternative case is a density-bounded HII region.)

Note that He^{++} is usually absent in HII regions ionized by MS stars. He^O is difficult to observe and therefore one usually resorts to the observations of other ions. For example O^+ should be found in an H^+-region and S^{++} in a He^+-region. However, the presence of O^+ does not allow the conclusion that He is neutral, since the ionization potential of O^+ is considerably higher than that of He^O. Likewise the presence of S^+ does not allow the conclusion that He is neutral but H is ionized, since S^+ is also expected to exist in the C^+-region (which, however, may be too cool for collisional excitation of forbidden $[S^+]$ lines).

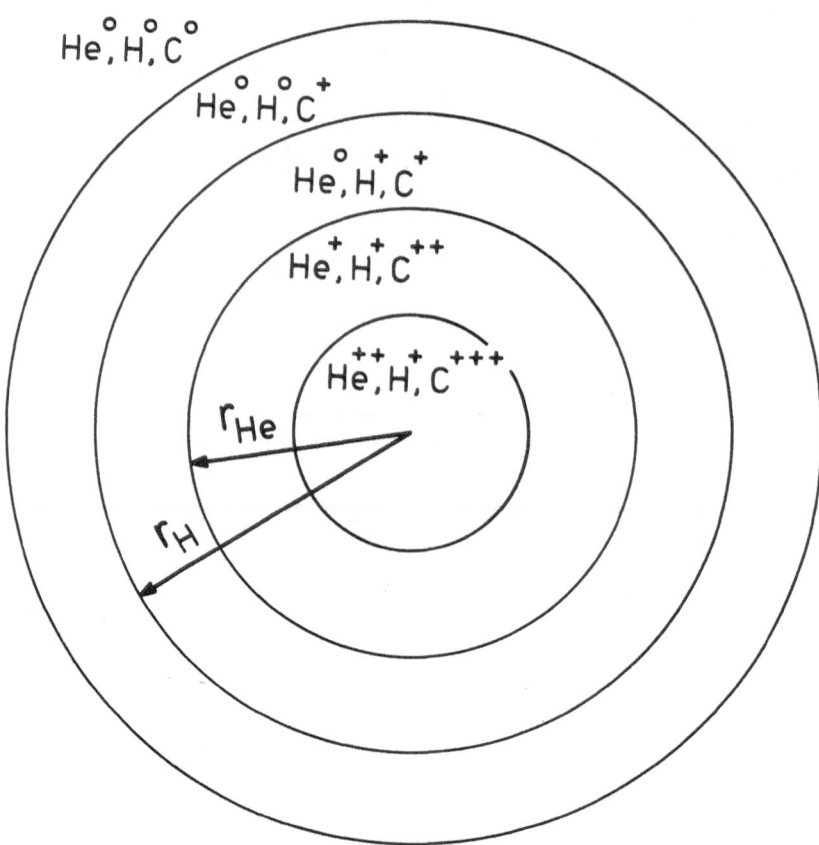

Figure III.1 Schematic representation of the ionization structure of an HII region. In HII regions ionized by
early O-stars the He^{++}-Strömgren sphere is usually
absent while He$^+$- and H$^+$-Strömgren spheres coincide.
A distinct, ionization-bounded C$^+$-Strömgren sphere
exists only for ionization-bounded HII regions surrounded by a dense shell of neutral gas.

The electron temperature of an HII region is $T_e \simeq 10^4$ K, apparently with only small deviations (Sect. 1.4 and Seaton, 1974). Observed electron densities are $n_e \overset{<}{\sim} 10^5$ cm^{-3}.

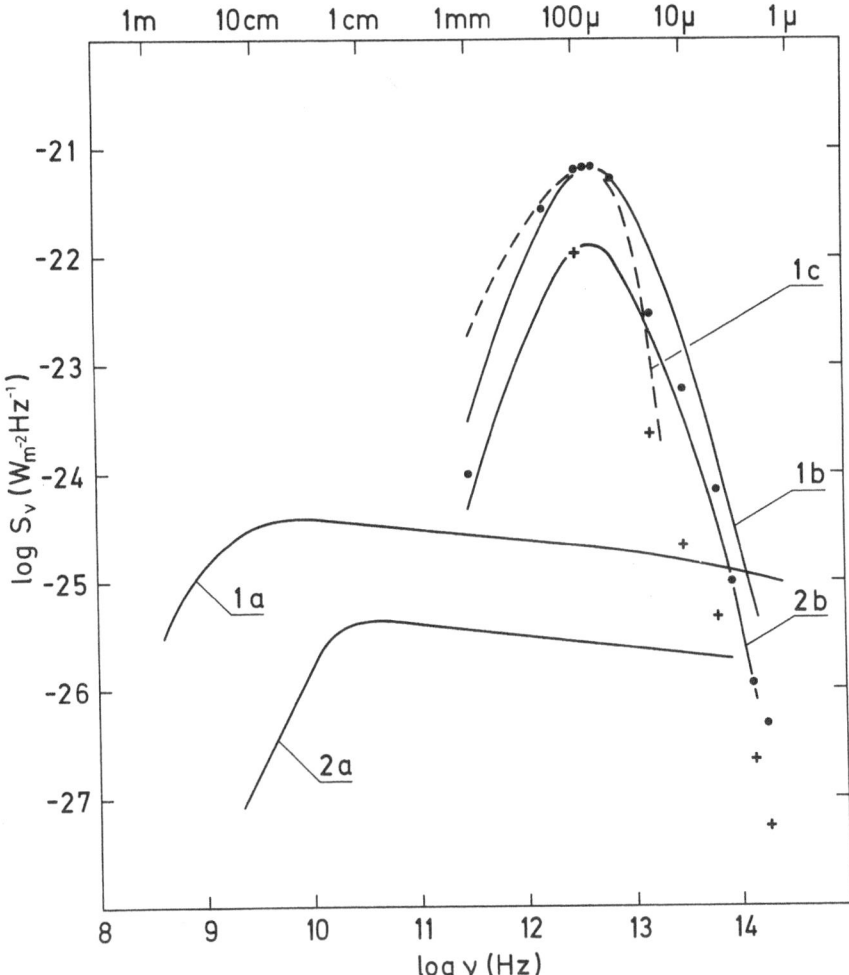

Figure III.2   Radio- and IR-continuum radiation of W3 (main component; curve 1a and ●●●) and of W3 (OH; curve 2a and +++).

Fig. III.2 shows the radio- and IR-continuum spectrum which is typical for compact (i.e. high density) galactic HII regions. The radio emission is free-free Bremsstrahlung; its spectrum is flat if the free-free optical depth $\tau_c \ll 1$. The IR-emission is thermal (black-body) radiation from dust grains. For details, I refer to my review in Wickramasinghe et al. (1972) and references therein.

3.2 Dust-free HII Regions

In dealing with the ionization structure of HII regions, one has to consider only the two most-abundant elements H and ^{4}He. It is important to realize that practically every recombination of a He$^+$-ion leads - through photon emission - to the ionization of a H-atom. Therefore, the size of an H$^+$-region is not affected if part of the stellar Lyc-photons are absorbed by He.

The number of stellar photons, N_c', absorbed per sec by the hydrogen gas must equal the number of recombinations per sec, in all but the ground state (since the latter recombinations produce another Lyc-photon).

$$N_c' = \int\limits_{V(H^+)} n(H^+) \; n_e \; \alpha(T_e) \, dV$$

α, the recombination coefficient to all states $n \geqslant 2$, depends on the electron temperature $\sim T_e^{-0.8}$. The free-free flux-density of an optically thin H$^+$ region is

$$S_\nu = \text{Const} \int n_e \; n(H^+) \; T_e^{-0.35} \, dV$$

Therefore, if $T_e \simeq \text{Const}$, $S_\nu \propto N_c'$. The exact relation, including the presence of He$^+$ and He^0 in the HII region is

$$N_c' = (1 + yR)^{-1} \; 7.55 \; 10^{46} \; \left[\frac{\nu}{\text{GHz}}\right]^{0.1} \; \left[\frac{S_\nu}{\text{f.u.}}\right] \; \left[\frac{D}{\text{kpc}}\right]^2$$

This and the following relations are derived or discussed by Mezger et al. (1974), if not noted otherwise. Here 1 f.u. = 10^{-26} W m^{-2} Hz^{-1}, D is the distance of the HII region and $T_e = 10^4$ K has been adopted. $y = 0.1$ is the number abundance of ^{4}He and

$$R = \int\limits_{V(He^+)} n(H^+)^2 \, dV \; / \; \int\limits_{V(H^+)} n(H^+)^2 \, dV$$

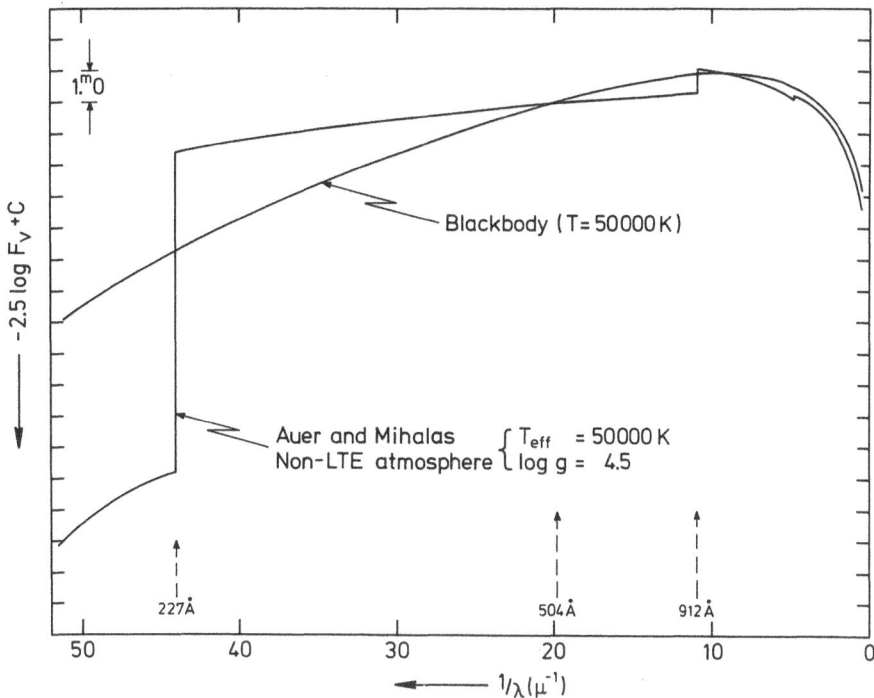

Figure III.3   Comparison of the non-LTE radiation spectrum of an
               O4-star; T_{eff} = $5 \cdot 10^4$) with the corresponding
               black-body spectrum.

is the ratio of He^+- and H^+-Strömgren spheres weighted with the
square of the proton density. If R = 1, the two Strömgren
spheres coincide. Otherwise R < 1. R is determined by one single
stellar spectral parameter, viz. the ratio, γ, of the number of
He-photons (504 < $\lambda/\text{Å}$ < 228) N_{He} to the number of H-photons
(912 < $\lambda/\text{Å}$ < 504) N_H

$$\gamma = N_{He}/N_H$$

If $\gamma \gtrsim 0.20$, R = 1. For black-body radiation, this would mean
that only for temperatures $T_{bb} \gtrsim 4.7 \ 10^4$ K are H and He ionized
throughout the same volume. For Auer and Mihalas non-LTE atmospheres, however, R = 1 for $T_{eff} \gtrsim 3.7 \ 10^4$ K (i.e. for spectral
types O9 and earlier). In Fig. III.3, the spectrum of an O4-star
($T_{eff} \simeq 5 \cdot 10^4$ K) is compared with the corresponding black-body
spectrum. One sees that non-LTE effects in the atmospheres of

O-stars (Auer and Mihalas, 1972) tend to increase the number of
He-photons. It should be noted, however, that line blanketing
is not included in these model calculations.

3.3 HII Regions with Dust

Dust in an HII region will absorb part of the intrinsic
stellar Lyc-radiation. Consequently, the number N_c' of Lyc-photons absorbed per sec by the gas will be smaller than the number
N_c of Lyc-photons emitted by the star. The fraction $f = N_c'/N_c$
depends on the optical absorption depth of the dust τ_{Lyc} only;
it can be conveniently approximated by $f \simeq \exp\{-\tau_{Lyc}\}$.

As shown in Sect. 2.2 and Fig. II.2, the absorption cross
section for He-photons, σ_{He}, is larger than the absorption cross
section for H-photons, σ_H. As a result of this selective absorption of Lyc-photons by dust, the intrinsic stellar Lyc-photon
radiation will be depleted of He-photons and the ratio of He-
to H-photons available for ionization of the gas $\gamma' = N_{He}'/N_H'$,
will be smaller than the stellar ratio γ. One finds

$$\gamma' = \gamma \exp\{-(\tau_{He} - \tau_H)\}$$

The total depletion of Lyc-photons by dust absorption is

$$N_c'/N_c = \{\exp(-\tau_H) + \gamma \exp(-\tau_{He})\}/(1 + \gamma)$$

As a result of absorption of Lyc- and less energetic photons,
the dust grains will be heated and become strong IR-radiators
(Fig. III.2). The total heating (here I use the term heating in
the sense of energy absorbed per sec) can be described by four
terms: L_α, L_U, L_H, L_{He}. $L_\alpha \simeq N_c' h\nu_\alpha$ is the heating due to absorption of Lyman alpha photons. Approximately, each Lyc-photon
absorbed by the gas is degraded into a Lyman alpha photon; therefore $N_\alpha \simeq N_c'$.

L_U is the heating due to absorption of photons with wavelength $\lambda > 912$ Å. L_H and L_{He} are the heating due to H- and He-
photons, respectively. These various contributions to the heating, normalized to the luminosity L_* of the ionizing star, are
shown in Fig. III.4.

The following conclusions are important in the context of
the ionization of IM and star formation:

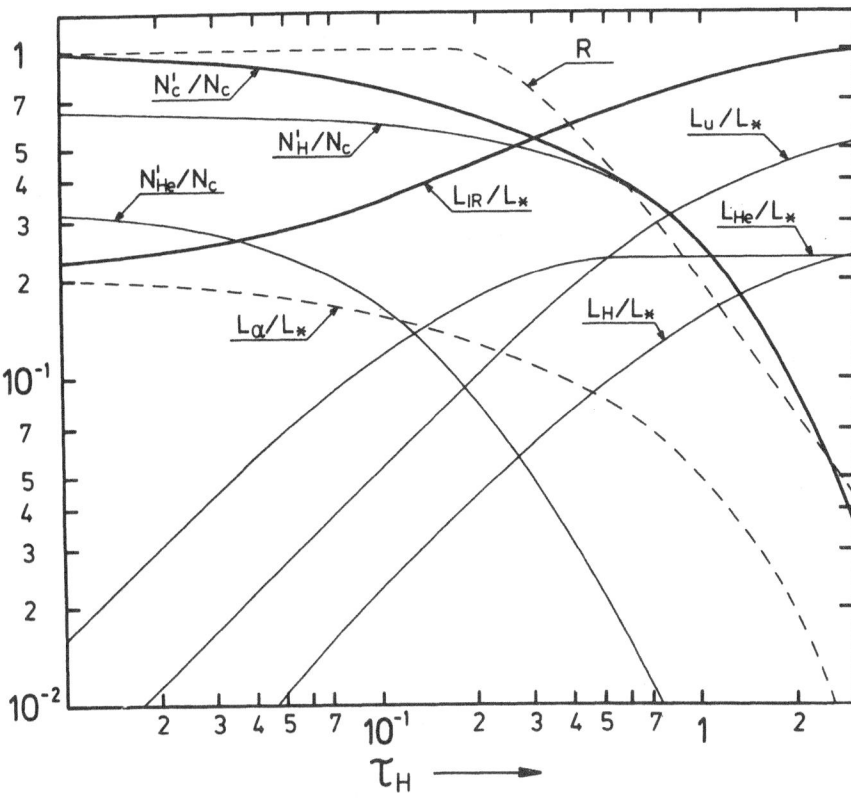

Figure III.4   The ratio of Lyc-photons available for ionization
               of the gas, N_c', to the stellar Lyc-photon flux,
               N_c; and the various contributions to the heating
               of dust grains expressed as fractions of the total
               stellar luminosity L_*. All quantities are given as
               function of the optical depth τ_H for H-photons.

   i) For $\tau_H \gtrsim 4.2$, $L_{IR} \simeq L_*$ and $N_c'/N_c < 0.01$. Since $\tau_H \propto$
$(n_e\,N_c')^{1/3}$, this means that, at high densities ($n_e > 10^5$ cm^{-3}),
an HII region becomes unobservable as a radio source, and the
total stellar luminosity is converted directly into IR-radiation.

   ii) If the HII region is surrounded by a sufficiently dense
shell of neutral gas, photons with $\lambda > 912$ Å, which can escape
the HII region, will be absorbed by dust in the neutral gas and
$L_{IR} \simeq L_*$ still holds. Observations of the IR- and radio continuum
thus allow us to estimate both the total luminosity L_* and the
total Lyc-photon luminosity N_c of the ionizing star(s) of an HII

region. This method can be used to estimate, in turn, the spectral type of the ionizing star(s). (Panagia, 1974; Mezger and Wink, 1974). Stellar parameters N_c, L_* and γ are given in Fig. III.5 as functions of stellar mass.

iii) If the HII region is ionization-bounded, we expect a strong CII region to form around it; this region is probably the source of the Carbon radio recombination lines (observed in emission).

iv) If the HII region is density-bounded and Lyc-photons can leak out into the general dilute IM, they should be depleted of He-photons. If density-bounded HII regions account for the ionization of the diffuse HII gas in interarm regions (Sect. 2.3.c.iv), one would expect He to be neutral in these regions.

Model calculations of the IR brightness distribution of HII regions as a function of wavelength have been made by Wright (1973), who concludes that the far IR radiation comes from the neutral gas surrounding the HII region because he believes that the HII region is depleted of dust. However, the result of these computations depends so strongly on the spectral distribution of the stellar radiation and on the absorption characteristics of dust (Krügel, private communication), that any quantitative conclusions drawn from Wright's model computation are very doubtful. In fact, Panagia (1974) concludes that dust may even be overabundant in HII regions. However, in two points Wright's results are qualitatively correct: The diameter of the IR-source increases with wavelength and the integrated IR spectrum cannot be approximated by a black-body spectrum. Therefore, the derived color temperatures are rather meaningless.

3.4 Star Formation

HII regions are the best indicators for star formation. However, their properties (determined from IR and radio continuum radiation) depend only on the most luminous stars. However, in an OB-star cluster, these luminous stars account for only a few per cent of the total stellar mass content. Any conclusions pertaining to the star formation rate which are based on observations relating to OB-stars, thus depend very strongly on the adopted luminosity function. We adopted Salpeter's (1955) luminosity function for unevolved stars (SLF) (see Sect. 1.3) to compute a Lyc-photon-to-stellar mass ratio of (Mezger et al., 1974)

$$\frac{<N_c>}{<M>} = 2.2\times10^{46} \text{ Lyc-photons } M_o^{-1}$$

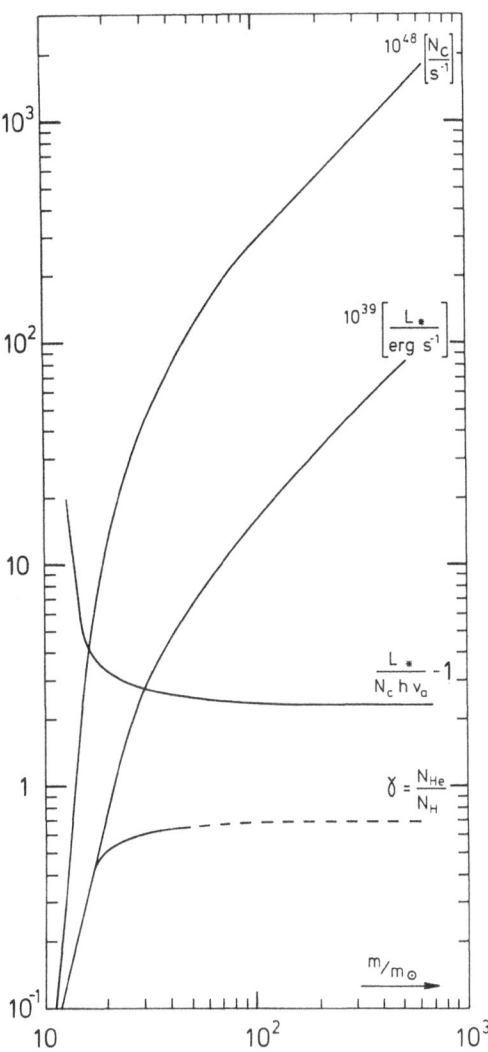

Figure III.5   Stellar parameters as a function of stellar mass
(from Mezger and Wink, 1974). N_C is the total
number of Lyc-photons emitted per sec, L_* is the
total stellar luminosity, $\gamma = N_{He}/H$ is the ratio
of He- to H-photons $(L_*/N_C h\nu_\alpha) -1$ is the IR-excess
of a dust-free HII region, where all photons are
absorbed in the neutral gas surrounding the HII
region.

I have summarized optical and radio observations pertaining to
star formation in material arms such as the Orion arm (Mezger,
1971). It appears that, in dust clouds, low-mass stars form over
extended periods of time, while O-stars form (if at all) only in
the densest parts of these clouds and only in the final stages
of the formation of a star cluster.

In density-wave spiral arms, the formation at least of
massive stars is initiated by compression of the gas when it
flows through the density wave. One of the best-studied giant
HII regions in a density-wave spiral arm is W3, located at a
distance of about 3 kpc in the Perseus arm. We have recently re-
analyzed various radio and IR observations of this HII region
(Mezger and Wink, 1974). It consists of a number of IR sources
and compact HII regions which represent various evolutionary
stages of O-stars. It appears that massive stars reach the MS
embedded in a dense cloud of neutral gas and dust. In the pre
MS stage radiation pressure pushes out the dust which subsequently forms a shell (cocoon). In the early stages, all Lyc-
photons are directly absorbed by dust which thus is observable
as a strong IR source. Subsequently, a compact HII region forms
inside the dust cocoon and eventually destroys the cocoon. In
its early stages, the compact HII region is still embedded in
a dense shell of neutral gas and dust and thus forms a sharply
defined Strömgren sphere of typical radii 10^{17} to 10^{18} cm.
Eventually, all of this neutral shell is ionized and the Strömgren spheres of the individual O-stars begin to overlap, thus
forming an extended low-density HII region, in which the remnants of the compact HII regions are still visible. This implies
that the neutral cloud out of which the O-stars form is not of uniform density but that O-stars form in the center of the densest
condensations in the cloud. Typical masses of these condensations are some 100 $M_\odot$.

Most giant HII regions are located in density-wave spiral
arms and can be studied by means of their radio and IR radiation,
from which one estimates: N_C, the intrinsic stellar Lyc-photon
flux; the total mass contained in stars (by multiplication of
N_C by $\left[<N_C>/<M_*>\right]^{-1}$); M_{HII}, the total mass of ionized hydrogen.
Giant HII regions are usually associated with massive molecular
clouds, whose mass M_{gas} can be estimated from observations of
molecular radio emission lines (see Sect. 2.3.a). A comparison
shows that, usually, $M_* \simeq 10\ M_{HII}$ but, on the other hand, $M_* \lesssim$
$10^{-1}\ M_{gas}$.

The total star formation rate can be estimated on the basis
of the total number of O-stars contained in our Galaxy. Terzian
(1974) has estimated that all O-stars contained in the galactic
disk between galactic radii 4 and 15 kpc emit $\sum N_C = 7.5\ 10^{52}$ Lyc-

photons s^{-1}. This corresponds to a total amount of $3\ 10^6\ M_\odot$ of gas transformed into stars. The main contributors to stellar Lyc-radiation are stars of spectral type O6 (Mezger et al., 1974) whose lifetime $t_{MS} \simeq 3\ 10^6$ yr. This yields a rate of star formation of 1 $M_\odot$ per year for the whole spiral arm region of the Galaxy.

Let us assume that 50 % of the total gas content of the Galaxy, i.e. $\simeq 10^{10}\ M_\odot$, passes through density-wave spiral arms. The maximum star formation occurs between galactic radii 4 and 8 kpc (Fig. I.2). Gas at these distances rotates around the galactic center in $\simeq 2\ 10^8$ yr and thus - for a two-arm pattern - $10^{10}\ M_\odot$ of gas passes every 10^8 yr through the compression shock of a density-wave spiral arm. The efficiency of star formation through compression by density waves thus is estimated to be about 1 %.

Although only 1 % of the total amount of gas passing through density-wave spiral arms eventually forms stars, a much larger fraction of this gas is expected to form clouds which may subsequently enter the interarm region.

4. THE GALACTIC CENTER

This lecture summarized a paper given by me at the ESO symposium on "Research Programs for large Optical Telescopes", Geneva, Switzerland, in May/June 1974. This paper is published in the proceedings of this symposium (E. Reiz, editor, CERN, Geneva, 1974, page 79) and therefore will not be included here.

It is my pleasure to thank U. Mebold, L.F. Smith and J. Wink for critical suggestions and help in preparing this manuscript, and Miss E. Dietz for her careful typing.

REFERENCES

Andriesse, C.D. and de Vries, J. (1974) Astron. & Astrophys. 30, 51

Arp, H. (1965) Ap. J. 141, 43

Auer, L.H. and Mihalas, D. (1972) Ap. J. Suppl. 24, 193

Barsuhn, J., Walmsley, C.M. (1974) in Proc. of the 2nd Europ. Astron. Meeting in Triest (to be published in Mem.Soc. Astr. Italiana)

Barsuhn, J. (1974) Mitt. Astron. Gesellsch. 35, 197

Brandt, J.C., Stecher, T.P., Crawford, D.I. and Maran, S.P. (1971) Ap. J., 163, L99

Brown, R. (1973) Ap. J. 184, 693

Cameron, A.G.W. (1973) Space Sci. Rev. 15, 121

Churchwell, E., Mezger, P.G., Huchtmeier, W. (1974) Astron. & Astrophys. 32, 283

Comte, G. and Monnet, G. (1974) Astron. & Astrophys. 33, 161
Dalgarno, A. and McCray, R.A. (1972) Ann. Rev. Astr. Astrophys.
    10, 375
Field, G.B. (1974) Ap. J. 187, 453
Georgelin, Y.P. (1971) Astron. & Astrophys. 11, 414
Goldreich, P. and Kwan, J. (1974) Ap. J. 189, 441
Grasdalen, G.L., Strom, K.M. and Strom, S.E. (1973) Ap. J. 184,
    L53
Greenberg, J.M. (1974) Ap. J. 189, L81
Grewing, M. and Walmsley, C.M. (1974), Astron. & Astrophys. 30,
    281
Grewing, M. (1974) Proc. of the IAU Symp. No. 60
Grewing, M. and Jenkins, E.B. (1974) Astron. & Astrophys. (in
    press)
Jenkins, E.B. and Meloy, D.A. (1974) Ap. J. (in press)
Jenkins, E.B. and Savage, B.D. (1974) Ap. J. 187, 243
Kaplan, S.A. and Pikelner, S.B. (1970) "The Interstellar Medium",
    Harvard Univ. Press, Cambridge, Mass.
Mebold, U. (1972) Astron. & Astrophys. 19, 13
Mezger, P.G. (1971) Highlights of Astronomy (Ed. C. de Jager)
    Vol. 2, P. 366
Mezger, P.G., Smith, L.F. and Churchwell, E. (1974) Astron. &
    Astrophys. 32, 269
Mezger, P.G. and Wink, J. (1974) Proc. 2nd Europ. Astron.
    Meeting, Trieste, to be published in Mem. Soc.Astr. Italiana
Monnet, G. (1971) Astron. & Astrophys. 12, 379
Morris, M., Zuckerman, B., Palmer, P., Turner, B.E. (1973) Ap.
    J. 186, 501
Morton, D.G., Drake, J.F., Jenkins, E.B., Rogerson, J.B.,
    Spitzer, L., York, D.G. (1973) Ap. J. 181, L103
O'Donnel, E.J. and Watson, W.D. 1974, Ap. J. 191, 89
Panagio, N. (1974) Proc. of the 8th ESLAB Symp. (in press)
Peebles, P.J. (1971) Phys. Cosmology, Princeton Univ. Press,
    Princeton, N.J.
Pinkau, K. (editor) (1974) "The Interstellar Medium" D. Reidel
    Publ. Comp., Doordrecht, Holland
Purcell, E.M. (1969) Ap. J. 158, 433
Radhakrishnan, V., Murray, J.D., Lockhardt, P. Whittle, R.P.J.
    (1972) Ap. J. Suppl. 24, 15
Reynolds, R.J., Scherb, F., Roesler, F.L. (1973) Ap. J. 185, 869
Rogerson, J.B. and York, D. (1973) Ap. J. 186, L95
Rogerson, J.B., York, D.G., Drake, J.F., Jenkins, E.B., Spitzer,
    L. (1973) Ap. J. 181, L110
Salpeter, E.E. (1955) Ap. J. 121, 161
Schmidt-Kaler, Th. and Schlosser, W. (1973) Astron. & Astrophys.
    29, 409
Schwartz, P.R., Wilson, W.J., Epstein, E.E. (1973) Ap. J. 186,
    529
Scoville, N.A., Solomon, P.M., Ap. J. 187, L67

Searle, L. (1971) Ap. J. 168, 327

Seaton, M. (1974) Quarterly Journal RAS (in press)

Simon, M., Righini, G., Joyce, R.R. and Gezari, D.Y. (1973)
    Ap. J. 186, L127

Simpson, J.A., Tuzzolino, A.J. (1973) Ap. J. 185, L149

Smith, L.F. (1974) Space Research XIV, ed. A.C. Strickland,
    Akademie Verlag, Berlin, in press

Spitzer, L. (1968) "Diffuse Matter in Space", Intersc. tracts
    on Phys. Astron., J. Wiley and sons, New York

Solomon, P.M. and Stecker, F.W. (1974) paper presented at the
    ESRO Symposium on Gamma Ray Astronomy, Frascati (to be
    published)

Talbot, R.J. Jr., Arnett, W.D. (1973) Ap. J. 186, 51

Terzian, Y. (1974) Ap. J. (in press)

Turner, B.E., Zuckerman, B., Palmer, P. and Morris, M. (1973)
    Ap. J. 186, 123

van den Bergh, S. (1968) J. Roy. Astron. Soc. of Canada, 62, 149

Wagoner, R.V. (1973) Ap. J. 179, 343

Wickramasinghe, N.C., Kahn, F.D., Mezger, P.G. (1972) "Inter-
    stellar Matter" Geneva Obs., CH-1290 Sauverney, Switzerland

Winnewisser, G., Mezger, P.G. and Breuer, H.D. (1974) in "Topics
    in Current Chemistry" Vol. 44, Springer Verlag, Berlin

Witt, A.N., Lillie, C.F. (1973) Astron. & Astrophys. 25, 397

Wright, E.L. (1973) Ap. J. 185, 569

Yentis, D.J., Novick, R., Vanden Bout, P., Ap. J. 177, 365

Zuckerman, B. and Palmer, P. (1974) Ann. Rev. Astronomy & Astro-
    phys., in press.

DYNAMICS OF INTERSTELLAR MATTER

F.D. Kahn

University of Manchester, Manchester, U.K.

1. THE BASIC EQUATIONS OF FLUID FLOW

Three basic equations describe the flow of a fluid. They are the equation of continuity

$$\frac{\partial \rho}{\partial t} + \bar{\nabla} \cdot (\rho \bar{u}) = 0 \ , \tag{1.1}$$

the equation of motion

$$\frac{\partial \bar{u}}{\partial t} + (\bar{u} \cdot \bar{\nabla}) \bar{u} = - \frac{\bar{\nabla} P}{\rho} \ , \tag{1.2}$$

and the equation for energy flow

$$\frac{\partial S}{\partial t} + (\bar{u} \cdot \bar{\nabla}) S = 0 \ . \tag{1.3}$$

In these relations ρ = density, P = pressure, $\bar{u}$ = velocity, S = entropy per unit mass. The equations are derived in any good text-book on hydrodynamics, and apply to any fluid in which transport phenomena (that is viscosity and heat conduction) are negligible everywhere. But as they stand they make no provision for various effects which are likely to be important in the interstellar medium.

For example, take the equation of continuity. There are likely to be regions in interstellar space where an appreciable interchange of mass takes place between the gas and the stars either by star formation, or as a result of stellar mass loss due to winds, or to supernova explosions. An appropriate source term must be added to the equation to describe these processes.

Next, according to equation (1.2), the motion of the gas is en-

G. Setti (ed.), Structure and Evolution of Galaxies, 179–195. All Rights Reserved.
Copyright © 1975 by D. Reidel Publishing Company, Dordrecht-Holland.

tirely determined by the pressure gradient. But there are several
other reasons why the interstellar gas may be accelerated. Almost
certainly there will be a gravitational field acting on the gas,
so that the equation of motion must include a term $\bar{\nabla}V$ (V is the
gravitational potential) on its right-hand side. If the interstellar
fluid is an electrical conductor then there may be hydromagnetic
forces acting on it. Alternatively the non-ionized fluid is an HI
region may co-exist with a plasma, which itself experiences hydromagnetic effects. The coupling between the plasma and the non-ionized fluid takes place via atom-ion interactions. In this case it
might be better to describe the interstellar gas as a two-fluid
medium, with friction between the two components.

Another modification is possible when radiation pressure is
important, for example in cocoon stars. Here it is believed that the
central star attracts towards itself a mixture of gas and dust from
interstellar space. Radiation from the star is repeatedly absorbed
and re-emitted by the dust which has collected around the star. The
trapped radiation exerts pressure on the dust grains and they, in
turn, drag the gas along. Thus the system really contains three
distinct fluids: the photon gas, the collection of dust grains and
the actual interstellar gas.

Finally the energy equation needs to be supplemented, as a
rule, in cases of astrophysical interest. The description of the
energy balance must take account of the rate G, at which a unit
mass of gas picks up energy from the surrounding medium, and the
rate L, at which it loses energy by radiation. Contributions to G
come chiefly from photons of UV radiation and of X-rays, usually
after ionizing a constituent atom (or ion) in the gas. Another
possible contribution may come from cosmic ray particles or "suprathermal" particles, either after they have caused ionization or
accelerated the ambient interstellar electrons.

The loss of energy from the interstellar gas is due to radiation by atoms or ions, such as C^+, Si^+, Fe^+, H or O^+ after collisional excitation, usually by electron impact. One species of coolant
dominates, as a rule, at any given temperature. A thorough description of these effects can be found in the article by Dalgarno and
McCray (1972). When the gain and loss processes are allowed for,
the equation for energy balance (1.3) changes to

$$T\left\{ \frac{\partial S}{\partial t} + (\bar{u}\cdot\bar{\nabla})S \right\} = G-L \qquad\qquad (1.4)$$

Very often the time for thermal adjustment is much shorter than
the characteristic time scale of the flow. The equation of energy
balance is then simply

$$G - L = 0; \qquad\qquad (1.5)$$

if so the thermal balance takes place in the same way as in a
static gas.

Interesting effects may be associated with peculiar forms of

dependence of the cooling rate on the gas temperature. In particular there may be thermal instabilities. They will be dealt with
later in these lectures.

2. PLANE WAVES

There are some classical results, due to Riemann, which describe
the progress of a plane wave of finite amplitude through a compress-
ible fluid. To obtain them, assume that the condition of energy
balance in the gas is replaced by a pressure-density relation of
the form

$$P = k\rho^{\gamma} \tag{2.1}$$

to begin with. This restriction will be replaced by a more general
relation later.

For a one-dimensional disturbance, the equation of continuity
becomes

$$\frac{\partial \rho}{\partial t} + u \frac{\partial \rho}{\partial x} + \rho \frac{\partial u}{\partial x} = 0 \tag{2.2}$$

and the equation of motion

$$\frac{\partial u}{\partial t} + u \frac{\partial u}{\partial x} + \frac{1}{\rho} \frac{\partial P}{\partial x} = 0. \tag{2.3}$$

With the pressure-density relation of equation (2.1), the
equation of motion becomes

$$\frac{\partial u}{\partial t} + u \frac{\partial u}{\partial x} + \frac{a^2}{\rho} \frac{\partial \rho}{\partial x} = 0. \tag{2.4}$$

Here a is the adiabatic sound speed, defined by

$$a^2 \equiv \frac{dP}{d\rho} (=k\gamma\rho^{\gamma-1}).$$

Note that with a pressure-density relation of this form,

$$2 \frac{da}{a} = (\gamma-1) \frac{d\rho}{\rho} \tag{2.5}$$

and that equation (2.2) and (2.3) therefore transform to give

$$\frac{2}{\gamma-1} \frac{\partial a}{\partial t} + \frac{2}{\gamma-1} u \frac{\partial a}{\partial x} + a \frac{\partial u}{\partial x} = 0 \tag{2.6}^{\cdot}$$

$$\frac{\partial u}{\partial t} + u \frac{\partial u}{\partial x} + \frac{2}{\gamma-1} a \frac{\partial a}{\partial x} = 0. \tag{2.7}$$

It then follows that

$$\{\frac{\partial}{\partial t} + (u \pm a) \frac{\partial}{\partial x}\}(u \pm \frac{2a}{\gamma-1}) = 0 \tag{2.8}$$

and this result shows that two wave systems exist. Some important consequences follow, but first I shall show that an equivalent result can be found for a gas which satisfies a pressure-density relation of the more general form

$$P = P(\rho). \tag{2.9}$$

Now we have to multiply equation (2.2) by $\frac{a}{\rho}$ and add it to (subtract it from) equation (2.3) to get

$$\{\frac{\partial}{\partial t} + (u \pm a) \frac{\partial}{\partial x}\} u \pm \frac{a}{\rho} \{\frac{\partial}{\partial t} + (u \pm a) \frac{\partial}{\partial x}\} \rho = 0, \tag{2.10}$$

in place of equation (2.8). Now define the auxiliary quantity q by the relation $dq \equiv ad\rho/\rho$, and we get

$$\{\frac{\partial}{\partial t} + (u \pm a) \frac{\partial}{\partial x}\}(u \pm q) = 0. \tag{2.11}$$

This relation is similar in form to that of equation (2.8) and will be used later; first let us deal with the consequences of equation (2.8). Here we see that the forward-going wave propagates along the trajectory of an r-characteristic

$$\frac{dx}{dt} = u + a \tag{2.12}$$

and that the quantity

$$r = u + \frac{2a}{\gamma-1} \tag{2.13}$$

remains constant along this trajectory. The backward-going wave propagates along the s-characteristics, given by

$$\frac{dx}{dt} = u - a, \tag{2.14}$$

and the quantity

$$s = u - \frac{2a}{\gamma-1} \tag{2.15}$$

remains constant along each of these paths. When a wave is sent forward into an undisturbed medium, then every s-characteristic starts from a region where the gas is at rest; hence

$$s = -\frac{2a_o}{\gamma-1} \tag{2.16}$$

on every s-characteristic. Here a_o = speed of sound in the undisturbed gas. Consequently

$$u - \frac{2a}{\gamma-1} = - \frac{2a_o}{\gamma-1} \qquad (2.17)$$

throughout the flow region, since an s-characteristic passes through
every point in the gas. The speed with which the r-characteristics
carry the pressure wave forward is therefore

$$u + a = (1 + \frac{2}{\gamma-1}) a - \frac{2a_o}{\gamma-1} = \frac{\gamma+1}{\gamma-1} a - \frac{2a_o}{\gamma-1} . \qquad (2.18)$$

It follows from this that, provided γ is not negative, the
characteristics propagate faster in regions where the density and
pressure are high and slower where the density and pressure are
low. In other words the crests of the wave catch up on the troughs.
This result comes from differentiation of equation (2.18), which
shows that

$$\frac{d(u+a)}{d\rho} = \frac{\gamma+1}{\gamma-1} \frac{da}{d\rho} = (\gamma+1) \frac{a}{2\rho} \qquad (2.19)$$

with the help of relation (2.5). In fact the crests overtake the
troughs in any case which is physically reasonable; if γ were
negative the speed of sound a would be imaginary, and therefore
sound waves with a real wavelength would be unstable. A gas cannot
continue to exist in such a state.

These results are readily extended to include gases which
have a more general form of pressure-density dependence $P = P(\rho)$.
This time the s-characteristics propagate the quantity $s = u-q$.
In the undisturbed gas $q = q_o$ and $u = 0$; therefore $u = q-q_o$ on
any s-characteristic, and consequently throughout the flow. The
r-characteristics therefore propagate forward with a speed $u+a=$
$=a+q-q_o$ and once again the crests will catch up the troughs provided that

$$0 < \frac{d(u+a)}{d\rho} = \frac{d(a+q)}{d\rho} = \frac{1}{\rho} \frac{d(\rho a)}{d\rho} . \qquad (2.20)$$

Thus the compression waves in the gas will steepen if and
only if the quantity ρa increases with density. An important
consequence is that sharp features - that is shocks - tend to
develop in any flow in which the gas is being pushed, <u>provided
condition (2.20) is satisfied</u>.

3. SHOCK WAVES, WITH AN APPLICATION TO THE SPIRAL PATTERN OF THE
GALAXY

The steepening of the pressure waves continues until the gradients
of pressure, density and velocity are so large that the usual
equations of hydrodynamics can no longer describe the flow. Eventually a shock must form, with a thickness of the order of a mean
free path in the gas.

Shock transitions are best described with respect to a frame of reference that travels with the shock. The conditions at shocks also are derived in any reasonable textbook, and they too relate to the conservation of mass, momentum and energy. If the gas velocity is perpendicular to the shock then the first two of the conditions are conservation of mass

$$\rho_1 v_1 = \rho_2 v_2 \qquad\qquad (3.1)$$

and conservation of momentum

$$P_1 + \rho_1 v_1^2 = P_2 + \rho_2 v_2^2. \qquad\qquad (3.2)$$

Here v is the gas speed relative to the shock. The suffices 1 and 2 refer respectively to the upstream and the downstream side of the shock. Shock waves are believed to be produced in the interstellar gas by the density wave associated with the spiral arms in the disk of the Galaxy. For a discussion of this effect we only need the shock relations for mass and momentum. The energy relation is superfluous.

According to the description developed by Lin and his associates, the spiral pattern in the galactic disk maintains an unchanging form, when viewed from a frame of reference rotating with angular velocity $\bar{\Omega}_p$ about an axis through the galactic centre and perpendicular to the galactic plane. The gas dynamics of spiral arms is generally presented in two-dimensional form. The standard equations of fluid dynamics are therefore re-written in terms of a surface density $\sigma \equiv \int \rho dz$, and a two-dimensional pressure $\Pi \equiv \int P dz$. In both cases the integration extends through the galactic disk, in a direction parallel to the rotation axis.

In the rotating frame the equation of motion of the gas becomes

$$(\bar{v} \cdot \bar{\nabla})\bar{v} = -\frac{\bar{\nabla}\Pi}{\sigma} + \bar{\nabla}V + \Omega_p^2\ \bar{r} - 2\bar{\Omega}_p \times \bar{v}. \qquad\qquad (3.3)$$

The second, third and fourth terms on the right-hand side are, respectively, the gravitational acceleration, the centrifugal acceleration, and Corioli's force. There is no time derivative of the velocity on the left-hand side, since the motion of the fluid is steady, by hypothesis.

In the Lin picture there are two spiral shock waves, symmetrically placed with respect to the centre of the Galaxy, and the gas moves forward with respect to the pattern along closed streamlines. In discussing this structure I shall apply an argument, due in its original form to George Field.

First integrate equation (3.3) along a streamline, from A on the downstream side of one shock to B on the upstream side of the next shock; this gives

$$\int_A^B (\bar{v}\cdot\bar{\nabla})\bar{v}\cdot d\bar{s} = -\int_A^B \frac{d\Pi}{\sigma} + V_B - V_A + \tfrac{1}{2}\Omega_p^2(r_B^2 - r_A^2) =$$

(3.4)

$$= \tfrac{1}{2}(v_B^2 - v_A^2)$$

But a shock is very thin, and therefore B and A are equivalent positions in the gravitational and centrifugal force fields. Therefore $V_A = V_B$ and $r_B = r_A$, so that

$$v_B^2 - v_A^2 = -2\int_A^B \frac{d\Pi}{\sigma} .$$

(3.5)

Since B is on the upstream side and A on the downstream side of a shock, we get immediately that

$$v_1^2 - v_2^2 = 2\int_1^2 \frac{d\Pi}{\sigma} .$$

(3.6)

Now resolve $\bar{v}_1$ into components $v_{1\parallel}$, and $v_{1\perp}$, which are respectively parallel and perpendicular to the shock, and do the same for $\bar{v}_2$. Since $v_{1\parallel} = v_{2\parallel}$ at the shock, it follows from (3.6) that

$$v_{1\perp}^2 - v_{2\perp}^2 = 2\int_1^2 \frac{d\Pi}{\sigma} .$$

(3.7)

The shock conditions require that $\sigma_1 v_{1\perp} = \sigma_2 v_{2\perp} = \Phi$, and so the momentum condition becomes

$$\Pi_2 - \Pi_1 = \Phi^2\left(\frac{1}{\sigma_1} - \frac{1}{\sigma_2}\right),$$

(3.8)

while (3.7) can be written

$$\int_1^2 \frac{d\Pi}{\sigma} = \tfrac{1}{2}\Phi^2\left(\frac{1}{\sigma_1^2} - \frac{1}{\sigma_1^2}\right).$$

(3.9)

According to (3.8) and (3.9), therefore

$$\int_1^2 \frac{d\Pi}{\sigma} = \tfrac{1}{2}(\Pi_2 - \Pi_1)\left(\frac{1}{\sigma_1} + \frac{1}{\sigma_2}\right).$$

(3.10)

To satisfy this relation requires either that $\Pi_2 = \Pi_1$, so that there is equal pressure on both sides of the shock. But this means that the shock is infinitesimally strong, and has no mechanical significance. Alternatively let

$$\frac{1}{\sigma} = A + B\Pi + f(\Pi),\tag{3.11}$$

where A and B are chosen in such a way that

$$f(\Pi) = 0 \quad\text{for}\quad \Pi = \Pi_1 \text{ and } \Pi_2.\tag{3.12}$$

It immediately follows from (3.10) and (3.11) that

$$\int_1^2 f(\Pi)\ d\Pi = 0\,.\tag{3.13}$$

Thus f vanishes for Π_1 and Π_2, and its integral vanishes over the range from Π_1 to Π_2. Therefore the function f has at least one point of inflection within the range and there

$$\frac{d^2 f}{d\Pi^2} \equiv \frac{d^2}{d\Pi^2}\ (\frac{1}{\sigma}) = 0.\tag{3.14}$$

Unless f vanishes identically there must also be a range of values for which

$$\frac{d^2}{d\Pi^2}\ (\frac{1}{\sigma}) < 0.\tag{3.15}$$

Now consider what this inequality implies for the process of shock formation by the steepening of pressure waves. We postulate the existence of a relation

$$\Pi = \Pi(\sigma),\tag{3.16}$$

which uniquely relates the surface pressure Π to the density σ. The speed of sound is then given by

$$a^2 = \frac{d\Pi}{d\sigma},\tag{3.17}$$

and

$$\frac{d}{d\Pi}\ (\frac{1}{\sigma}) = -\frac{1}{\sigma^2 a^2}\tag{3.18}$$

Inequality (3.15) is thus equivalent to

$$\frac{d}{d\Pi} \left(\frac{1}{\sigma^2 a^2}\right) > 0 \text{ , or to } \frac{d(\sigma a)}{d\Pi} < 0 \tag{3.19}$$

Since $d\Pi/d\sigma > 0$ if the system is to be stable, the inequality implies that somewhere in the gas

$$\frac{d(\sigma a)}{d\sigma} < 0 \text{ .} \tag{3.20}$$

Compare this result with that of relation (2.20). If a shock is to form at all, it does so by the steepening of pressure waves. In the present application this occurs only for those regions in the gas where

$$\frac{d(\sigma a)}{d\sigma} > 0 \text{ ,} \tag{3.21}$$

if the speed of sound is real, as it must be. But condition (3.20) was derived from the requirement that there is a unique pressure-density relation, and that the shock pattern is steady. Obviously conditions (3.20) and (3.21) contradict one another. The shock pattern therefore cannot exist with our assumptions.

The most reasonable conclusion to be drawn from these results is that the shock pattern is more complicated. If Π depends not only on the density σ, but also on the position in the Galaxy, then our discussion no longer applies. Physically the change means that the pressure depends not only on how much gas there is in a given position, but also on how far we are from the regions where energy is injected into the interstellar medium by early type stars or supernovae. Any reasonable account of spiral structure must allow for the fact that these energy sources are located preferentially in particular parts of the spiral pattern. The injection of energy is probably an important part of the process which maintains the spiral pattern in the gas.

4. THERMAL INSTABILITIES

Consider now the thermal balance of a region in H I space. There are various ways in which the gas may gain energy: two favourite suggestions are that the heat input comes from soft X-rays or from ionization losses of suprathermal particles. For the present it will be assumed that the rate of input of heat G, per unit mass of gas, remains fixed.

The gas can cool by a variety of processes. They usually involve the excitation of some higher level of an atom or ion by electron (or possibly atom) impact, and the subsequent radiation by the excited particle. The student is referred to the article by Dalgarno and

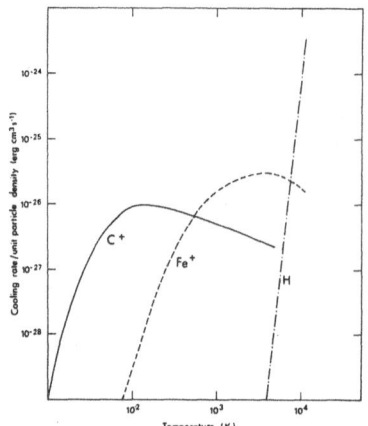

Fig. 1. Contribution to the cooling rate of the interstellar gas, made by various coolants. The fractional ionization below 10^4 K is assumed to be 1%. (after Dalgarno and McCray 1972).

McCray for a full discussion. When the fractional ionization x exceeds 1%, electron impact dominates.

Characteristically the rate of cooling due to any one species of ion goes like the solid curve sketched in figure 1, as the temperature is varied. On the high temperature side the dependence is like $T^{-\frac{1}{2}}$, and the Boltzmann factor makes little difference to the rate at which energy is communicated to the ion. But when the temperature is low, the factor $\exp(-\chi/kT)$ clearly dominates the cooling

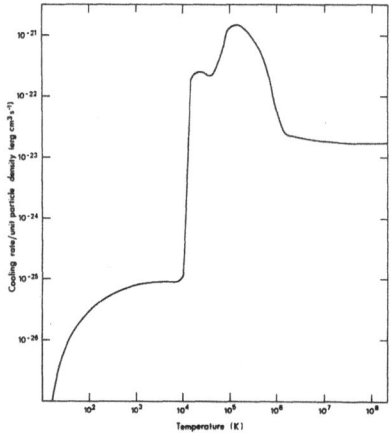

Fig. 2. The total cooling rate of the interstellar gas, with allowance made for contributions from all likely coolants. (after Dalgarno and McCray 1972).

rate. As one goes to higher temperatures other excitations, with
larger values of χ, switch in and so the Fe^+ ion (broken curve in
figure 1) eventually takes over. Ultimately, for $T \gtrsim 7000$ K,
cooling on H atoms becomes dominant.

The resultant combined cooling rate, for the interstellar
mixture, is sketched in figure 2.

The rate per atom present can be expressed as

$$L = n \, \Lambda(x,T) \tag{4.1}$$

where n is the number of atoms per cm^3. When x exceeds 1% the
rate is proportional to the fractional ionization and therefore

$$\Lambda(x,T) \equiv x \, \lambda(T). \tag{4.2}$$

A thermal instability characteristically takes place when
λ only varies slowly with T. This applies in the temperature
range below that at which the steep rise due to hydrogen cooling
sets in. The characteristic time for energy loss at, say, 5000 K
is

$$t_c = \frac{3kT}{2\lambda nx} \sim \frac{10^{-12}}{\lambda nx} \sim \frac{2 \cdot 10^{11}}{nx} \; \text{sec},$$

since λ equals about $5 \cdot 10^{-24}$ erg cm^3/s there. The recombination
time t_R is rather longer. With a recombination coefficient
typically equal to $B \sim 2$ or $3 \cdot 10^{-13}$ cm^3 s^{-1} we get $t_R \sim (Bnx)^{-1} \sim$
~ 3 or $5 \cdot 10^{12}/(nx)$ sec. In our approximation we neglect changes
in fractional ionization, since they take place rather more slowly
than the thermal adjustments in the gas.

To show that the gas is thermally unstable, under certain
conditions, consider a small amplitude, one-dimensional disturbance
in a gas. Magnetic fields are ignored, for the moment.

The linearization of the equations allows for small fractional
changes in pressure, density and temperature, so that these quantities can respectively be expressed in terms of their equilibrium
values by

$$P = P_o(1+\varpi), \quad \rho = \rho_o(1+s), \quad T = T_o(1+\theta), \tag{4.3}$$

and ϖ, s and θ are small in comparison with unity. The gas also
is assumed to move slowly, and therefore only linear terms in u
are retained. The linearized equations of motion and continuity
become

$$\frac{\partial u}{\partial t} = -\frac{P_o}{\rho_o} \frac{\partial \varpi}{\partial \zeta} \tag{4.4}$$

$$\frac{\partial s}{\partial t} + \frac{\partial u}{\partial \zeta} = 0 \tag{4.5}$$

The condition of thermal balance is

$$T \frac{\partial S}{\partial t} = G - L \tag{4.6}$$

Our assumption is that G remains fixed and equal to G_0. A standard formula gives the specific entropy in a monoatomic gas as

$$S = \frac{k}{m} \log \frac{P^{3/2}}{\rho^{5/2}} \tag{4.7}$$

so that the perturbed value of the specific entropy is

$$S = S_0 + \frac{k}{m} \left(\frac{3}{2}\vartheta - \frac{5}{2} s\right). \tag{4.8}$$

The energy loss rate is

$$\begin{aligned} L &= \lambda\{T_0(1+\theta)\}n_0(1+s)x = \\ &= \lambda(T_0)n_0 x\{1+ \frac{T_0\lambda'(T_0)}{\lambda(T_0)} \theta + s\}. \end{aligned} \tag{4.9}$$

In first order the equation of thermal balance becomes

$$\frac{kT_0}{m} \left(\frac{3}{2} \frac{\partial \vartheta}{\partial t} - \frac{5}{2} \frac{\partial s}{\partial t}\right) = - \frac{\lambda(T_0)n_0 x}{m} \frac{T_0\lambda'(T_0)}{\lambda(T_0)} \theta + s . \tag{4.10}$$

Finally the equation of state $P = (k/m)\rho T$ linearises to give

$$\vartheta = s + \theta. \tag{4.11}$$

In terms of the characteristic cooling time t_c the equation of energy balance becomes

$$\frac{\partial \vartheta}{\partial t} - \frac{5}{3} \frac{\partial s}{\partial t} = - \frac{1}{t_c} \{\alpha \vartheta + (1-\alpha)s\} \tag{4.12}$$

where $\alpha \equiv T_0\lambda'(T_0)/\lambda(T_0)$.

Equations (4.4), (4.5) and (4.12) together describe small amplitude waves in the gas, when the effects of the thermal balance are allowed for. To see whether the system is stable, let an infinitesimal disturbance have the spacetime dependence $\exp(\sigma t + iK\zeta)$ where K is real. Then a little algebra leads to the dispersion relation

$$\sigma^3 + \frac{\alpha}{t_c} \sigma^2 + \frac{5}{3} K^2 c_0^2 \sigma - (1-\alpha) \frac{K^2 c_0^2}{t_c} = 0, \tag{4.13}$$

with $c_o = (kT_o/m)^{1/2}$, the isothermal sound speed in the medium.
The wave will grow, and therefore the medium is unstable, if values
of σ can be found which satisfy relation (4.13) and have positive
real parts. Such disturbances grow without limit, in the linearized approximation. The interstellar medium cannot of course continue to exist in a state where it is subject to unstable disturbances. It is clear from the relation that there will be at least
one positive root for σ if

$$1 - \alpha > 0, \text{ or } \frac{T_o \lambda'(T_o)}{\lambda(T_o)} < 1. \tag{4.14}$$

In small wavelength disturbances, that is for K large, the growth
rate becomes, approximately,

$$\sigma = \frac{3(1-\alpha)}{5t_c} \tag{4.15}$$

and the growth rate is thus rather smaller than the cooling rate.
Large K, in this connection, clearly means that

$$Kc_o \gg \frac{1}{t_c} \text{ or } \lambda_w \equiv \frac{2\pi}{K} \ll 2\pi c_o t_c. \tag{4.16}$$

The wavelength of these disturbances must be small in comparison
with the distance which a sound wave can travel in a few cooling
times.
     For large wavelength disturbances, or small K, the relation
leads to the approximate result

$$\sigma = (\frac{1-\alpha}{\alpha})^{1/2} Kc_o. \tag{4.17}$$

In this case small K means $Kc_o \ll t_c$ so that these disturbances
grow on a time scale considerably longer than the cooling time
unless α is very close to zero. In fact instability occurs in
about the same period as it takes a sound wave to cross the region
concerned. In general therefore the growth of a thermal instability
is usually rather slow.

5. THERMAL INSTABILITIES IN THE PRESENCE OF A MAGNETIC FIELD

The presence of a magnetic field, in a conducting gas, alters the
nature of the waves that the gas can carry, and so the discussion
of thermal instabilities has to be reconsidered.
     In general, if a gas is a (nearly) perfect conductor, there
are three possible kinds of wave. The first kind is a purely
transverse Alfven wave. Here the disturbance velocity and the
disturbance magnetic field are both perpendicular to the plane
containing the wave-vector $\bar{K}$ and the unperturbed magnetic field
$\bar{H}_o$. No pressure or density changes are associated with such a
wave provided the amplitude remains small. But thermal instabilities arise just because the thermal balance of a gas depends on

its pressure and density. For the transverse Alfven waves there can be no such effect.

The remaining possible waves couple together to form a system of combined sound and MHD waves. The disturbance magnetic field and the velocity are now coplanar with $\bar{K}$ and $\bar{H}_o$, the perturbation vector potential is perpendicular to that plane. We shall briefly derive the properties of the wave in a convenient form. Let $\bar{A} = (o, A_2, o)$, $\bar{u} = (u_1, o, u_3)$ and let the perturbed physical parameters be pressure $P = P_o(1+)$, density $\rho = \rho_o(1+s)$. The wave is assumed to have a space-time dependence $\exp\{i(k_1 x_1 + k_3 x_3) + + \sigma t\}$. Then we get from the equation of continuity

$$\frac{\partial \rho}{\partial t} + \bar{\nabla} \cdot (\rho \bar{u}) = 0 \quad \rightarrow \quad i\sigma s = k_1 u_1 + k_3 u_3; \quad (5.1)$$

from the perfect conductivity condition

$$\bar{E} + \frac{1}{c} \bar{u} \times \bar{H} = -\frac{1}{c} \frac{\partial \bar{A}}{\partial t} + \frac{1}{c} \bar{u} \times \bar{H} \quad \rightarrow \quad \sigma A_2 + u_1 H_o = 0; \quad (5.2)$$

from the equation of motion

$$\frac{\partial \bar{u}}{\partial t} + (\bar{u} \cdot \bar{\nabla})\bar{u} = -\frac{\bar{\nabla}P}{\rho} + \frac{1}{\rho c} \bar{j} \times \bar{H}$$

we have

$$i\sigma \rho_o u_1 = k_1 P_o \vartheta + \frac{iH_o}{4\pi} (k_1^2 + k_3^2) A_2 \quad (5.3)$$

$$i\sigma \rho_o u_3 = k_3 P_o \vartheta. \quad (5.4)$$

In each case the full equation turns into the equations (5.3) and (5.4) when the substitutions given above are made for $\bar{A}, \bar{u}, P$, and ρ, and the resulting forms are linearised. It follows from these equations that

$$\{\frac{v_A^2}{\sigma^2} (k_1^2 + k_3^2) + 1\}(s + \frac{k_3^2}{\sigma^2} c_o^2 \vartheta) + \frac{k_1^2}{\sigma^2} c_o^2 \vartheta = 0, \quad (5.5)$$

where $v_A = (H_o^2/4\pi\rho_o)^{1/2}$ is the Alfven speed.

A further relation between ϑ and s comes from equation (4.12) for the energy balance. This shows that

$$\vartheta = s \frac{5\sigma t_c - 3(1-\alpha)}{3(\sigma t_c + \alpha)} \quad (5.6)$$

From (5.5) and (5.6) it follows, after some re-arrangement, that the dispersion relation with allowance made for cooling becomes

$$\sigma^5 + \frac{\alpha}{t_c} \sigma^4 + \sigma^3(v_A^2 + \frac{5}{3} c_0^2)K^2 + \sigma^2 k^2\{\frac{\alpha}{t_c} v_A^2 - \frac{(1-\alpha)}{t_c} c_0^2\} +$$

$$+ \frac{5}{3} \sigma v_A^2 c_0^2 k_3^2 K^2 - \frac{(1-\alpha)}{t_c} v_A^2 c_0^2 k_3^2 K^2 = 0. \qquad (5.7)$$

In this formula $K^2 \equiv k_1^2 + k_3^2$; clearly there will be a thermal instability if the dispersion relation has at least one positive root for σ. The necessary and sufficient condition for this is that $\alpha < 1$ (unless $k_3 = 0$, but that is an exceptional case). The growth rate is now approximately given, in the case of short wavelengths, by setting the last two terms in relation (5.7) equal to zero. This leads to

$$\sigma = \frac{3(1-\alpha)}{5t_c}, \qquad (5.8)$$

the same growth rate as was found in the absence of a magnetic field.

For waves with small K (and large wavelength) we again expect that the growth rate will be determined by the speed with which a sound wave can cross a wavelength. Therefore the growth rate will be small compared with the cooling rate. Picking out the dominant contributions from relation (5.7) then gives that

$$\alpha\sigma^4 + \sigma^2 K^2\{\alpha v_A^2 - (1-\alpha)c_0^2\} - (1-\alpha)v_A^2 c_0^2 K^2 k_3^2 = 0, \qquad (5.9)$$

and again there will be a growing mode, with $\sigma > 0$, provided $\alpha < 1$. (The case $k_3 \equiv 0$ is once more exceptional). The presence of a magnetic field does nothing therefore to alter the thermal stability of the medium.

If the magnetic field is strong, the Alfven speed much exceeds the speed of sound, and it then follows that the unstable mode has a growth rate σ given by

$$\sigma^2 = \frac{1-\alpha}{\alpha} c_0^2 k_3^2 \qquad (5.10)$$

approximately. In this case the term $\alpha\sigma^4$ makes only a negligible contribution. Compare this result with the corresponding result, in equation (4.17), when there is no magnetic field. It seems that there has been not much change. The growth rate is now proportional to k_3, the component of K along the zero-order magnetic field. Previously it was proportional to K. The main consequence is that among instabilities with a given wave length the ones that grow best are those whose wave-vector is parallel to the magnetic field.

To summarize then: thermal instabilities occur in principle

whenever the cooling rate of the interstellar gas does not rise
fast enough with the temperature, so that $\alpha \equiv T_0 \lambda'(T_0)/\lambda(T_0) < 1$.

But the growth time is never less than the cooling time t_c,
or the time that it takes a sound wave to travel one wavelength,
whichever is the longer. Therefore thermal instabilities grow rather
slowly. This conclusion is hardly affected by the presence or
absence of a magnetic field.

The whole discussion concerning these instabilities is based
on the assumption that the background pressure and the heating rate
are maintained constant while the instability grows. But suppose
that we wish to study the stability of a region, say 30 pc across,
and that the speed of sound is 3km/s. Unless the gas density is very
low, we have here a case in which the long wavelength limit applies.
The time scale for an instability to grow is at least of the order
of 10^7 years. It may well be that the background conditions do not
remain steady for so long a period. Therefore the instability has
no chance to develop, and we can no longer insist that the interstella
medium must always be found in one of its stable phases. The two-
phase theory of the interstellar medium should consequently be applied with some reservations.

6. COCOONS AROUND EARLY STAR [†]

A cocoon will form around any star which is accreting interstellar
matter at a large enough rate. The interstellar grains in the cocoon
absorb the direct starlight, and re-emit in the infra-red. In many
cases the cocoon is also opaque to this reradiated radiation, and
therefore a gradient of radiation pressure builds up and retards
the infalling material. As the mass of the central star grows, and
its luminosity to mass ratio increases, this effect becomes more
important. Eventually the retardation becomes so severe that the
infalling gas and dust mixture cannot cross the surface where the
grains melt. Once this has happened the flow is turned back and a
dense shell of interstellar mixture begins to be driven away from
the star. However the shell is unstable, and will probably break
up into fragments.

This process sets a limit to the luminosity to mass radio
L/M of those stars which can accrete interstellar gas. Under present
day conditions in our part of the Galaxy the limit appears to be
$L/M \leq 10600$ erg/gm s. This ratio is appropriate to a star of about
40 $M_\odot$.

[†]   Only the abstract of this lecture is given here. It was based
on the text of a paper, to be published soon in Astronomy and Astrophysics.

REFERENCES

Section 1.
An excellent introduction to fluid mechanics is given by L.D. Landau
and E.M. Lifshitz, "Fluid Mechanics", London, Pergamon Press 1959.
The cooling processes in interstellar space are comprehensively
described by A. Dalgarno and R.A. McCray, Annual Reviews in As-
tronomy and Astrophysics, 10, 375, 1972.
Section 2.
Landau and Lifshitz, Chapter X, contains a description of plane waves
of finite amplitude.
Section 3.
The spiral density wave theory, and its verifications, are treated
by C.C. Lin in this volume.
Section 4.
An early discussion of thermal instabilities is given by G.B. Field
in his article in "Interstellar Matter in Galaxies" (ed. L. Woltjer)
New York, W.A. Benjamin 1962.
Dalgarno and McCray summarize later developments, and also the theory
of the two-phase model of the interstellar medium.
Section 5.
For MHD waves see V.C.A. Ferraro and C. Plumpton "Magneto-Fluid
Mechanics" Oxford 1961.
Section 6.
The paper on Cocoon Stars will be published soon in Astronomy and
Astrophysics.

ATOMS AS MONITORS OF GALACTIC EVOLUTION

H. Reeves

Centre d'Etudes de Saclay - GIF 91190,
Institut d'Astrophysique de Paris

The evolution of the chemical composition of the interstellar
matter (gas and dust) is intimately related to the history of
stellar formation and destruction throughout the galactic life.
If we could know the abundance of elements and isotopes as a
function of time, we could go a long way in understanding galactic
evolution. Each isotope has potentially something to say, and it
is our job to extract this information from it. To extract this
information, we first need to identify the nuclear processes by
which this isotope is made and by which it can be destroyed, and
then the plausible astrophysical circumstances in which these
processes did take place.

1. FATE OF INDIVIDUALS OR GROUPS OF ISOTOPES

Let us first consider the case of Deuterium: it appears very likely
that, by far, the largest source of D is the Big-Bang itself and that
only very small contributions were made later either by Shock-Wave in
Supernova envelopes or by Galactic Cosmic Ray bombardment of interstellar Helium. Furthermore, because Deuterium atoms are very fra-
gile (they have very short life time against thermonuclear destruction
at typical Main-Sequence stellar temperature), they will be entirely
destroyed in any mass of interstellar matter which has been astrated
(that is: been involved in a star and later been ejected back in
space). In other words the fraction of original (Big-Bang) Deuterium
still present in the interstellar gas is expected to be equal to the
fraction of galactic gas which has never been involved in stars.
(The "unastrated" mass fraction of the gas).
    Next we consider ^{4}He. Again we have good reasons to expect that
this isotope was made to a very large extent in the Big-Bang, mainly

G. Setti (ed.), Structure and Evolution of Galaxies, 197–215. All Rights Reserved.
Copyright © 1975 by D. Reidel Publishing Company, Dordrecht-Holland.

because its abundance appears to be so uniform throughout quite a number of galaxies, (while from "local" origin we would have no reason to expect such a uniformity) and also because Big-Bang nucleosynthesis very nicely predict the correct abundance.

However some ^{4}He should also come from stellar synthesis since it is through the $4H \rightarrow ^4$He burning that Main-Sequence stars get their light. This extra contribution of ^{4}He is not expected to be more than one or two He atoms per one hundred H atoms (while the observed abundance is about ten He atoms per one hundred H atoms). Several attempts (Mezger 1973, Peimbert 1974) have been made at separating this contribution from the Big-Bang contribution. It is clear that its magnitude would give us important information on the integrated amount of stellar activity in any given neighborhood.

The isotopes of ^{3}He and ^{7}Li appear also to be of Big-Bang origin, although this origin can be disputed since other mechanisms could have contributed importantly. Here a clearer determination of origin is needed before the analysis can proceed any further.

The isotopes of ^{6}Li, ^{9}Be, ^{10}B, ^{11}B appear to have been created by the bombardment of the interstellar gas by the Galactic Cosmic Rays throughout the life of the galaxy. As for Deuterium, the abundance of these elements is altered by astration since they are hardly more resistent to stellar heat. We note that both the formation rate of these elements and also their destruction rate will be affected by the rate of astration: first it is generally believed that cosmic rays originate in Supernova which are believed to come from large stars of short duration; second the abundance of target elements (CNO) in the gas increases with the amount of past astration; third their rate of destruction by astration is evidently related to the rate of astration.

The elements from Carbon to Uranium (except perhaps ^{50}V which appears also to come from the Galactic Cosmic Rays) are believed to be the result of stellar cooking at various stages of stellar evolution. Some elements like C, O and the s elements (heavy isotopes, with A>60, lying at the bottom of the valley of nuclear stability) are probably made in quiescent periods of more or less isothermal nucleosynthesis. Other elements like the iron peak elements and the r-elements (heavy isotopes lying on the neutron-rich side of the valley of nuclear stability) are probably made during explosive stages, presumably taking place during supernova explosions. The stellar cooking is returned to the galactic gas through stellar winds, planetary nebulae phenomenon or supernova explosions and, in this way, the interstellar matter is gradually enriched in these elements. Some of these atoms are later taken back from the gas by further astration, etc. Hence again, we expect their evolutionary abundance curve (abundance as a function of galactic age) to be related to astration, but in a way different from the way of the previously discussed atoms.

2. THE FRACTIONAL GAS MASS IN THE GALAXY

Most models of galactic evolution have assumed that, at its "birth",
the galaxy was entirely gaseous and that, since then, it has gradually
transformed its gas in stars, with some stars going partially back
in gas. Furthermore, it is generally assumed that the galaxy is a
"closed" system which is not gaining or losing matter to its extra-
galactic environment. There is not much foundation for these assumptions except that they keep the problem reasonably simple. But
before we understand anything about galactic formation, it is very
imprudent to believe too strongly in these assumptions. It is wiser
to leave the situation open since we have there a chance of learning
something about galactic origin.
     For the present lectures, we shall adopt these conventional
assumptions and derive a simple formalism mainly for illustrative
purposes. It will be interesting to see, from this formalism, how
the monitor-atoms can tell their story.
     The mass of the galaxy M_G is constant in the present context.
It is made of a mass $S^*(t)$ of stars and a mass $U(t)$ of gas and dust.
Then, clearly $dU/dt = -dS^*/dt$. Let us define $\tau^*(t)$ the lifetime of the
gas against condensation into stars:

$$\frac{dS^*}{dt} \equiv \frac{U(t)}{\tau^*(t)} = -\frac{dU}{dt} \tag{2.1}$$

Then, at time t after the birth of the galaxy we have

$$U(t) = M_G \exp(-t/<\tau^*>) \tag{2.2}$$

where[†]

$$\frac{1}{<\tau^*>} = \frac{1}{t} \int_0^t dt'/\tau^*(t') \tag{2.3}$$

The present age of the galaxy is evaluated as $\sim 11 \times 10^9$y, from studies
of old clusters, and the present fractional gas mass is ~ 0.1 in the
solar neighborhood, yielding a value $<\tau^*> \approx 5 \times 10^9$y. For the galactic
center where the fractional gas mass is $\sim 10^{-2}$ one would have $<\tau^*> =$
$= 2.5 \times 10^9$y and at the galactic limb, the fractional gas mass is
~ 0.5, hence $<\tau^*> = 15 \times 10^9$y. Thus it appears that the lifetime $<\tau^*>$
is a function of the radial distance in the galaxy. However, this
tells us nothing of the time history of $\tau^*(t)$.

3. THE ACT OF ASTRATION

We define $M_i(t)$ as the mass of nuclide i in gas and dust, and $Z_i(t)=$
$= M_i(t)/U(t)$ as the mass fraction of i. Also $Z(t) = \sum_i Z_i(t)$ is

---

[†]I have been rather careless in the definitions here. I confuse
such things as $<1/\tau^*>$ and $<\tau^*>^{-1}$. In practice the confusion is not
very serious.

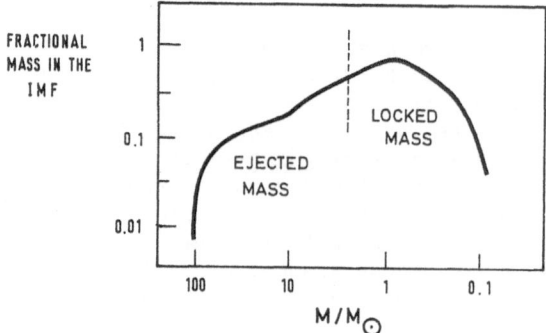

Fig. 1. The Initial Mass Function is the distribution of stars of
mass M in a cluster as a function M. From the graph it appears that
most of the mass is in the form of low mass stars. This distribution
applies to stars formed recently. It is of interest to know if it
has varied with time throughout the galactic life.

fractional mass of metals (A>4) in gas and dusts.

The elementary act of astration consists in the following: an
interstellar cloud of unit mass (for convenience), and of metal
abundances Z_i, collapses somewhere. After multiple fractionations
and condensations, there is left a number of stars with various
masses. The distribution of stellar masses as a function of the
mass is called the initial mass function (IMF) (Salpeter 1959).
The Fig. 1. shows this distribution as observationally determined
on young clusters (star formation in very recent time).

Clearly, most of the mass of the gas goes into rather small
stars (M $\leqslant$ 2 $M_\odot$) and a rather small fraction goes into the stars
which, from stellar evolution calculations (Truran and Cameron 1971,
Talbot and Arnett 1973), we expect to be the main producers of nucleosynthesis. Since, on the other hand, these big stars have a rather

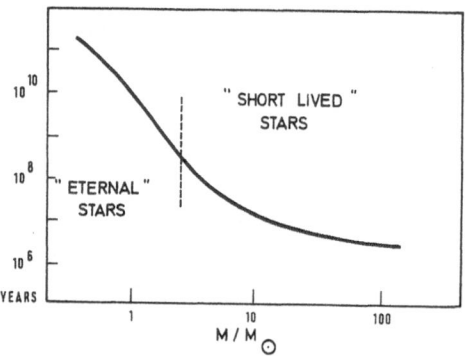

Fig. 2. Lifetime of the stars as a function of their masses. Stars
with more than a few solar masses have lives considerably shorter
than the age of the galaxy.

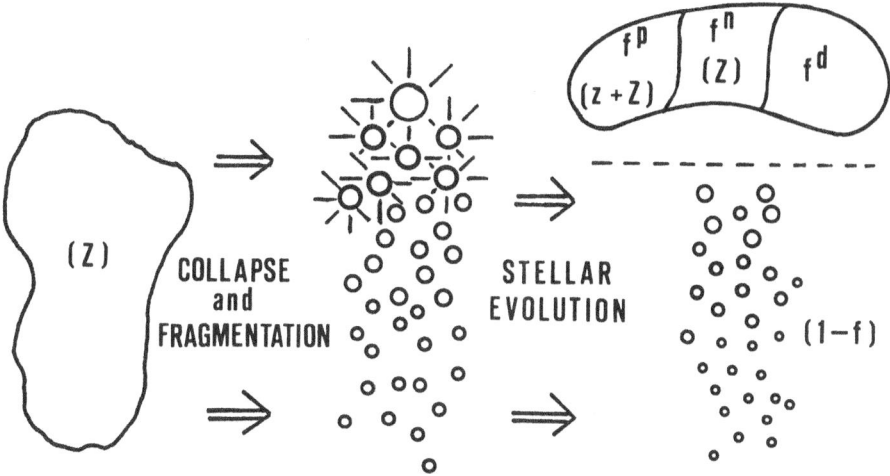

Fig. 3. Sketch of the elementary act of astration: an interstellar
cloud with metal abundance Z collapses and fragments in a number of
stars with various masses. The stars of large masses evolve rapidly
and shed back into space a fraction f of the cloud's initial mass.
For each element Z_i this ejected fraction can further be divided
in three parts. One part f^d where this element is destroyed; one part
f^n where it is restituted with its initial value Z_i (no alteration);
and one part f^p where a new amount Z_i is added.
The rest of the cloud (with fractional mass (1-f)) remains locked in
small stars with very large lifetime.

short lifetime ($\leqslant 10^8 y$), one may introduce very conveniently the
"Instant Cycling Approximation" which consists in breaking the distribution in two parts, one part which is locked eternally in small
stars (with lifetime comparable to the age of the galaxy), and
another part which is returned immediately to the interstellar
medium with its load of heavy elements cooked in stars(Fig. 2.).
Let us call f the mass fraction returned into space by these big
stars (Fig. 3.). To compute f we need to know that when a large
star dies, it probably does not shed its entire mass back into space,
but keeps a remnant of approximately one solar mass under the form
of a white dwarf or a neutron star or perhaps even a black hole. The
value of f turns out to be ~ 0.2 from present day stellar data (Tinsley
1974). We do not know if this value has always been the same. Some
people believe that in the early days of the galaxy the fraction
of large stars in the IMF was larger, resulting in a larger f (Schmidt
1963).

An estimation of the value of f can be obtained also through a
statistical study of the pulsar numbers and a plausible interpretation of their origin. Much of the present material comes from various
papers by Ostriker (see in particular Ostriker, Richstone and Thuan
1974).

Pulsars are believed to be rotating neutron stars. Neutron stars cannot result from the quiet contraction of a small star - which would rather lead to the white dwarf configuration - but more likely from the violent collapse of the core of a large star accompanying the explosion and ejection of the outer layers. Hence, qualitatively, we should expect a relationship between the ejected mass fraction in a generation of stars and the rate of pulsar formation.

The rate of pulsar formation can be obtained from data on the pulsar density and pulsar age, determined by combining measurements of their rotation period (Ω) and their rate of deceleration ($\dot{\Omega}$).(The age is then given by ($\Omega/\dot{\Omega}$)).

It has been customary to discuss this data in a cylindrical volume of one parsec square around the sun, in the galactic plane, and extending up and down on both sides of the galactic plane until the border of the Galaxy. This is usually called our "local pool". For our discussion we need to know that the total mass of matter in this pool is $\sim$75 $M_\odot$ including $\sim$7.5 $M_\odot$ of interstellar gas.

The rate of pulsar birth in our pool is $\sim$5x10^{-11}y^{-1}pc^{-2}. This number can be compared to rate of star formation as a function of stellar mass. This rate is obtained in a way which is quite similar to the way used for the pulsar formation rate. One counts the number of big stars with mass M in our pool and divides this number by the life-time of this star (essentially its life-time on the main-sequence For stars of M>3 $M_\odot$ these lifetimes are quite short (less than 3x10^{8}y) and we may assume equilibrium between stellar formation and destruction. From these statistics it appears that the stellar birth rate of stars of M>3 $M_\odot$ is about equal to the pulsar birth rate. This gives some support to the view that pulsars originate as cores of stars with M>3 $M_\odot$. Furthermore, from the number distribution of these stars of mass M, we may estimate the stellar mass ejected, and we get $\approx$2x10^{-10}(pc)$^{-2}$y^{-1} $M_\odot$ = $(dU/dt)_{ej}$. In terms of our previously defined terminology, it is clear that $(1/U)(dU/dt)_{ej} = f/\tau^*(1-f)$. could be called the rate of interstellar gas replenishment by stellar ejecta. With the numbers given previously, we obtain $f/\tau^*=3x10^{-11}$y^{-1}. As discussed previously, if τ^* is assumed to be time independent it takes the value 5x10^{9}y, hence f$\approx$0.15, in reasonable agreement with the previously quoted value. Actually, the two methods have large inherent uncertainties and they result in a total uncertainty of at least (but probably not much more than) a factor two each side.

The fact that some gas is "immediately" returned to space implies that the lifetime against gas depletion (τ^*) is somewhat larger than the lifetime against astration (τ). (For clarification: τ^* refers to condensation into long-lasting stars only, while τ refers to condensation into all stars). Let us define dS/dt as the rate of incorporation of matter in collapsing clouds (or the total mass of clouds collapsing per unit time). Then:

$$\frac{dS}{dt} = \frac{1}{1-f} \frac{dS^*}{dt} \qquad\qquad (3.1)$$

and the lifetime τ against astration is

$$\frac{dS}{dt} = \frac{U(t)}{\tau(t)}, \tag{3.2}$$

hence $\tau(t) = (1-f)\tau^*(t) \simeq 4\times10^9$ years today.

(To be coherent with this discussion we should redefine $S^*(t)$ as the mass of all stars with less than 2-3 $M_\odot$ in the galaxy at time t. However, the mass fraction neglected this way is very small, much smaller than f itself since for all previous generations only the small stars are leftover).

4. THE DEUTERIUM ABUNDANCE

One can now discuss the variation of the D/H ratio in the galactic gas as a function of these two lifetimes. Since the galactic gas is almost entirely made of H and He ($\sim$98%), and since the ratio of H to He has remained constant, it is clear that the lifetime for decrease of the total mass of H is the same as the lifetime τ^* of gas depletion into stars. But since, because of its fragility, Deuterium is entirely burned in the fraction f ejected from big stars (as well as, of course, in the fraction (1-f) locked in stars), its lifetime is $\tau(t)$. Hence if the galaxy started with a D/H ratio as given at the end of the Big Bang $(D/H)_{BB}$, we should find the ratio to decrease with:

$$(D/H)_t / (D/H)_{BB} = \exp(-t/<\frac{(1-f)\ \tau^*}{f}>) \tag{4.1}$$

where

$$<\frac{(1-f)\ \tau^*}{f}>^{-1} \equiv \frac{1}{t} \int_0^t \frac{f(t)\ dt}{(1-f(t))\ \tau^*(t)} \tag{4.2}$$

which we shall call the "average ejection rate". Assuming for the moment that f and τ^* are time independent, the ejection rate is $\sim 2\times10^{-11}y^{-1}$ and the D/H ratio in the gas is ~ 0.6 of its original value (or equivalently 60% of the present gas has never been astrated (although $\sim$94% of the initial gas has been astrated)).

A full history of these parameters could be drawn if we had measurements of D/H ratios at different moments of the galactic life. We have a value of D/H $\sim 1.5\times10^{-5}$ in interstellar matter (hence applying today) and a measurement of D/H $\sim 2\times10^{-5}$ in the atmosphere of Jupiter (hence applying some 5×10^9 years ago, at the birth of the solar system). The previous formula, assuming constant f and τ^*, would predict a decrease of 20% from the birth of the sun until today which, given the uncertainties, is quite consistent with the data. (Actually the data and uncertainties could be consistent with no decrease at all (see Fig. 4.)). Before we can obtain, from these Deuterium measurements, the Big Bang D/H ratio, we shall need more information on the past history of f and τ^*. (Reeves 1974).

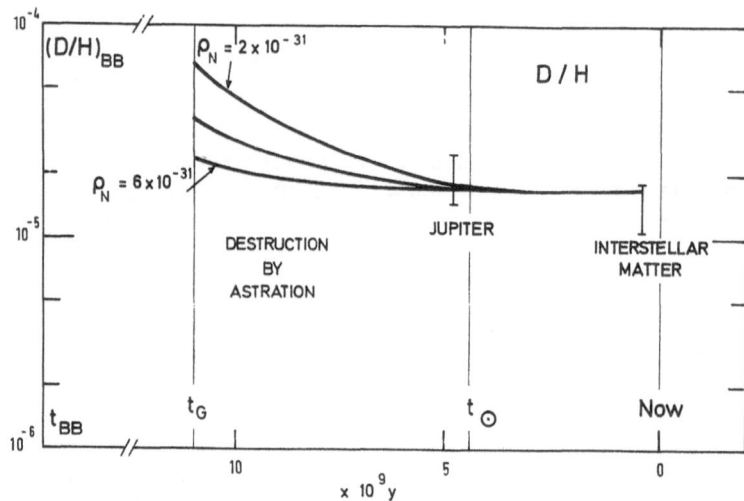

Fig. 4. The Evolutionary Abundance Curve of Deuterium. Measurements
of D/H in Jupiter (corresponding to the interstellar value at the
birth of the sun) and in the present interstellar space give the
value $\sim 2 \times 10^{-5}$ within the uncertainties. We have no data for previous
time but we believe that during the galactic life D/H can only
decrease, and the amount of reduction is related to the time integrated history of astration (Reeves 1974). On the other hand,
the initial D/H after Big Bang depends strongly upon the present
universal density. This way, the history of astration can be related
to the universal density. For example, from the graph, a minimal
amount of astration (and consequent deuterium reduction) would
correspond to a density of $\sim 6 \times 10^{-31}$gr cm^{-3},while a reduction by
about a factor thirty in the deuterium abundance would correspond
to a density roughly equal to the density of the visible universe
1×10^{-31}gr cm^{-3} (hence this is an upper limit to the amount of astration which can have taken place) (Wagoner 1973).

5. THE METAL ABUNDANCE

Next we would like to know the distribution of chemical abundances
in the mass fraction of ejected matter. For each nuclide i, we may
define a mass fraction $f_i{}^d$ in which it is entirely destroyed, a
mass fraction $f_i{}^n$ in which it remains at its original cloud value
of Z_i and a last fraction $f_i{}^P$ in which its value is changed to
$(z_i + Z_i)$, where in general $z_i \gg Z_i$. Clearly $f_i{}^d + f_i{}^n + f_i{}^P = f$.
       To determine these mass fractions we would need to understand
stellar evolution from the beginning to the end for each stellar
mass. We would need to have reliable models of presupernova stars,
and we should know how to compute the effect of tne explosion and of
the expansion on the chemical abundances. Needless to say, we are
far from this goal. Even our belief that we understand at least the

Main Sequence (Hydrogen Burning) stage will not be justified before
we have clarified the mystery of the "missing solar neutrinos".

However, the situation may not be as dark as it seems since
the nuclear physics aspects of the problem do give us some interesting hints and boundaries. We know for instance that ^{12}C and ^{16}O
are made at the same times, by helium-burning processes. In the
same fashion, we may consider together certain groups of nuclides
which are generated in the same nuclear processes and in this way
have some fairly good knowledge of the ratio of their abundances at
formation although we may not know when and where they are formed.

Two main avenues of investigation are pursued. One consists
in grouping all the elements in three groups: Hydrogen, Helium and
the metals Z. The present studies indicate that metals are never
transformed back into H and He so that we can set $f_Z^d = \sum_i f_i^d = 0$.
Attempts at computing f_Z^n and $f_Z^p = \sum_i f_i^p z_i$ can be made by looking
at existing models of advanced stages of stellar evolution for various
stellar mass M and plotting the profile of the mass fractions in
metals as shown in Fig. 5. (It is difficult to estimate the uncertainties involved in this computation...). Then, by convolving
this graph with the initial mass function (Fig. 1) one could in
principle compute the required quantities (Talbot and Arnett 1973).

A more promising approach consists in looking directly in the
sky for the result, by making statistics of the number of stars
with metal abundance Z as a function of Z.

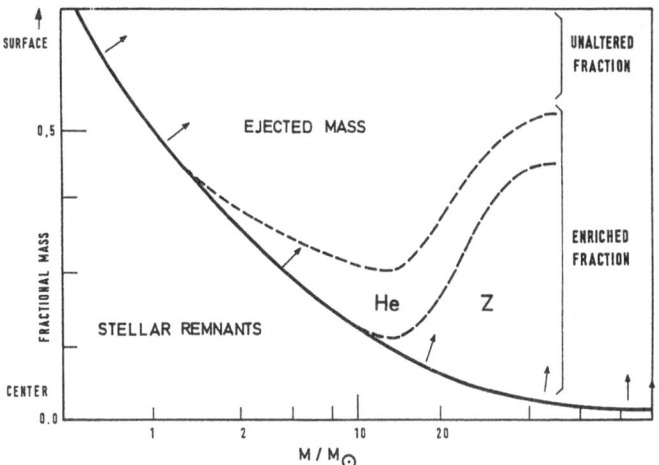

Fig. 5. This figure shows the fractional mass of a star of mass M
which is returned to interstellar space, and the fractional mass which
is left as a remnant (white dwarf, neutron star, etc...). The
ejected mass is subdivided into three fractions, a fraction where the
chemical composition is essentially unaltered (except for destruction
of D and other light elements), a fraction where the main product is
Helium (from Hydrogen burning), and a fraction where the main pro-
ducts are nuclei heavier than Helium.

Let us first consider the problem theoretically in the framework established previously. The total mass of metal in the galactic gas (UZ) varies with time in the following way (Pagel 1973;
Searle 1973):

$$\frac{d(UZ)}{dt} = -Z\frac{dS}{dt} + Zf_Z^n\frac{dS}{dt} + (Z+z)f_Z^p\frac{dS}{dt} = \frac{dU}{dt}(Z-y) \qquad (5.1)$$

where $y = (f_Z^p z)/(1-f)$ is usually called the "yield" of metals.
It is useful to eliminate the time variable and to replace it by
Z. Then

$$\frac{dZ}{dU} = -\frac{y}{U}, \qquad (5.2)$$

yielding

$$U(Z) = M_G \exp(-Z/<y>) \qquad (5.3)$$

where

$$\frac{1}{<y>} = \frac{1}{Z}\int_0^Z \frac{dZ}{y(Z)} \qquad (5.4)$$

This gives us a relation between the fractional gas mass and the
mean metal abundance Z. With the present values $U(t)/M_G = 0.1$ and
$Z \sim 0.02$ in the solar neighborhood, we obtain $<y> \sim 0.01$. Assuming,
again momentarily, that the astration parameters (zf_Z^p) and f are
independent of time, we get $(zf_Z^p) \simeq 0.008$, implying a stellar-
produced metal abundance of $zf_Z^p/f = 0.04$ in the whole ejected mass
fraction. This is, then, what is required from the nuclear cooking
of elements in big stars. One nice confirmation of the applicability
of these ideas is found in the metal variation from center to limb
of spiral galaxies (our's and others). The Z abundance varies by a
factor of four (decreasing outward), while fractional gas decreases
by about one hundred (in the same direction), as expected from the
previous formalism if $<y>$ is space independent but τ^* is space
dependent (as discussed previously).

The same approach can be used to test the time-independence
of y and in fact the matter turns out to be not so simple, as we
shall see shortly. The task consists first in selecting a class of
stars with Main-Sequence lifetime longer than the age of the galaxy,
for instance G type stars, and then, counting the number of stars
of this class which have a metal abundance smaller than or equal to
a given value Z, $N(\leqslant Z)$. From the total mass of all these stars one
can compute the fractional stellar mass at the moment the galactic
gas reaches the value Z, $(S^*(Z)/M_g)$, from which one can easily
obtain the fractional gas mass $U(Z)/M_g$ at the same time (there are
of course some technically difficult but conceptually simple problems of normalization into which we need not go at the present
time). Then one may draw a plot of $\ln(U(Z)/M_g)$ as a function of
Z, and one expects this plot to be a straight line if y is time
(hence Z) independent. This data is presented in Fig. 6, adapted

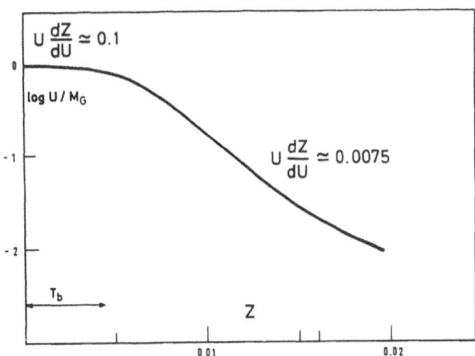

Fig. 6. Mass fraction in local galactic gas, as a function of the
metal abundance Z in the gas, obtained by counting the number of
stars with a given value of Z, and going through a simple formalism,
as described for instance in Searle 1973. The slope of the line
is directly related to the "yield" of astration at any one time.
One way of accounting for the data is to assume that the yield
started at the value of ∿0.1 and, when Z∿0.004, decreased rather
abruptly to 0.01. (see text).

from Pagel (1973). One can indeed draw a straight line, but it
does not extrapolate to a zero value of Z at $U(Z) = M_G$, the moment
the galaxy was entirely gaseous! That the difficulty lies at a
period corresponding to the birth of the galaxy may not be so surprising after all, since, as discussed before, we know nothing about
the formation of galaxies and the model used here represents probably a foolhardy extrapolation into these nebulous times which are
hardly disconnected from cosmology.
      In more visual terms the problem is the extreme paucity of
low-metal stars: less than one star in one thousand has less than
one sixteenth of the solar metal abundance while from a constant
y of ∿0.01 we should expect about ten percent. Several theories
have been put forward to account for this fact. One consists of saying
that the galaxy started with a metal abundance of about 0.004 (20%
of the solar value) which would explain why the straight line of
Fig. 6 does not extrapolate to zero. But then, we are faced with
the question: where do these metals come from? An origin in the
Big Bang seems excluded by the present calculations, and also
would not explain the very few stars which have <u>less</u> metals than the
20% solar value...
      Another theory consists of saying that the earliest astration
phases were much more effective at generating metals than today.
This could be due to the fact, for instance, that the Initial
Mass Function was, then, biased toward much larger stars, thereby
increasing the ejected fraction f and also the yield y. In fact,
if we were bold enough, we could read from Fig. 6 the value of y
required to match the data. We would need two phases, one with

y∿0.1, lasting from Z = 0 to Z = 0.004 and then a sudden drop to y = 0.007 lasting until today.

The time sequence of those events could in principle be studied by plotting the metal abundance of stars as a function of the galactic age at their birth. We may use equation (2.1) and (5.1) to obtain

$$\frac{dZ}{dt} = \frac{y(t)}{\tau^*(t)} \tag{5.5}$$

or

$$Z = <y/\tau^*>t \tag{5.6}$$

where

$$<y/\tau^*> = \frac{1}{t} \int_0^t \frac{y(t')dt'}{\tau^*(t')} \tag{5.7}$$

is the mean value of the rate of metal yield. If this parameter is constant, we should get a linear increase of Z with respect to time. The observational data has always been difficult to interpret, mostly in view of the fact that there is considerable scatter on the metal abundance at each age, as shown in Fig. 7 from Powell (1973), but also in view of the difficulty in assessing the uncertainties of these points. Nevertheless it is probably fair

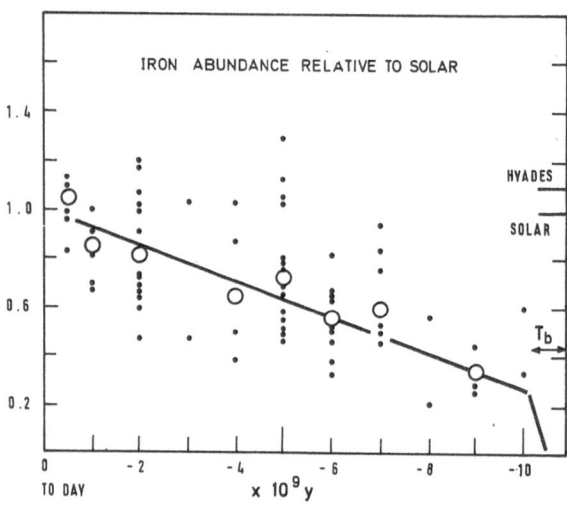

Fig. 7. Iron abundance in stars as a function of their age (Powell 1973). There is such a large dispersion for each time value that it is difficult to draw any conclusion. However, in combination with Fig. 6, the data may be understood in terms of two phases, one initial burst of duration T, in which we form ∿20% of the metals, and a period of continuous nucleosynthesis during the rest of the galactic life.

to say that the "mean" metal abundance has increased somewhat faster
than linearly with time. Adding the information obtained from
Fig. 6, the data is best explained by considering again two phases;
one initial burst of intense activity of duration T_b, where about
twenty percent of the metals are made, hence

$$Z(T_b) = 0.004 = <y/\tau^*>T_b \qquad\qquad (5.8)$$

From Fig. 6, again, we had y~0.1 in this period, hence, taking
averages over this short period, we have

$$T_b/\tau_b^* \simeq 0.04, \qquad\qquad (5.9)$$

where τ_b^* is average lifetime of the gas during this period. From
equation (2.2), we find that the gas fraction of the galaxy has
decreased by only 4% at the end of the burst period, as apparent
from Fig. 6. Then, nucleosynthesis continued through a second
phase with a quieter rhythm $<y/\tau> = 2x10^{-12}y^{-1}$ extending until
today. The important point to notice here is that the majority
of the metals present in the galactic gas today appear to have
originated through a phase of astration lasting only slightly less
than the galactic life and characterized by constant values of the
astration parameters. The initial burst is responsible for only
a small fraction of the metal abundance.

6. THE ABUNDANCE OF LONG-LIVED RADIOISOTOPES

More information, and of a different kind, on the time history
of astration parameters comes from a study of long-lived radioisotopes. To illustrate the matter we shall consider the case of
the two unstable isotopes of Uranium: $^{235}U(\tau = 10^9y)$ and $^{238}U(\tau =$
$= 6.5x10^9y)$. The present ratio in mines is $n(^{235}U)/n(^{238}U) = 0.007$,
from which we calculate that this ratio was 0.30 at the birth of the
solar system, some $4.7x10^9$ years ago. On the other hand, we may
compute from nuclear physics that when these isotopes are formed
in the r-process their formation abundance ratio is about 1.7.
Now, because an r-process involves a huge density of free neutrons
(grams to kilograms of neutrons per cm^3), it appears very likely
that its occurrence takes place in highly explosive situations,
usually identified with supernova explosions. Therefore, we expect
these uranium isotopes to start their interstellar life with an
isotopic ratio of ~1.7. If the turnover of gas in stars had
been very quick (that is if $\tau(t)$ is small), the ratio at the birth
of the sun would have been close to this formation ratio. This
appears not to be the case; a lot of ^{235}U appears to have decayed.
    We have several isotopes which can be analyzed in this way,
through the formalism we have set up earlier. These are $^{129}I(\tau =$
$= 2.3x10^7y)$, $^{244}Pu(\tau = 1.2x10^8y)$, $^{235}U(\tau = 10^9y)$, $^{238}U(\tau = 6.5x10^9y)$,
$^{232}Th(\tau = 2x10^{10}y)$ and $^{187}Rh(\tau = 4x10^{10}y)$. These lifetimes nicely

span the range from 10^7 to 10^{10} years. The first task is to obtain
the abundance ratios of all these isotopes at the birth of the sun.
For ^{244}Pu and ^{129}I this is done through studies of Xenon isotopes
resulting from their decay. For the other isotopes they are observed
directly and corrected back for the lifetime of the solar system.
One important difficulty comes from the possibility of elemental
fractionation of Pu, Th, and U during the formation of the solid
bodies of the solar system (Manhes, Minster, Allegre 1974). Although
this problem seems not to be too serious, we should not overlook
possible important "recalibration" of the ratios in the future.

Following our previous formalism let us define f_r^n the ejected
mass fraction in which the r elements remain unaltered (to their
initial cloud value) and f_r^P the mass fraction in which they were
changed by an r-process. Fortunately the ratio of production of
two isotopes i and j (z_i/z_j) depends almost solely on nuclear physics
parameters and very little on the astrophysical conditions, and can
be calculated once for all (B^2FH) (Schramm 1974). The values quoted
in the Table I, with uncertainties, come from a recent reanalysis of
the situation by Johns and myself (1974).

We write the time variation of the total mass (UZ_i) of isotope
i with radioactive lifetime τ_i as (Reeves and Johns 1974):

$$\frac{d}{dt}(UZ_i) = - Z_i \frac{dS}{dt} + Z_i f_r^n \frac{dS}{dt} + (Z_i+z_i) f_r^P \frac{dS}{dt} - \frac{UZ_i}{\tau_i} \qquad (6.1)$$

(One could debate whether the coefficient of ($f_r^P dS/dt$) should be
(Z_i+z_i) instead of z_i, however since $z_i >> Z_i$ the question is not
important). This can be written as:

$$\frac{d}{dt}(UZ_i) = \frac{y_i}{\tau^*} M_G \exp(-t/<\tau^*>) - UZ_i (\frac{1}{\tau^*} + \frac{1}{\tau_i}) \qquad (6.2)$$

where $y_i = z_i f_r^P/(1-f)$ is the yield of the isotope i.

The solution of this equation for the abundance $z_i(t)$ is

$$Z_i(t) = <\frac{y_i}{\tau^*}> \tau_i \{1 - \exp(-t/\tau_i)\} \qquad (6.3)$$

where

$$<\frac{y_i}{\tau^*}> = \frac{\int_0^t (y_i/\tau^*) \exp\{-(t-t')/\tau_i^*\} dt'}{\int_0^t \exp\{-(t-t')/\tau_i\} dt'} \qquad (6.4)$$

is the rate of yield of i averaged over a period τ_i just before
the observation period t. The interest of this averaging lies in
the fact that one can successively consider several isotopes with
different periods τ_i and hence obtain information on the process
of astration over various periods of the galactic life. The term
$\{1 - \exp(-t/\tau_i)\}$ is a correction term for the finite lifetime of the
galaxy. For a stable element it should be replaced by t (as in the
previous section). Furthermore, since the isotopes available to

Table I

Parameters for the Interpretation of
the Data on the Long-lived Radioisotopes

j/i	τ_j/τ_i (years)	Z_j/Z_i	z_j/z_i	$\dfrac{1-\exp(-T/\tau_j)}{1-\exp(-T/\tau_i)}$	η_{ij}
$\dfrac{^{235}U}{^{244}Pu}$	$\dfrac{10^9}{0.12 \times 10^9}$	21.0	2.0 ± 0.6	$\dfrac{0.99}{1}$	1.1
$\dfrac{^{235}U}{^{238}U}$	$\dfrac{10^9}{6.5 \times 10^9}$	0.31	1.9 ± 06	$\dfrac{0.99}{0.60}$	0.8
$\dfrac{^{232}Th}{^{238}U}$	$\dfrac{2 \times 10^{10}}{6.5 \times 10^9}$	2.4	1.9 ± 0.4	$\dfrac{0.26}{0.60}$	1.73
$\dfrac{^{244}Pu}{^{232}Th}$	$\dfrac{0.12 \times 10^9}{20 \times 10^9}$	0.006	0.5 ± 0.1	$\dfrac{1}{0.26}$	0.65
$\dfrac{^{129}I}{^{127}I}$	$\dfrac{2.3 \times 10^7}{\text{stable}}$	10^{-4}	$1.5^{+1.5}_{-0.7}$		0.009

Z_j/Z_i is the relative abundance of isotope j and i at the
birth of the meteorites (4.7×10^9 years ago), assuming that, in
spite of chemical fractionation, one may reconstitute the true
cosmic abundances through meteoritic studies. For Pu, U, and Th
this assumption may be hazardous (Manhes, Minster, Allegre 1974).
The ratios z_j/z_i are the ratios of formation in r-process synthesis.
These ratios and their uncertainties are reviewed in a companion
paper (Johns and Reeves 1974). The next to last column is a
correction term for the finite duration of solar system nucleo-
synthesis ($T \approx 6 \times 10^9 y$). The last column gives η_{ij}, the astration
probe discussed in the text.

us are solar system material, the appropriate value of t is T, the period extending from the birth of the galaxy until the birth of the sun. From data pertaining to the oldest clusters (Sandage 1970) and data pertaining to meteoritic material (Schramm 1974, Allegre et al. 1974) we estimate T to be $(6\pm2)\times10^9$y. (see also Fowler 1972).

Selecting two isotopes i and j, we define η_{ij} as

$$\eta_{ij} = \frac{Z_i \tau_j z_j}{Z_j \tau_i z_i} \times \frac{1-\exp(-T/\tau_j)}{1-\exp(-T/\tau_i)} = \frac{\left\langle \frac{y_i}{\tau^*} \right\rangle_{\tau_i} z_j}{\left\langle \frac{y_j}{\tau^*} \right\rangle_{\tau_j} z_i} \qquad (6.5)$$

Clearly, from the definition of η_{ij}, if the Initial Mass Function (and hence the f's), and also the lifetime τ^*, were constant through-out the life of the galaxy, we should get $\eta_{ij} = 1$ for all pairs. Conversely, any strong departure from one (outside of the uncer-tainties which are of about a factor of two each side) should indi-cate the time variation of at least one of these parameters. In this sense, η_{ij} is a "time probe of the rate of yield astration".

In the Table I, the value of η_{ij} is computed for five ratios of isotopes. The required parameters are also listed. The remarkable thing is that averages taken between the last 10^8y before the birth of the sun (^{244}Pu) and the age of the galaxy (^{238}U, ^{232}Th) show no significant departure from the value $\eta_{ij} = 1$ (within a factor of two). This is best explained by assuming that the IMF and τ^* were largely constant throughout the life of the galaxy (the case of ^{129}I will be discussed separately later).

The connection with the previous discussion on the metal abundances is best accomplished by stating that to account for the consistency of the η_{ij} we only need to say that most (more than fifty percent) of these long-lived isotopes were made in a phase characterized by constant IMF and τ^* Again we could well tolerate an initial burst of large yield of astration (which would hardly affect the present value of ^{238}U, ^{232}Th), if it involves only an extra contribution of $\sim20\%$ to the total metal abundance. Combining the information from the metal abundance to the information from the long-lived radioactivities we conclude that:
—most of the elements were made during a time independent astration regime lasting almost as long as the life of the galaxy;
—a small fraction was made during an intense period, lasting a rather short time and involving only a minimal depletion of the interstellar gas.

We may now try to evaluate quantitatively the resulting deu-terium depletion. At the end of the first period T_b we had from equation (4.1):

$$(D/H)_{T_b} / (D/H)_{BB} = \exp\{-T_b f_b / (1-f_b) \tau_b^*\} \qquad (6.6)$$

where f_b is the value of the ejected fraction appropriate for the initial burst period. The data on metal abundance give us the value $f_b{}^p z_b = 0.1$ for this period (remember that $f_b{}^p < f$ by definition), but don't give us f_b^P nor z_b. (The relationship between f^P and z could in principle be obtained through calculation of nucleo-synthesis throughout stellar evolution (see Fig. 3), but we still have little to go on). It is however reasonable to assume that to obtain such a metal enrichment the value of f must have been appreciably larger than the 0.2 of today, probably as large as 0.5.

From equation (6.1) and the value $T_b/\tau_b^* = 0.04$ it is clear that important deuterium depletion (more than twenty percent) could only occur if $f_b \geqslant 0.8$ which, at the present time, would appear to be somewhat extreme, in view of the still moderate value of $f_b^P z_b$. For the second period we have, as mentioned before, $f \sim 0.2$ and $\tau^*(t) \sim 5 \times 10^9$ years, thereby implying a further decrease of $\sim$ forty percent.

All in all then, it seems reasonable to conclude that about fifty percent of the original D is still around, leading as discussed before to a present universal density of $\sim 5 \times 10^{-31}$ g cm^{-3}. This density is about five times larger than the density of visible matter (galaxies) and hence leaves some room for intergalactic gas (needed to bind the clusters of galaxies?) or other forms of invisible matter. On the other hand, this density is about one order of magnitude less than the density required to close the universe. The corollary is that we live in an ever-expanding universe, never to contract anymore.

The same discussion can be applied to the history of other light isotopes. Isotopes issuing from the Big Bang would have been affected similarly by astration. If ^{7}Li originates in the Big Bang, its depletion fraction is the same. For ^{3}He and ^{4}He we have potential formation mechanism in stars which complicates the matter somewhat.

For the isotopes generated in the galactic cosmic rays the effect of astration is even smaller since they were formed only gradually and hence traversed only a fraction of the whole astration process in the galaxy. The fraction of ^{6}Li, ^{9}Be, ^{10}B, and ^{11}B destroyed in stars is quite insignificant in view of the uncertainties of the observations.

7. THE IODINE RATIO AND THE TIME-SCALE OF GALACTIC ROTATION

It is quite remarkable that the only strong departure from $\eta_{ij} = 1$ is found for the shortest time scale of $\sim 2 \times 10^7$y related to ^{129}I. The decrease of η_{ij} to 0.02 is most likely due not to a variation of the IMF during this short period, but to an increase in τ^* implying a decrease of the nucleosynthesis activity some 2×10^7y before the formation of the meteorites. The reason for this decrease is still a matter of speculation. It is of interest, of course, to notice that we are here entering in a time-scale shorter than the period of galactic rotation, and that, in our previcus formalism, we have, in defining τ^*, taken no account of the galactic rotation.

If the nucleosynthesis activity is, in any way, related to galactic rotation (as in the density-wave-model) the use of τ^* has a meaning only if averages over a period larger than the galactic rotation are considered. Otherwise, one should consider a series of "peaks" in the nucleosynthetic activity (occurring, in the frame of the density-wave-model, at the moment of passage through the arms), separated by $\sim 10^8 y$. The reason why the η_{ij} value for Iodine is low would, in this context, simply be due to the fact that, in the last $\sim 10^8 y$ before the birth of the sun, the protosolar cloud (or clouds) was cruising between two arms of the galaxy where very little astration takes place. The last contribution to solar system ^{129}I was incorporated into the protosolar cloud when it crossed the last arm before its arrival in the arm where the sun was born (Reeves 1972).

REFERENCES

Allegre, C.J., Birck, S., Fourcade, S. and Semet, M., 1974, "^{87}Rb -^{87}Sr Age of Juvinas Basaltic Achondrites and Early Igneous Activity in the Solar System". Science. (in press).
Fowler, W.A., 1972, in "Cosmology Fission and Other Matters". A Memorial to George Gamow (Edit. Reines F.), Colorado Assoc. Univ. Press, Boulder, Colorado
Johns, O. and Reeves, H., 1974, "Clock-Element Production Ratios", (in preparation).
Metzger, P., 1973, Lecture notes from the school on "Interstellar Matter", Schliersee, April 1973
Manhes, G., Minster, J.F. and Allegre, C.J., 1974, "U-Th-Pb Age of Saint Severin Meteorite. Fractionation of U-Th-Pb in the Formation of Meteorites and Limits on Nucleochronologies". (in preparation).
Ostriker, J.P., Richstone, D.O. and Thuan, T.X., 1974, Ap. J. L. 188, L87
Pagel, B., 1973, (unpublished).
Peimbert, M., 1974, Lecture notes from the school on "Stellar Populations", Erice, 1974
Powell, A.L.T., 1973, M.N.R.A.S., 155, 492
Reeves, H., 1972, "Some Unwritten Chapters of our Book", comptes rendus de la conference sur "l'origine du système solaire", Nice, 1972 Avril (Ed. H. Reeves) C.N.R.S., 15 quai Anatole France, Paris 75700
Reeves, H. and Johns, O., 1974, "The Long-Lived Radioisotopes as Monitors of Stellar, Galactic and Cosmological Phenomena" (in preparation).
Reeves, H., 1974, Annual Review of Astronomy and Astrophysics, 14, 437
Salpeter, E.E., 1959, Ap. J. 129, 608
Sandage, A., 1970, Ap. J. 162, 841
Schmidt, M., 1963, Ap. J. 137, 758
Schramm, D.N., 1974, Annual Review of Astronomy and Astrophysics, 14, 383

Searle, B., 1973, in "L'Age des Etoiles", Meudon. (G. Cayrel and
A. Delplace, Editors).
Talbot, R.J. and Arnett, W.D., 1973a, Ap. J. 186, 51
Talbot, R.J. and Arnett, W.D., 1973b, Ap. J. 186, 59
Tinsley, B., 1974, Lecture notes from the school on "Stellar Pop-
lations", Erice, May 1974
Truran, J.W. and Cameron, A.G.W., 1971, Ap. and Space Science,
14, 179
Wagoner, R.V., 1973, Ap. J. 179, 343

RADIO CONTINUUM EMISSION FROM GALAXIES

R.D. Ekers

Kapteyn Astronomical Institute, University of Groningen,
Groningen, The Netherlands

1. INTRODUCTION

The continuum radio emission detected from external galaxies is
produced either by thermal process or by the synchrotron process
with the latter dominating in most cases. For the synchrotron process the volume emissivity is given by

$$\epsilon(\nu) = n \, B^{\frac{1+\gamma}{2}} \, \nu^{\frac{1-\gamma}{2}} = n \, B^{1-\alpha} \, \nu^{\alpha}$$

where the number of relativistic electrons with energy E is $N(E) =$
$n \, E^{-\gamma}$, B is the magnetic field and ν the frequency. Hence an observation of the radio continuum is a measurement of the product
of the magnetic field strength and the relativistic particle density.
The general questions which arise in interpreting these radio continuum observations are: where do the magnetic fields and relativistic electrons come from and how are they contained?

For the spiral galaxies observations of the radio continuum
can be of considerable importance. They may give information on
conditions, such as compression ratios, in the interstellar gas.
Further, if supernovae are significant sources of particles or
magnetic field the present non-thermal emission may depend on the
evolutionary history of the galaxy. Finally, the regions of strong
non-thermal emission are often indicators for more violent events
in the galaxy.

For the elliptical galaxies we have strange morphologies and enormous powers still demanding explanation as well as the possibility of using these objects as probes both of the intergalactic medium and of the universe as a whole.

I will start in Sections 2 and 3 by looking in detail at the closest systems. These detailed observations then provide the framework for the discussion in Sections 4 to 9 of the integral properties of a larger sample of galaxies. Finally, the properties of the very powerful but much rarer radio galaxies which dominate the radio source catalogues are summarized in Section 10.

Throughout I have used distances based on a Hubble constant of 55 km sec^{-1} Mpc^{-1}. Flux densities at 1400 MHz have been used when available, otherwise they have been estimated using an assumed spectral index.

2. NEARBY SPIRAL GALAXIES

2.1 The Galaxy

To discuss our Galaxy in the context of external galaxies we are mostly interested in the large scale properties. The radio emission from discrete sources e.g. stars, planets, pulsars, etc. has a negligible integrated effect. Supernova remnants also make a negligible contribution to the total flux but some are strong enough to be individually detectable in the external galaxies. More importantly, they may play an indirect role as a possible source of relativistic electrons and also perhaps magnetic fields. Thermal emission is seen both from discrete HII regions and from the gaseous disk but even the biggest galactic HII regions would be detectable only in external galaxies closer than a few Mpc.

The dominant radio emission is the diffuse non-thermal component from the whole disk of the Galaxy. This can be separated from the thermal emission on the basis of spectra and polarization. It is almost certainly due to the synchrotron process as the cosmic ray electrons interact with an interstellar magnetic field of order 10^{-6} gauss. It is clear from the large scale angular distribution of this emission that most of it originates in the disk of the Galaxy and, with less certainty, it is also thought that this emission is strongest in the spiral arms. For a detailed review of the situation in our Galaxy the reader is referred to Baldwin (1967).

2.2 Disk emission in nearby spiral galaxies

The total radio emission from many of the large nearby spiral galaxies has been measured with pencil beam radio telescopes and a

few of these observations, together with values for our Galaxy,
are collected in Table 1. Data for the nuclei and the median are
discussed in later sections. The disk brightness is calculated
from

$$T_B = 189 \times S_{tot} / (D^2_{major} \times \frac{\pi}{4}) \text{ Kelvins}$$

where S_{tot} is the flux density in units of 10^{-26} W m^{-2} Hz^{-1} at
1400 MHz and D_{major} is the optical major axis diameter in arc
minutes. Simply by using D_{major} we are automatically correcting
the brightness to its face-on value. This data clearly indicates
the large range in brightness even for galaxies of similar type.

For the more powerful of these galaxies the integrated spectra are clearly non-thermal but in general the separation of the
thermal and non-thermal components on the basis of the radio spectrum is not easy because of the difficulties in obtaining good flux
density measurements. At low frequencies the observations are limited by confusion with unresolved background sources while at higher
frequencies correction for resolution is difficult. In M33 many
individual HII regions are detected on the high resolution map of
Israel and van der Kruit (1974) and, by extrapolating to the less
luminous HII regions, they estimate that all the radio emission at
1400 MHz could be thermal. In M101 Israel, Goss and Allen (in preparation) have identified a number of the radio sources in the Westerbork map with HII regions and have shown that they have an optically
thin thermal spectrum between 600 MHz and 10,000 MHz.

Our first knowledge of the brightness distribution in an ex-

Table 1. Radio emission from spiral galaxies

Galaxy	Type	Distance (Mpc)	Disk		Nucleus	
			Brightness ($^{\circ}K_{1400}$)	Power (1)	Power (1)	Linear size (pc)
The Galaxy	Sbc		0.6	–	0.5	180
M31	Sb	0.6	0.1	3.4	0.04	240
M33	Sc	0.7	0.2	1.6	<0.002	–
M81	Sab	3.5	0.3	5.5	1	<35
M101	Sc	7.0	0.2	22	0.2	~500
M51	Sbc	9.5	6.0	120	7	750
NGC6946	Sc	10.0	3.1	130	8	<200
Median			0.8[2]	16[2]	0.8	200

1. 10^{19} WHz^{-1} ster^{-1}
2. Estimated from the 408 MHz data of Cameron (1971) assuming
$S \propto \nu^{-0.7}$

ternal galaxy came with the observation of M31 at 400 MHz by Pooley
(1969) using the one mile telescope at Cambridge. Further maps have
now been obtained at 1400 MHz at Westerbork (van der Kruit 1972)
and at 2700 MHz in Bonn (Berkhuizen and Wielebinski 1974). They
show that the non-thermal emission is 2-3 times stronger in an area

Fig. 1. Isophotes of 1415 MHz emission from M51 and NGC5195. The
contour unit is 0.8 K brightness temperature. The contours drawn
are Nos. 1 to 15, then 18, 20, 25, 30, 40, 50 and 60.

∿1 kpc wide in the regions of the spiral arms.

The Westerbork observations of the spiral galaxy M51 (Mathewson, van der Kruit and Brouw 1972) show a far more spectacular concentration of radio emission in the spiral arms (Figure 1). This galaxy was already known to have a high non-thermal brightness (∿6 K) from single disk observations. The Westerbork observations show that this consists of an unresolved (≲300 pc) ridge of emission generally coinciding with the dust lanes along the inside of the spiral arms. The average brightness in the arms is 8 K compared with 2.5 K between the arms.

Mathewson et al (1972) suggested that this enhanced emission might be due to the compression in this region as predicted by the density-shock wave theory eg Roberts and Yuan (1970). In a very simple model in which both the magnetic field and the relativistic electrons are compressed we have

$$\frac{\varepsilon}{\varepsilon_o} = \frac{n}{n_o} \cdot \left(\frac{B}{B_o}\right)^{1-\alpha} = \left(\frac{\rho}{\rho_o}\right)^{2-\alpha}$$

where ε_o and ε are the volume emissivity corresponding to the densities ρ_o and ρ before and after compression. For M51

$$\alpha = -0.7 \text{ so } \frac{\varepsilon}{\varepsilon_o} = \left(\frac{\rho}{\rho_o}\right)^{2.7}$$

and hence the synchrotron emission is a very sensitive indicator for compression. In fact the observed ratio of 3:1 for the continuum is much less than predicted by this model since the Westerbork observations of the neutral hydrogen distribution (Shane private communication) give

$$\frac{\rho}{\rho_o} = 3 \text{ implying } \frac{\varepsilon}{\varepsilon_o} \sim 20.$$

This simple model is almost certainly incorrect for at least two reasons.

1) As noted by Mathewson et al (1972) the non-thermal emission region in our Galaxy is thicker than the gas layer and if this is the case in M51, the non-thermal emission region may not be all compressed.

2) Parker instabilities may cause the magnetic field to buckle up out of the gas layer. This will allow a redistribution of relativistic electrons in the field and cause a reduced continuum emission.

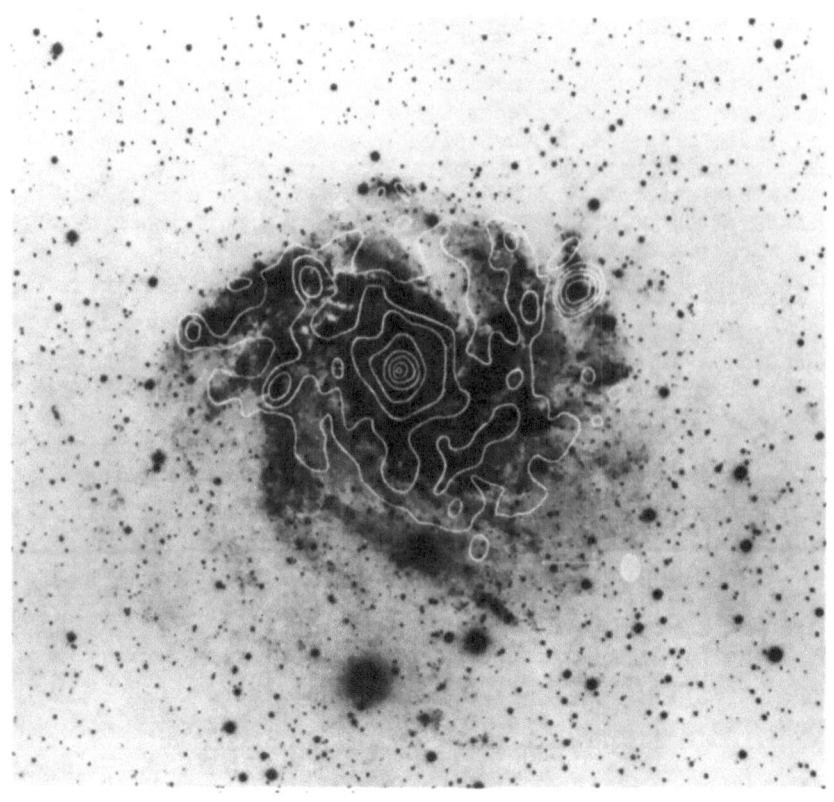

Fig. 2. Isophotes of the 1415 MHz emission from NGC6946. The outer
contour corresponds to T_B = 2.3 K; the other contours are for 4.6,
6.9, 11.6, 23.1, 46.2, 69.3, and 92.4 K. The horizontal line is
1'. The optical picture is from Arp's Atlas.

How common is the enhanced spiral arm emission? NGC6946 is
another spiral galaxy with very similar high brightness non-thermal
emission. The Westerbork map at 1415 MHz (Rots private communication) shows a complex distribution of non-thermal emission (Figure
2). Some of this is associated with HII complexes in the arms but
there are no well defined spiral ridges such as seen in M51. If we
look at M81, which is a "grand design" spiral like M51, we see
sharp ridges of neutral hydrogen (Rots 1974) lying along the arms
indicating compression $\geqslant 3$ but no associated non-thermal ridge. Unfortunately M81 has much weaker average non-thermal emission
($T_B \sim 0.2$ K) and although the observations would exclude non-thermal emission at the level predicted by the simple compression model,
the limits can only just exclude compression comparable to that seen
in M51. In the higher sensitivity lower resolution Westerbork data
obtained after smoothing,van der Kruit (1973a) finds a broad (~ 2 kpc)

region of enhanced emission in the region of the spiral arms. This
may then be more comparable to the broad arms seen in M31.

In fact, of the 33 galaxies which have now been mapped with
high resolution there is only one other clear case in which spiral structure is seen in the continuum emission. That is NGC4258
but here the most prominent radio arms are at quite a different
phase from the main optical arms and van der Kruit, Oort and
Mathewson (1972) suggest quite a different explanation for this
object. This will be discussed in the lecture by Professor Oort.

In our Galaxy it has been suggested that the "steps" in the
continuum emission can be associated with the spiral arms. In a
recent evaluation of this evidence Price (1974) concludes that
the data is consistent with an enhancement of a factor of four
in the spiral arms. Although this is comparable to what is seen
in M51 it is probably much greater than the average enhancement
in the spiral arms of other galaxies.

A number of possible theories can be considered which might
explain the distribution of the non-thermal emission. It has often been argued in the past e.g. by Chandrasekhar and Fermi (1953),
Hoyle and Ireland (1961), and Piddington (1972) that magnetic
fields play an important role in maintaining the spiral structure
and hence the enhanced non-thermal radio emission from spiral arms
could also be explained naturally in this way. The difficulty with
this, as with the density wave hypothesis, is that the enhanced
emission is not seen in all spirals. Another possibility is that
the supply of relativistic electrons determines the distribution.
If the supernovae are the main source of relativistic electrons a
higher supernovae rate in the region of the spiral arms and the
HII complexes could result in enhanced emission. However, in the
case of M51 this could not explain the positional agreement with
the dust lanes. Finally, the density wave compression first discussed may still be the most interesting possibility but further
work on the balance between the thin gas disk and the magnetic
field and relativistic plasma is needed.

2.3 The distribution of emission in z - the radio haloes

To study the distribution in z we need to observe an edge-on system.
This can be done in our Galaxy e.g. Baldwin (1967) estimates an
equivalent width of 750 pc, but the results are somewhat model dependent.

In a recent study (Allen private communication) of the edge-
on galaxy NGC891 with the Westerbork telescope at 1400 MHz the
disk emission is clearly resolved in z and has a full width at half
intensity of ∿4 kpc. This can be compared with the gas layer in the

same galaxy which has a width of <400 pc (Sancisi private communication). It is interesting to note that the ratio in brightness
between our Galaxy and NGC891 is similar to the ratio of the widths
in z.

As well as the general broadening of the disk seen in NGC891,
weaker non-thermal emission can be traced to ∿7 kpc above the plane.
This large z extent is of considerable interest for the problem
of galactic haloes. The existence of a halo of radio emission surrounding our Galaxy has been a controversial matter since it is
difficult to separate the more-or-less isotropic halo component
from the disk emission and even more difficult to separate it from
local irregularities.

For M31 the case for a radio halo seems reasonably well established. Pooley (1969) finds that the total flux density from
the disk, nucleus and field sources is still much less than the
total flux density observed in the direction of M31 with large
beams. The difference could be explained by a halo of radius 24
kpc and average brightness temperature of 3 K at 408 MHz (∿0.1 K
at 1400 MHz). However, no further cases had been found until the
recent 600 MHz Westerbork observation of the edge-on galaxy NGC4631
by Sancisi and Ekers. Emission is seen extending to 12 kpc above
the plane, giving a slightly flattened halo of ratio 3:2. The
brightness at this height is 2 K at 600 MHz (∿0.2 K at 1400 MHz).
This galaxy also has a strong disk and a strong nuclear source.
Interesting questions for the future are whether the presence of
halo emission is related to either the strength of the disk or
the nuclear emission, or whether it is just a peculiarity in some
galaxies.

2.4 Nuclear emission

A clear component of the Galactic continuum radiation is the complex distribution seen in the direction of the Galactic centre.
This emission has three main components: an extended source about
$1^{\circ}.0 \times 0^{\circ}.4$ (180 x 70 pc) elongated along the Galactic plane which
is probably non-thermal, a complex of giant HII regions in a similar
area and another non-thermal source, Sgr A, about $3'.0 \times 2'.4$
(9 x 7 pc) very close to the centre of the Galaxy. Recent high resolution observations (Ekers et al, in preparation) show that Sgr A
is also complex. It consists of two main features: Sgr A East which
has a steep non-thermal spectrum and is of low surface brightness,
and Sgr A West which is a centrally peaked source of linear size
∿3 pc. Sgr A West has a flat spectrum and could be thermal. Its
peak coincides with a much smaller source of size <0.01 pc (Balick
and Brown 1974). Although some external galaxies may be more interesting because of their stronger nuclear activity such high linear
resolution can only be obtained in our Galactic centre.

The radio power and linear dimensions for the nuclear sources in a number of nearby galaxies that have been measured with
synthesis telescopes are given in Table 1. These parameters for
the most extended component of our Galactic centre are also given
since this would be the dominant component at similar resolution.
These examples clearly illustrate the enormous range in nuclear
emission from M33 for which only a lower limit has been obtained
to M51 and NGC6946.

The nuclear source in M31 was detected by Pooley (1969). It
is about 10 times weaker than our Galactic centre but has comparable linear dimensions. Observations by van der Kruit (1972) show
that it is resolved into a complex of sources but contrary to the
situation in our Galaxy they do not appear to lie in the galactic
plane. The nuclear source in M101 is more nearly comparable to our
Galactic centre. It agrees in position with a small optical HII
region and may also be partly thermal. The nuclear source in M51
is ten times brighter than our Galactic centre and can be clearly

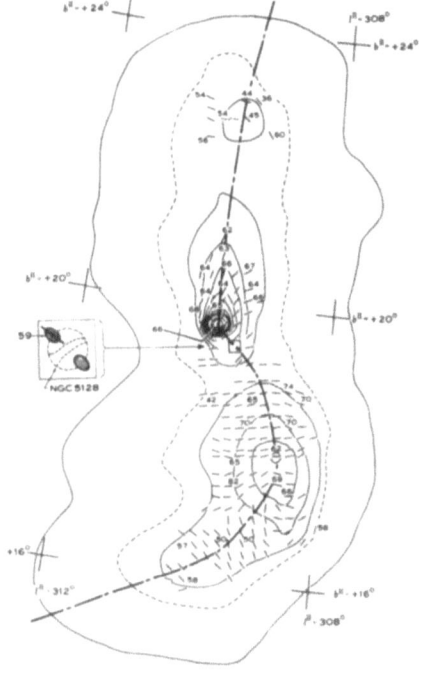

Fig. 3. Isophotes of the radio emission from Centaurus A. Contours
are at equally spaced intervals in brightness. Enlarged inset shows
the position of the inner components and the galaxy NGC5128. Vectors indicate the direction of the magnetic field and numbers give
the total Faraday rotation measures.

seen in the map of Mathewson et al (1972) (Figure 1). Higher resolution observations at 5000 MHz (de Bruyn private communication)
show that this is also resolved into at least two components.

3. NEARBY ELLIPTICAL GALAXIES

The nearest elliptical galaxies are the dwarf systems in the local
group but no radio emission has been detected from any of these.
For example, the companions of M31, NGC205 and NGC221, have radio
power less than 2.10^{16} W Hz^{-1} ster^{-1} (van der Kruit 1972). The
nearest giant elliptical galaxy is NGC 5128 at a distance of 9 Mpc
and here we encounter one of the strongest radio sources in the
Southern Hemisphere, Centaurus A. It has an integrated radio power
of 5.10^{23} W Hz^{-1} ster^{-1} – thousands of times more powerful than the
most powerful spiral galaxy! The radio brightness distribution from
Cooper, Price and Cole (1965) is shown in Figure 3. The radiation
comes predominantly from two giant lobes almost 1 Mpc in size lying

Fig. 4. 1400 MHz continuum map of the inner components of Cen A
superimposed on an optical photograph from the Hubble Atlas. This
map has been obtained with the Fleurs Synthesis Radio Telescope
and has been kindly supplied by Prof. W. Christiansen.

well outside the optical image. The central region of the radio
source is resolved into two lobes and a compact nuclear source
that can be seen in the new Fleurs synthesis telescope map in
Figure 4. Both these inner lobes and the outer lobes are in the
same position angle and this is also the position angle of the
rotation axis as determined from optical spectroscopy. The nuclear source has a radio power of 10^{21} W Hz^{-1} $ster^{-1}$ which is
comparable to the brightest spiral galaxy nuclei but it is more
compact than most of the spiral galaxy nuclei.

Because of their lower space density there are only a few
more giant elliptical galaxies before we reach the distance of
the Virgo cluster. At this distance we do not have intrinsic sensitivity or linear resolution comparable to that obtained for
nearby spirals, so further discussion of the elliptical galaxies
is left until we discuss radio galaxies in Section 10.

4. SAMPLE FOR STATISTICAL STUDY OF PROPERTIES

In order to sensibly discuss the distribution functions for the
properties of galaxies or the correlations between different
properties it is necessary to use complete samples of galaxies
or at least samples for which the incompleteness is known. This
is especially important for radio observations since most of the
galaxies are near the sensitivity limit of modern radio telescopes. The selection of the sample of nuclei of spiral galaxies is
described in detail as an example of the way in which these difficulties can be overcome.

4.1 Spiral nuclei

In order to clearly separate the nuclear radio sources from the
sometimes complex emission from the galactic disks high resolution observations with good spatial coverage must be used. Van
der Kruit (1973b) discusses the properties of a sample of 45
galaxies observed with the Westerbork array and to this can be
added further observations from the Cambridge, Greenbank and
Westerbork arrays. This sample is not complete to a given optical magnitude since a variety of selection criteria have been
used in the various programmes. In such a heterogeneous sample
it is especially important to avoid spurious correlations
arising because of different sensitivities and differing distances of various types of galaxies. From this data an unbias-
sed sample is made as follows. For each galaxy the distance
and the sensitivity of the observations have been used

to calculate the minimum absolute radio power level at which it
would be detectable. All galaxies for which the detection limit
is greater than the median power of the total sample are then excluded and the median power of the new sample is computed. By continuing to iterate in this way a final sample is obtained which
includes only galaxies which could be detected if they were stronger than the medium of the sample. The final median value obtained
is 8.10^{18} W $Hz^{-1}ster^{-1}$ at 1400 MHz and above this value there is
no bias in the sample with respect to radio power. The use of the
median in this method has no significance other than to maximise
the size of the final sample. If this procedure is not used the
sample will be dominated by very strong sources just because they
are visible to greater distances. There may still be bias in the
sample with respect to other properties e.g. in the numbers of
each optical type but these have less serious consequences. The
sample found in this way and used in the following discussion is
tabulated in Ekers (1974).

4.2 Disk emission

The sample of galaxies used to investigate the disk emission is
taken from the 408 MHz pencil beam observations of Cameron (1971)
which are complete to a given optical magnitude. These observations
have insufficient resolution to separate the disk from the nucleus
so I have assumed that all the observed emission comes from the
disk. The average fraction of the flux density in the nuclear components for the sample of galaxies discussed in the previous section is ⩽17%, so this assumption does not introduce a serious error. The average brightness temperatures are calculated by using
the major axis optical diameter from the Reference Catalogue of
Bright Galaxies (De Vaucouleurs and de Vaucouleurs 1964).

4.3 Elliptical galaxies

There are few detected elliptical galaxies in the sample of Cameron
(1971) so the results of the surveys of elliptical galaxies by
Heeschen (1970) and Ekers and Ekers (1973) are also used. These
have higher sensitivity and also separate the nuclei and extended sources.

    The more powerful elliptical galaxies, the radio galaxies,
are mainly found in radio source surveys and are discussed separately in Section 10.

5. CORRELATION BETWEEN HUBBLE TYPE AND RADIO PROPERTIES

For different Hubble types Table 2 gives the median disk brightness

Table 2. Disk Brightness v. Hubble type

Type	Number in sample	Median brightness (K_{1400})	Range of brightness (K_{1400})
Sd	17	0.5	<0.2 - 4
Scd	12	0.7	<0.3 - 8
Sc	28	1.4	<0.2 - 6
Sbc	23	1.8	<0.2 - 9
Sb	18	1.8	<0.4 - 4
Sa	18	0.6	<0.4 - 5
S0	4	(1.0)	-
E	36	2.8	<1.6 - 100

and the range of brightness that includes 90% of the sample. For the spiral galaxies there is relatively small variation of average brightness with Hubble type compared with the range of brightness.

The tendency for intermediate types to be brighter is probably significant. For the S0 class the numbers are too small to give a reliable result. As has already been noted, the situation changes completely with the elliptical galaxies, the main difference being the enormous increase in their range of radio brightness. In the optically complete sample of Cameron (1971) the brightest elliptical galaxies are ten to one hundred times brighter than the spiral galaxies although the majority of elliptical galaxies remain undetected. In more sensitive observations some nearby elliptical galaxies are detected and they have luminosities comparable with that of the spiral galaxies. However, even these low luminosity elliptical galaxies still have brightness distributions similar to that already described for Centaurus A.

In Figure 5 the power of the nuclear sources is plotted against type. Again no obvious correlation is seen for the normal spiral galaxies but the numbers are now too small to see a small effect such as noted for the disk brightness. Because of their greater average distance the elliptical and Seyfert galaxies should not be compared directly with the spiral galaxies but even so it is immediately clear that these two classes contain more powerful radio nuclei than seen in any of the normal spiral galaxies.

For one barred system, NGC5393, emission is seen from the bar itself as well as from the nucleus. Other barred systems, however, are either weak or undetected and in general no clear comparison between either disk brightness or nuclear power with the presence of a bar is seen.

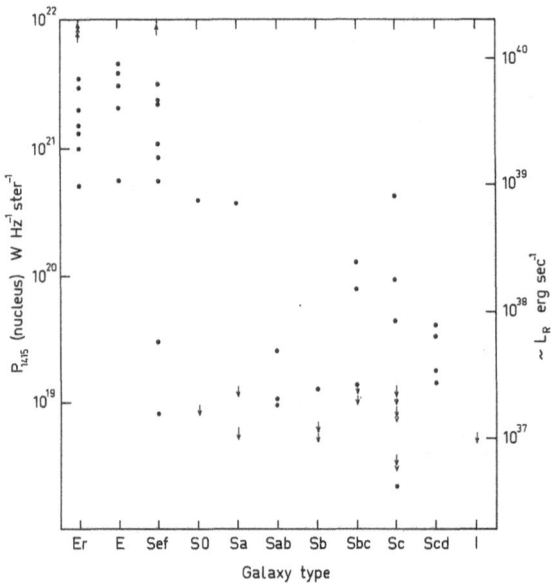

Fig. 5. Plot of the monochromatic radio power at 1415 MHz for different Hubble types and for the Seyfert galaxies, Sef, and the radio galaxies, E_R. The radio luminosity scale, L_R, is shown for a power law spectrum of index -0.7 integrated from 10^7 to 10^{11} Hz.

6. PROPERTIES OF THE SPIRAL GALAXIES

6.1 Luminosity and Spectra

The median disk brightness and integrated luminosity from the disks of the spiral galaxies and the median power for the nuclear sources are given in Table 1. It is interesting to note that our own Galaxy falls very close to these median values. Hence calculations of conditions in our Galaxy can be taken as typical for bright spiral galaxies. The distribution of spectral indices, α, where $S \propto \nu^\alpha$, for the integrated spiral emission between 408 and 5000 MHz has been determined by Wright (1974) and is shown in Figure 6(a). The mean of this distribution is $\alpha = -0.72$. For the nuclei some high resolution measurements are available at other frequencies (de Bruyn private communication and Wade private communication) and the distribution and spectral indices for these galaxies shown in Figure 6(b). The mean spectral index for the nuclei is $\alpha = -0.6$.

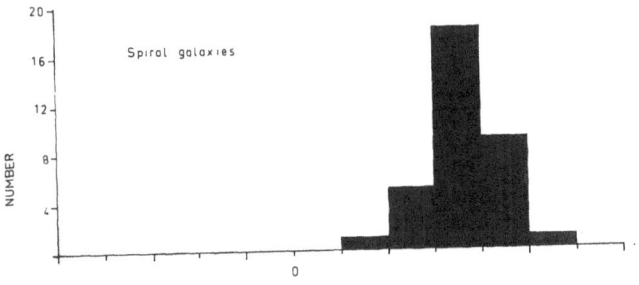

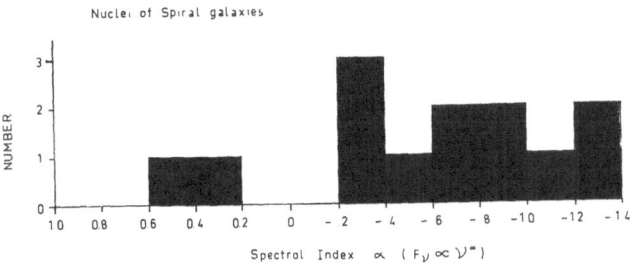

Fig. 6. Distribution of radio spectral index; (a) for the total
radio emission and (b) for the nuclear radio emission of spiral
galaxies.

In general, the nuclei seem to have a similar spectral index to
the disk emission. The distribution is somewhat broader but this
will be partially due to the relatively larger errors in the flux
density from the nuclear components. Two spiral galaxies, M81 and
NGC4544 (Sombrero), stand out with inverted radio spectra.

6.2 Brightness Distribution

The complete sample used for the statistical discussion so far was
obtained with insufficient resolution to use for a discussion of
the brightness distribution. The largest sample for which the brightness distributions are known is that discussed by van der Kruit
(1973b) but even this data is limited by the low signal to noise
in many cases so that at the present time most of our knowledge
of the brightness distribution comes from the nearby systems already discussed in Sections 2 and 3.

The nuclei in many of the spiral galaxies are resolved and

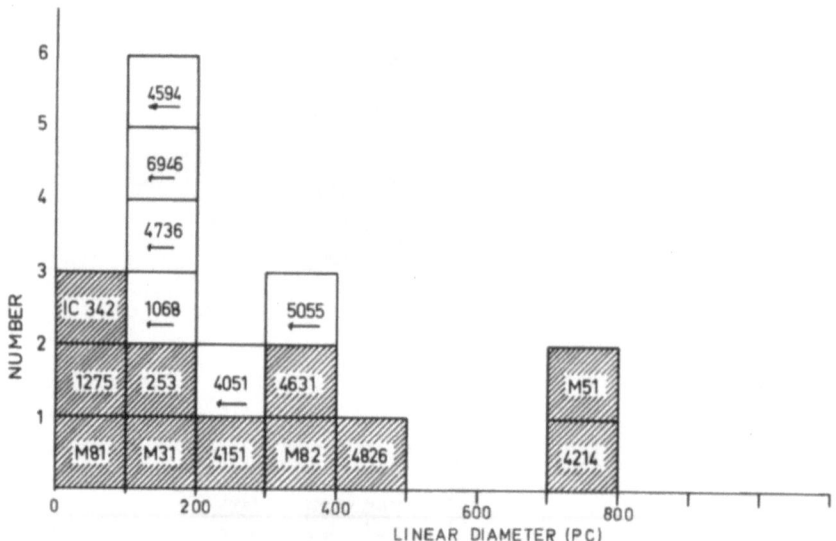

Fig. 7. Distribution of the major axis diameters of the nuclear
radio sources in spiral galaxies. All limits for diameters are
included if they are <500 pc.

the distribution of linear diameters is shown in Figure 7. All unresolved nuclei in the sample with upper limits less than 500 pc
are also included. The median diameter for this sample is 200 pc,
quite similar to the diameter of the most extended component of
the Galactic centre. A few of the nuclei are much smaller, e.g.
M81, one of the spiral nuclei in Figure 8(b) with an inverted
radio spectrum, has diameter ∿2 arc sec (∿35 pc) and is the smallest spiral nuclei known (Wade 1968). But in general the nuclei
of the spiral galaxies do not resemble either in linear size or
in spectra the compact radio sources with synchrotron self absorption type spectra which are commonly found in the more active N
galaxies and quasars.

    For M82 (Kleinmann and Low 1970) and NGC253 (Becklin et al.
1973) both the radio and infrared nuclei are extended and have comparable dimensions. In M82 the radio and the infrared structure
have been resolved into bright knots (Kronberg et al. 1972) but
there is no detailed correlation between the positions of the infrared and radio knots.

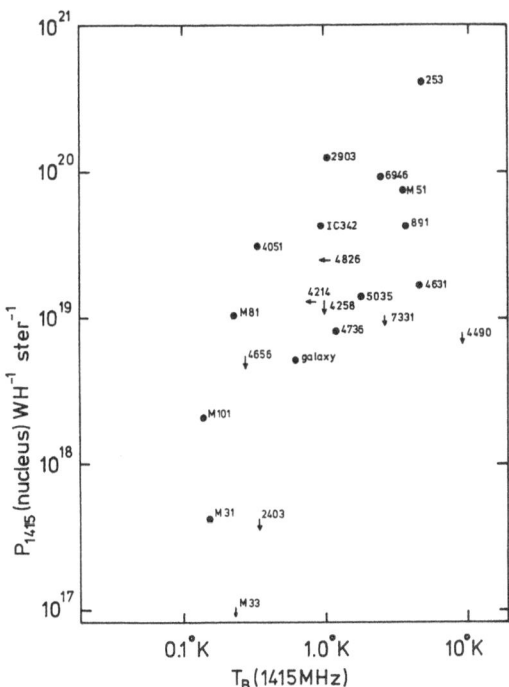

Fig. 8. Plot of the monochromatic radio power of the nuclear sources against the disk brightness in Kelvins at 1415 MHz for the
spiral galaxies.

6.3 Correlation between disk and nuclear emission

Figure 8 is a plot of nuclear power, P_{nuc}, against disk brightness,
T_{disk}. Again the large range of values of these parameters is noticeable – of order 10^2 for the disk brightness and 10^4 for the
power of the nuclear sources. Although at first sight it looks as
though these two parameters may be well correlated we must remember that the sample of galaxies with nuclear emission is only complete above 8.10^{18} W Hz^{-1} ster^{-1} and above this level little correlation is evident. Furthermore, there are galaxies such as NGC
4258, NGC7331 and NGC4490 with relatively strong disk emission and
no distinguishable nuclear sources which could occupy the lower right
part of the diagram. Because of the weakness of this correlation
we cannot at present make a strong case for the nucleus providing
a significant supply of relativistic particles to produce the disk
emission.

For both the nuclear and the integrated radio emission there
is some correlation with absolute optical magnitude. This is to be

expected for the integrated emission but that it also occurs for
the nuclear emission implies that the nuclear activity is not in-
dependant of the rest of the galaxy.

7. PROPERTIES OF ELLIPTICAL GALAXIES

7.1 Luminosity function

One of the most striking properties of the elliptical galaxies is
the very strong dependance of radio emission on absolute optical
magnitude. This property results in the well known effect that
galaxies identified with radio sources are intrinsically very
bright, the dispersion in their brightness being sufficiently
small that their distance can be estimated directly from their
magnitude.

The combination of this simple distance estimate and the po-
pularity of observing quasars resulted in a dearth of redshift
determinations that is only just now being remedied. As a consequence we can only study the actual dependence of radio emission
on absolute optical magnitude for a relatively small number of
nearer systems.

Table 3 gives the probability of radio emission stronger than
10^{21} W Hz^{-1} ster^{-1} at 1400 MHz for a range of absolute optical
magnitudes.

A new complete sample of elliptical galaxies identified with
radio sources from the Bologna Catalogue of Radio Sources now
have measured redshifts (Colla et al. 1974) and a bivariate optical
radio luminosity function has been determined for the luminosity
range 10^{22} - 10^{25}. W Hz^{-1} ster^{-1} (corrected to 1400 MHz and H = 55
km sec^{-1} Mpc^{-1}). From this data they deduce the relation

(radio luminosity) $\propto$ (optical luminosity)$^{1.5 - 2.0}$

Table 3. Probability of radio emission from elliptical galaxies

Mpg	Number detected	Number sampled	%
<-20.5	5	9	55
-20	8	41	19
-19	4	31	13
>-18	2	17	11

7.2 Nuclear Sources

Heeschen (1968) found that two elliptical galaxies, NGC1052 and
NGC4278, contained small diameter sources with radio spectra typical of the spectra of the optically thick variable radio sources
frequently identified with quasi stellar objects. One of these,
NGC 1052, was strong enough to be subsequently detected in a very
long baseline interferometer experiment and found to have a diameter <0".001, corresponding to a linear diameter of <0.1 pc
(Cohen et al. 1971). Further surveys of elliptical galaxies
(Heeschen 1970 and Ekers and Ekers 1973) have yielded an additional 9 nuclear sources.

The distribution of the spectral indices between 1400 and 5000
MHz for these sources are shown in Figure 9(b). The distribution
is strikingly different from the distribution of spectral indices
for the extended emission from elliptical galaxies shown in Figure
9 (a) and is also clearly different from the distribution for the
nuclei of the spiral galaxies shown in Figure 6(b).

Unfortunately,very high resolution observations are only available
for a few of these nuclear sources. Three, in NGC1052, M87 and

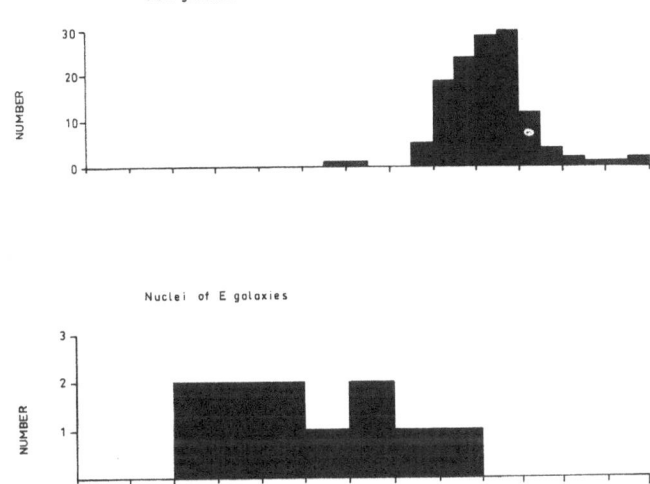

Fig. 9. Distribution of radio spectral index; (a) for the total
radio emission from 3CR radio galaxies and (b) for the nuclear
radio emission from elliptical galaxies.

Centaurus A, are known to be smaller than a few tenths of a parsec. Furthermore, some of the elliptical galaxy nuclei flux densities are now known to vary with time (Sramek private communication). This is in striking contrast to the nuclei in the spiral
galaxies which have typical sizes of a few hundred parsecs.

8. SO SYSTEMS

The SO systems form a transition between the elliptical and spiral
galaxies so since we have such a striking change in radio properties between these classes it would be of interest to consider the
radio properties of the SO galaxies. If the greater sphericity of
the elliptical galaxies causes them to behave differently from the
spiral galaxies, then the radio properites of the SO class would
resemble those of the spiral galaxies. If it is the lack of gas in
the elliptical galaxies which either causes or results from their
more violent radio activity, then the SO class would resemble the
elliptical galaxies. The radio data for a good sample of SO galaxies is lacking, and furthermore the classification is not always
unambiguous, so a definite answer is not yet possible, however it
seems reasonably sure that few if any SO galaxies have strong extended radio sources. No careful analysis has been made of the
nuclear sources in SO galaxies but it is interesting to note that
the Sombrero nebula is one of the galaxies in Figure 6(b) with an
inverted spectrum and this is more like the nuclear sources in
elliptical galaxies.

9. SEYFERT GALAXIES

The nuclear radio emission from Seyfert galaxies is stronger than
that from normal spiral galaxies (Wade 1968, van der Kruit 1971)
and shows a good linear correlation with the 10μ infrared emission
(van der Kruit 1973b). As can be seen from Figure 5 most of the
Seyfert galaxies have radio luminosity within a factor of ten of
the luminosity of the normal spiral galaxies. For these Seyfert
galaxies de Bruyn and Willis (1974) have measured diameters between 300 and 1500 pc and spectral indices near -0.8. Thus in these
respects the Seyfert galaxies seem similar to the nuclei of normal
spiral galaxies. But a few of the Seyfert galaxies are much more
powerful, e.g. NGC1275 (3C84) is more than a thousand times stronger than the brightest spiral nuclei and it contains a flat spectrum
nuclear component with very small linear size (∿1 pc). Thus in
these respects it resembles the nuclei of the elliptical galaxies.
In addition to this nuclear component NGC1275 contains components
of sizes 15 pc, 15 kpc and 200 kpc. This complex morphology is quite
different from anything seen in the spiral galaxies nor is it typical of the extended components of the strong elliptical galaxies.

A recent puzzling observation of NGC 1275 is the narrow (6 km/sec wide) HI absorption line seen in the radio continuum at the redshift of the higher of the two optical redshift systems (de Young, Roberts and Saslaw 1973). Since this line is seen in absorption it must be produced by HI in front of the radio source and yet it is at +3000 km/sec with respect to the systematic velocity, implying an implosion rather than an explosion! Possible explanations might be that the absorption is only in a part of the complex of radio emission which lies behind the centre of the galaxy, or that the higher velocity system is an unrelated (?) foreground galaxy.

The general impression obtained of the Seyfert class is that if the Seyfert characteristic is seen in an otherwise normal spiral galaxy we see significant but not enormous enhancement of the radio emission, but if the Seyfert characteristic is seen in an elliptical like galaxy e.g. NGC 1275, 3C 120, then very strong emission is seen.

10. RADIO GALAXIES

10.1 Definitions

Historically the term "radio galaxy" has been applied to the relatively faint galaxies identified with radio sources found in radio surveys. When bright nearby galaxies are detected in surveys or in special searches for their radio emission they are called "normal galaxies". In order to understand the distinction between these two types of galaxies it is useful to see how the separation arises. In a radio source catalogue complete to a given flux density, s, we can see a source of luminosity L out to a distance

$$R = \left(\frac{L}{s_{min}}\right)^{1/2} , \text{ i.e. in a volume } V = \frac{4}{3}\pi\left(\frac{L}{s_{min}}\right)^{3/2} .$$

If the density of sources of luminosity L is $\rho(L)$ then the number seen above the flux density limit s_{min} is,

$$N_{obs}(L) = \rho(L)\left(\frac{L}{s_{min}}\right)^{3/2} = K L^a \left(\frac{L}{s_{min}}\right)^{3/2}$$

if we assume a power law form for the radio luminosity function. Hence we have three possibilities:

(i) $a > -\frac{3}{2}$. Then as L increases N_{obs} <u>increases</u>.

(ii) a = -3/2. Then any L is equally probable.

(iii) a < -3/2. Then as L increases N_{obs} decreases.

For radio sources identified with elliptical galaxies a $\simeq$ -2.2 above 10^{24} W Hz^{-1} ster^{-1} (Colla et al. 1974) but becomes > -1.5 at lower luminosities. As a result, most of those found in radio source catalogues are above this knee in the luminosity function. The relative number of low luminosity sources will not increase until catalogues reach a flux density sufficiently low that the cosmological effects at this luminosity significantly increase the geometric factor of -3/2. For the spiral galaxies a knee in the luminosity function occurs at a much lower luminosity of $\sim 10^{19}$ W Hz^{-1} ster^{-1} (Cameron 1971) and above this knee it is much steeper with a = -2.8. The result is that although the luminosity functions overlap for these two classes of objects there is still a clear separation of the observed distribution of sources into a low and a high luminosity group corresponding to the spiral and the elliptical galaxies. These two groups are called the "normal" and the "radio galaxies". In some cases where a more exact definition of a radio galaxy has been required a lower limit in luminosity is set at a few x 10^{23} W Hz^{-1} ster^{-1} (10^{42} ergs sec^{-1} integrated over the radio spectrum). However such a definition is clearly quite artificial as radio sources identified with elliptical galaxies and with similar morphology to the radio galaxies can be found right down to the luminosities of the "normal" spiral galaxies. A more suitable and a simple defining characteristic of the class would be the identification with an elliptical galaxy.

10.2 Morphology

The most striking features of the radio galaxies is the basically double structure such as already seen in Centaurus A. Sixty five percent of the resolved radio galaxies are double (including variations such as multiple and nonlineal doubling). By comparison, only six percent of the resolved radio galaxies contain solitary components and these can be entirely explained as doubles viewed end on. The doubles seen are rather symmetrical in their large scale structure, seventy five percent have component intensity ratios < 2.1 and the optical galaxies frequently lie very near the centre of the doubles. However, higher resolution observations, e.g. that of Cygnus A (Figure 10), show considerable differences in fine scale structure and polarization.

Despite the relatively simple morphology almost no other correlations are seen between observed parameters. For example, in a plot of radio power versus linear separation (open circles in Figure 12) it is only for the lowest of the six orders of magnitude in power that we see any systematic change in linear size and even this may be due to selection against weak sources of large angular

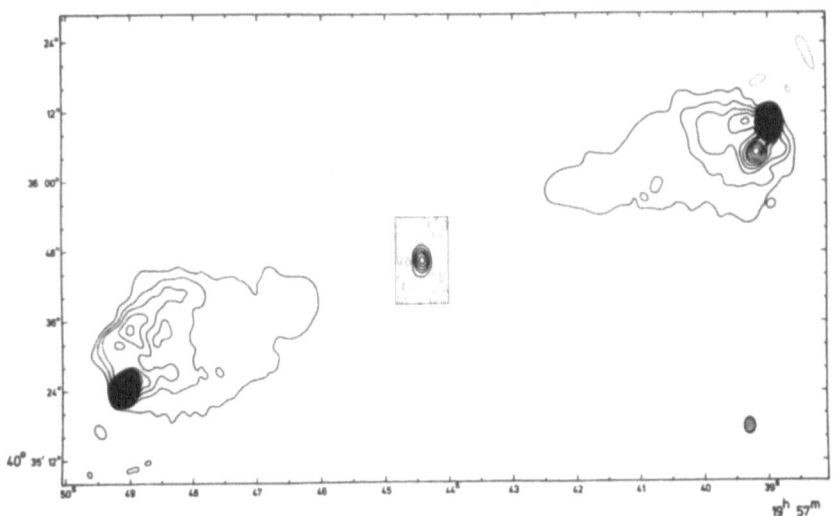

Figure 10. Contours of brightness temperature in Cygnus A at
5000 MHz. The contour interval in the area surrounding the central
component is $\frac{1}{5}$ of that for the outer components. (Reproduced from
Hargrave and Ryle 1974).

size. The effect of this selection can be seen in the recent discovery of the giant radio galaxies 3C236 and DA240 (Willis, Strom
and Wilson 1974). DA240 is a 34' arc double with an integrated
flux density of 22×10^{-26} W m^{-2} Hz^{-1} at 178 MHz. However, because
of its low brightness, it is missed in the 3C Catalogue which has
a nominal completeness limit of 9×10^{-26} W m^{-2} Hz^{-1}.

10.3 Energetics

As an example of the physical conditions in the radio galaxies
parameters are given in Table 4 for the most powerful, the largest
and the closest known system. The energetics and the problem of
confining the radiating blobs are discussed by Professor Rees.

10.4 Lifetimes

A minimum age of the radio galaxy is given by the light travel
time from the galaxy to the outer components. For 3C236 this is
2×10^7 years and if the expansion rate is <0.1 c, as is required
to keep the observed degree of symmetry (Mackey 1973), the light
travel time is $>2 \times 10^8$ years. Another estimate can be obtained
by comparing the luminosity with the total energy in relativistic
electrons and this is the value given in Table 4. However, if the
conditions are far from equilibrium or if new ejection of parti-

cles is occurring this estimate is not meaningful. An estimate
of the minimum time that a galaxy is a radio galaxy can be obtained from the fraction of a given type of a galaxy that we
see as radio sources now. Ten percent of the giant elliptical
galaxies have radio luminosity $>10^{42}$ erg sec^{-1} hence galaxies of
this luminosity spend ten percent of their life as radio galaxies

Table 4. Properties of Radio Galaxies

	Cyg A	3C236	Cen A
Distance (Mpc)	290	503	9
Total radio luminosity (erg/sec)	3×10^{45}	2×10^{43}	10^{42}
Linear size (Mpc)	0.14	5.2	1.4
Minimum total energy * (ergs)	6×10^{58}	10^{60}	2×10^{59}
Radiative life time (years)	6×10^{5}	2×10^{9}	6×10^{9}

* relativistic electrons and magnetic field only

i.e. 10^9 years (Schmidt 1966). It is possible that this time is
spent in a number of shorter periods.

10.5 Nuclei of radio galaxies

Recent high frequency radio observations show that the presence
of a compact radio nucleus such as discussed in Section 7.2., is
a common phenomenon in the radio galaxies. Previously these
compact nuclei had eluded detection since their flat spectra and
weakness relative to the extended sources have made them incon-
spicious in lower frequency maps.

    Most of the galaxies with radio nuclei are relatively nearby
radio galaxies and in most cases the compact nuclear sources are
near the radio detection limit. This suggests that in the more
distant radio galaxies there may be many more nuclear sources
which are below the present detection limit.

    Compact nuclear sources have now been detected in 15 out of
25 extended 3C sources identified with galaxies brighter than
m_p = 16 which have been observed with either the Cambridge or
Westerbork synthesis telescopes at 5000 MHz (Ekers 1974). In comparison, out of a sample of 114 E Galaxies brighter than m_p = 16
without extended radio sources, Ekers and Ekers (1973) found no
nuclear components as strong as any of those in the radio galaxies.
The presence of compact neclear sources in such a large fraction
of the radio galaxies, and only in the radio galaxies, is of considerable interest since it implies that the compact and extended

components commonly occur concurrently. This could happen if (i)
both types of radio emission are continuous for most of the lifetime of a unique type of elliptical galaxy or (ii) both types of
radio emission are transient but switch on and off together. Both
these cases argue that the extended component is maintained by a
continuous injection of energy - in (i) because the present
energy in relativistic particles in the extended sources is insufficient to enable it to continue radiating for the life of the
galaxy and in (ii) because the information about whether the
nucleus is on or off has to be transmitted to the extended component and the only plausible way to do this is to switch on or off
the energy supply.

10.6 Radio sources with tails

Ryle and Windram (1968) found two galaxies in the Perseus cluster,
NGC 1265 and IC 310, with morphology different from the usual double
sources. These two radio sources have extended tails trailing away
from the parent galaxy. Further observations (Hill and Longair
1971, Miley and van der Laan 1973) have given more detailed maps
of these objects and added further examples. The tails are up to
300 kpc long and trail away from the galaxy in a slightly curved
trajectory. Some appear to be double in cross section. Especially
interesting is the high resolution 5000 MHz observation of NGC
1265 (Wellington, Miley and van der Laan 1973) which is shown in
Figure 11. One obtains the strong impression that blobs of radiating plasma are ejected in pairs from the nucleus of the galaxy
and that these are forming a trail as the galaxy moves with respect to the intergalactic medium. Whether or not the details of
this picture are true it would seem that this observation provides
some of the strongest evidence available for the presence of an
intergalactic medium. It also strongly supports the idea of
continual ejection suggested by the statistical argument of the
last section.

11. SUMMARY

Given the general lack of correlation between observed parameters
for the radio galaxies, and in particular the lack of any obvious
evidence for the most popular theory in which the double sources
themselves were initially ejected and are expanding with time, it
might be wise to seek guidance from the strong pieces of evidence
that we do have.

1) Only the elliptical galaxies, including N and D systems,
have this strong radio emission while the spiral galaxies are
never observed in this state. Is it their higher mass, their
sphericity or their lack of gas?

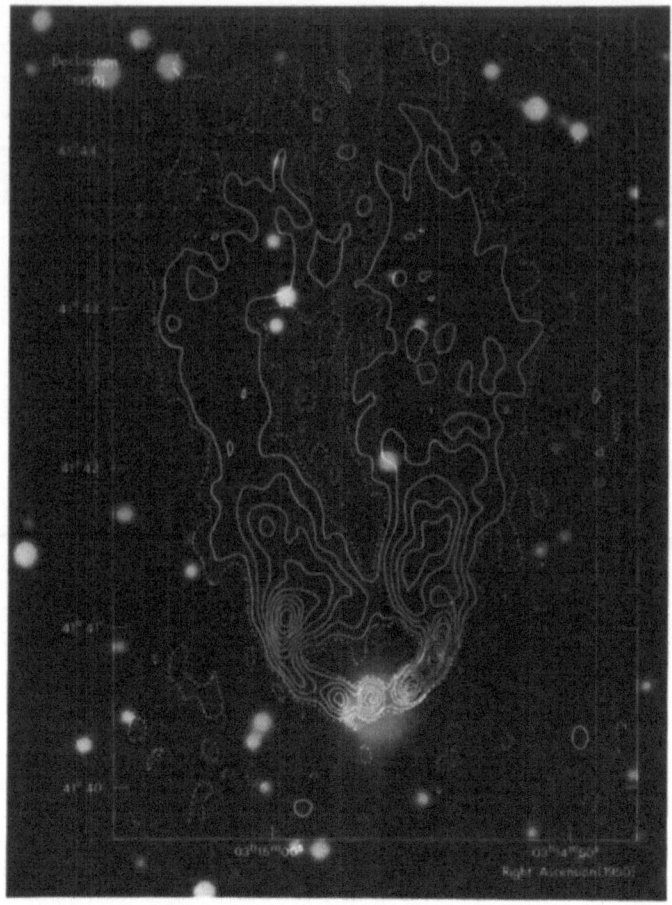

Figure 11. Contours of brightness temperature for NGC 1265 superimposed on the blue Palomar Sky Survey Print. (Reproduced from
Wellington et al. 1973).

2) Over a very large range of power and linear size the
radio galaxies are basically symmetric doubles. Is this caused
by the generation mechanism or by the external medium?

3) The probability of radio emission depends strongly on
absolute optical magnitude. Is it the greater mass of the giant
galaxies or a different evolutionary history?

Finally, to give the overall picture for the extragalactic
radio sources I have plotted all types together on a linear size
versus radio luminosity plot (Figure 12). For clarity, many objects
with only an upper limit on their radio diameter are not included

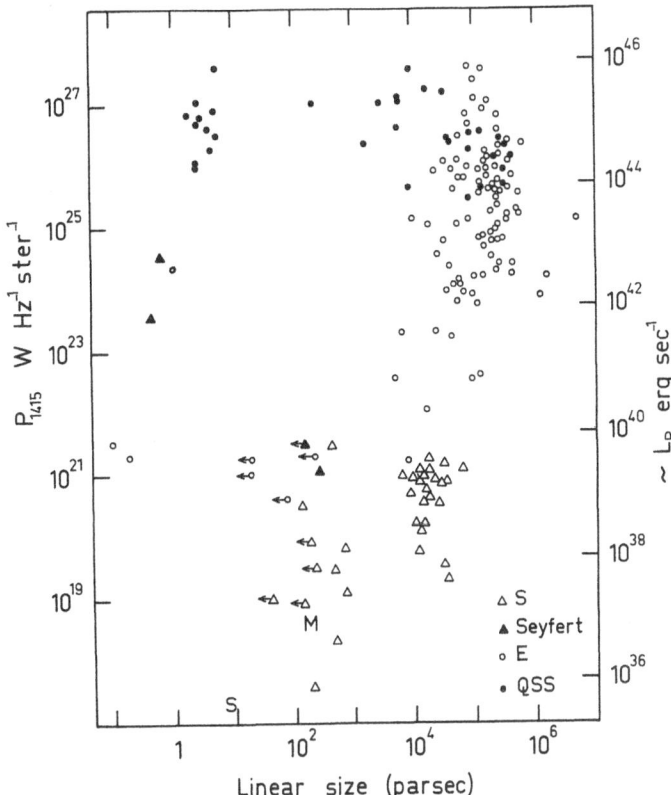

Figure 12. Plot of the monochromatic radio power at 1400 MHz against linear diameter for all known classes of extragalactic radio sources. For clarity most objects with upper limits to the diameter are not plotted. The radio luminosity scale, L_R, is defined in Figure 5. The extended source in our galactic centre is indicated by M and the Sgr A source by S.

and galaxies are plotted twice if they contain distinct nuclear and extended components. It is striking that the spiral galaxies are completely restricted to the lower part of this diagram. Even so they still have considerable variation in radio output which does not correlate particularly well with any of their properties. This may mean that some attribute, e.g. the magnetic field, which is important for the non-thermal radiation, is not an important factor in determining other properties of galaxies. There is much greater scatter for the elliptical galaxies which span the six orders of magnitude in luminosity between the spiral galaxies and the quasars. At the upper luminosity they become indistinguishable from the quasars (assuming cosmological distances) but at the lower luminosity, although they overlap with the spiral galaxies, they remain morphologically distinct.

On the left at P $\simeq 10^{21}$ W Hz^{-1} ster^{-1} we find the two elliptical galaxies M87 and NGC 1052, which contain the smallest extra-
galactic radio sources known. Unfortunately, the angular size of
most of the elliptical galaxy nuclei are not known and they do not
appear in Figure 12. If they are just as compact as those of M87
and NGC 1275 they will form a continuous distribution running up
to the N galaxy 3C371 and the two Seyfert galaxies 3C120 and
NGC 1275 at P $\simeq 10^{24}$ W Hz^{-1} ster^{-1}. These two Seyfert galaxies
have nuclei quite different from the other Seyfert galaxies which
lie much closer in radio properties to the nuclei of ordinary
spiral galaxies. The sequence of compact sources may then continue
on up to the quasars, the small number of radio galaxy nuclei in
this region being at least partly due to selection in favour of
quasars in the very long baseline interferometry observations.

REFERENCES

Baldwin, J.E., 1967, in Radio Astronomy and the Galactic System
     (IAU Symp. No. 31), ed. H. van Woerden (Academic Press,
     New York), p. 337.
Balick, B. and Brown, R.L. 1974, Astrophys. J. 194 (in press).
Becklin, E.E., Fomalont, E.B., and Neugebauer, G. 1973,
     Astrophys. J. 181, L27.
Berkhuijsen, E.M. and Wielebinski, R. 1974, Astron. Astrophys.
     34, 173.
de Bruyn, A.G., and Willis, A.G. 1974, Astron. Astrophys. 33, 351.
Cameron, M.J. 1971, Monthly Not. Roy. Astr. Soc. 152, 403.
Chandrasekhar, S. and Fermi, E. 1953, Ap.J. 118, 113.
Cohen, M.H., Cannon, W., Purcell, G.H., Shaffer, D.B., Broderick,
     J.J., Kellermann, K.I., and Jauncey, D.L. 1971, Astrophys.
     J. 170, 207.
Colla, G., Fanti, R., Fanti, L., Gioia, C., Lari, C., Lequeux, J.,
     Lucas, R., Ulrich, M.H. 1974, submitted to Astron. Astrophys.
     Supplements.
Cooper, B.F.C., Price, R.M. and Cole, D.J. 1965, Aust. J. Phys.
     18, 589.
Ekers, R.D. 1974, IAU Symposium No. 58 "Formation and Dynamics
     of Galaxies", ed. J. Shakeshaft.
Ekers, R.D., and Ekers, J.A. 1973, Astron. Astrophys. 24, 247.
Hargrave, P.J., and Ryle, M. 1974, Monthly Not. Roy. Astr. Soc.
     166, 305.
Heeschen, D.S. 1968, Astrophys. J. 151, L135.
Heeschen, D.S. 1970, Astron. J. 75, 523.
Hill, J.M., and Longair, M.S. 1971, Monthly Not. Roy. Astr. Soc.
     154, 125.
Hoyle, F. and Ireland, J.G. 1961, Monthly Not. Roy. Astr. Soc.
     122, 35.

Israel, F.P., and van der Kruit, P.C. 1974, Astron. and
    Astrophys. 32, 363.
Kleinmann, D.E., and Low, F.J. 1970, Astrophys. J. 161, L203.
Kronberg, P.P., Pritchet, C.J., and van den Bergh, S. 1972,
    Astrophys. J. 173, L47.
van der Kruit, P.C. 1971, Astron. Astrophys. 15, 110.
van der Kruit, P.C. 1972, Astrophys. Letters 11, 173.
van der Kruit, P.C., Oort, J.H., and Mathewson, D.S. 1972,
    Astron. Astrophys. 21, 169.
van der Kruit, P.C. 1973(a), Astron. Astrophys. 29, 231.
van der Kruit, P.C. 1973(b), Astron. Astrophys. 29, 263.
Mackay, C.D. 1973, Monthly Not. Roy. Astr. Soc. 162, 1.
Mathewson, D.S., van der Kruit, P.C., and Brouw, W.N. 1972,
    Astron. Astrophys. 17, 468.
Miley, G.K., and van der Laan, H. 1973, Astron. Astrophys.
    28, 359.
Piddington, J.H. 1972, Cosmic Electrodynamics 3, 129.
Pooley, G.G. 1969, Monthly Not. Roy. Astr. Soc. 144, 101.
Price, R.M. 1974, Astron. Astrophys. 33, 33.
Roberts, W.W., and Yuan, C. 1970, Ap.J. 161, 887.
Rots, A.H. 1974, Distribution and Kinematics of Neutral Hydrogen
    in the Spiral Galaxy M81, University of Groningen,
    Ph.D. Thesis.
Ryle, M., and Windram, M.D. 1968, Monthly Not. Roy. Astr. Soc.
    138, 1.
Schmidt, M. 1966, Astrophys. J. 146, 7.
de Vaucouleurs, G., and de Vaucouleurs, A. 1964, Reference
    Catalogue of Bright Galaxies, University of Texas Press,
    Austin.
Wade, C.M. 1968, Astron. J. 73, 876.
Wellington, K.J., Miley, G.K., and van der Laan, H. 1973,
    Nature 244, 502.
Willis, A.G., Strom, R.G., and Wilson, A.S. 1974, Nature 250, 625.
Wright, A.E. 1974, Monthly Not. Roy. Astr. Soc. 167, 273.
de Young, D.S., Roberts, M.S., and Saslaw, W.C. 1973, Astrophys.
    J. 185, 809.

THE DETERMINATION OF THE HUBBLE PARAMETER: AN INTERIM REPORT

S. van den Bergh

David Dunlap Observatory, University of Toronto,
Richmond Hill, Ontario.

1. INTRODUCTION

The present paper should be regarded as an interim report on efforts
to derive the numerical value of the Hubble parameter. A review of
the extra-galactic distance scale seems appropriate because the
present epoch marks the culmination of a number of beautiful investigations of the members of the Local Group. The distances to these
galaxies, which are our nearest neighbours in extra-galactic space,
are now relatively secure. Changes of more than say 15 or 20 percent
in the distance scale within the Local Group appear quite improbable.
The second stage in the quest for the Hubble constant is now well
underway. At the Hale Observatories the 200-inch telescope is being
used in a determined effort to push accurate distance determinations
beyond the limits of the Local Group to the small nearby clusters
that are associated with the supergiant spirals M81 and M101.

The present paper presents a rediscussion of the observational
material on the extra-galactic distance scale that is currently
available. This discussion falls naturally into three parts:

1. The determination of the distances to certain key members
of the Local Group.

2. A discussion of the distance scale beyond the Local Group.

3. An investigation of the possible large-scale anisotropy
and/or inhomogeneity of the Hubble flow.

Needless to say the second and third sections of this discussion are based on observational material that is far less secure
than is that on which the distance scale within the Local Group
is based. All determinations of the extra- galactic distance scale
are ultimately based on the assumption that recognizable types of
distant objects are similar to nearby objects of the same type.
Due to possible differences in age, evolutionary history and abun-

G. Setti (ed.), Structure and Evolution of Galaxies, 247–259. All Rights Reserved.
Copyright © 1975 by D. Reidel Publishing Company, Dordrecht-Holland.

dance of the elements this assumption may not be correct for any
particular type of distance indicator. The basic philosophy adopted
in this review is that systematic errors resulting from differences
in chemical abundances etc. will be minimized if the largest
possible number of methods of distance determination is employed.
Admittedly this procedure has the drawback that a bad apple might
be included in the pie.

2. DETERMINATION OF THE DISTANCE SCALE WITHIN THE LOCAL GROUP

Hubble and Humason (1931) used the mean absolute magnitude of the
seven brightest members of the Local Group to estimate the distances
to a number of distant clusters of known redshift. The best Local
Group distance estimates available at that time gave $<M_{pg}>=-13.5$
for the seven brightest members of the Local Group. This value
yielded a Hubble constant $H=630$km $sec^{-1}Mpc^{-1}$. Using the most recent
distance estimates to members of the Local Group gives $<M_B>=-17.7$
for the seven brightest Local Group members. This yields $H=90$km
$sec^{-1}Mpc^{-1}$. The difference between these two values of the Hubble
constant is entirely due to an increase in the adopted distance
scale within the Local Group. This shows how vitally important
it is to obtain an accurate distance calibration within the Local
Group.

3. THE DISTANCE SCALE BEYOND THE LOCAL GROUP

In this section the current observational status of twelve distinct
methods of measuring the Hubble constant is discussed. Each one of
these methods is itself quite uncertain. The degree of internal
agreement between the distances obtained by different methods may,
however, provide some indication of the degree of confidence which
can be placed in the resulting distance scale.
    1. Supernova expansion. Kirschner and Kwan (1974) have recently
applied the "Baade-Wesselink method" to the expansion of supernovae.
From observations of two type II supernovae they obtain $H=60\pm15$km
$sec^{-1}Mpc^{-1}$. This result is consistent with a somewhat less accurate
value of H that Branch and Patchett (1973) have obtained from two
supernovae of type I. This type of determination of the Hubble
parameter is to be preferred to all others because it avoids reliance
on the long and uncertain chain of arguments that start with the
distance modulus of the Hyades.
    2. Diameters of HII regions. Hot stars that are imbedded in
interstellar gas ionize the hydrogen in their vicinity. A number
of investigators have compared the diameters of the largest HII
regions in members of the Local Group with the diameters of HII
regions in distant galaxies to obtain an estimate of the Hubble
constant. Such comparisons are fraught with a number of dangers. In
the first place the sizes of the largest HII regions will depend

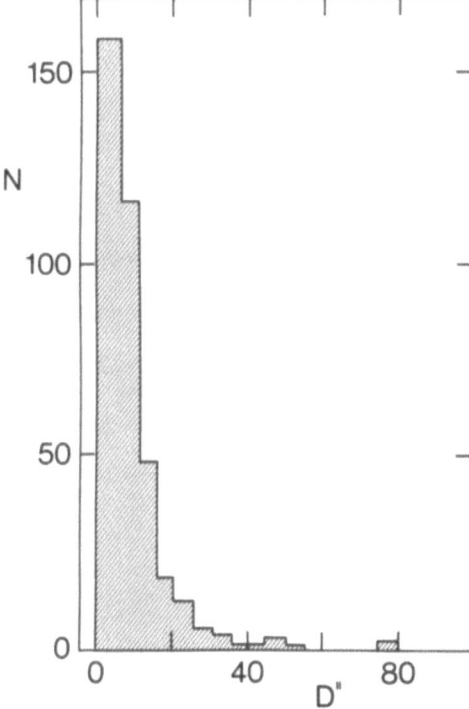

Fig.1. Histogram showing the observed frequency of HII region diameters in M33 according to Sérsic (Sandage 1962).

on the total number of hot young stars in a galaxy. This implies that the diameters of the largest HII regions will depend on both the luminosity and on the Hubble type of a galaxy. Even within a given classification type and luminosity class there are large differences between individual galaxies. The second difficulty results from the fact that only the very largest HII regions, which occur at the thinly populated upper end of the size-spectrum, can be measured at large distances. Because we are dealing with the statistics of small numbers, the mean diameter of the largest three HII regions will not be a very stable quantity (see Fig.1). This is particularly obvious in the case of the Large Magellanic Cloud. According to Sérsic (1959) the largest HII regions in the LMC has a diameter which is 2.5 times greater than that of the third largest emission region. A third difficulty arises from the fact that individual HII regions in the most active areas of star formation will frequently overlap. It is, therefore, often difficult to decide what is the largest HII region and what is, in fact, a complex of smaller but partly overlapping ionized hydrogen zones. This difficulty becomes particularly acute in the most luminous galaxies of Hubble type Sc which are critically important for the calibration of the

Table I

Distance Modulus of the LMC

	Author	Method	$(m-M)_V$
i	RR Lyrae Variables		
	Graham (1973)	Field variables	18.4
ii	Globular Cluster Giants		
		NGC1466,1978,2257	
iii	Novae		
	van den Bergh (197?)		19.3
iv	Supernovae		
	Mathewson and Clarke (1973)		18.4
v	Cepheids		
	Sandage and Tammann (1968)	PLC	18.6
	Sandage and Tammann (1971)	PLA	18.9
vi	Hydrogen Line Strengths		
	Hutchings (1966)	H_γ B stars	17.9
		H_γ A stars	18.5
vii	Spectral Gradients		
	Divan (1973)		18.6
viii	Luminosity Classification		
	Feast et al (1960)		18.7
ix	Horizontal Branches		
	of Clusters		18.9

$<m-M>_V = 18.6 \pm 0.1$

extra-galactic distance scale. In such supergiant Sc galaxies overlapping HII regions often produce almost completely ionized spiral arm segments. In view of these difficulties it is, perhaps, not surprising that the two most recent attempts to use HII regions for the calibration of the extra-galactic distance scale do not show very good agreement. Sérsic (1960) obtains H=125±5 km sec^{-1} Mpc^{-1} whereas Sandage and Tammann (Babcock 1971) obtain H=53±5 km sec^{-1}Mpc^{-1}. The size of the quoted mean errors gives some indication of the optimism with which astronomers tend to embark on voyages through the uncharted regions of extra-galactic space.

From a rediscussion of Sérsic's observational material (van den Bergh 197?) and the Local Group distance moduli given in Tables I-IV, a value H=99^{+16}_{-13} km sec^{-1}Mpc^{-1} is obtained. A harmonic mean of this value and Sandage and Tammann's preliminary results yield H=72^{+30}_{-20}km sec^{-1}Mpc^{-1}. (The quoted mean error represents an estimate of the uncertainty of this result).

Compared to photometric parameters, which scale as H^{-2}, HII

Table II

Distance Modulus of the SMC

	Author	Method	$(m-M)_V$
i	RR Lyrae Variables		
	Graham (1973)	Field variables	18.9
ii	Globular Cluster Giants		
	Tifft (1963)	NGC121 vs. NGC362	18.9
iii	W Virginis Stars		
	Payne-Gaposchkin and Gaposchkin (1966)	Field variables	19.7
iv	BL Herculis Star		
	Tifft (1963)		19.4
v	Novae		
	van den Bergh (197?)		19.4
vi	Supernovae		
	Mathewson and Clarke (1972,1973)		18.8
vii	Cepheids		
	Sandage and Tammann (1968)	PLC	19.0
	Sandage and Tammann (1971)	PLA	19.2
	Gascoigne (1974)	PLC+Z/4	18.6
viii	Strength of Hydrogen Lines		
	Buscombe and Kennedy (1962)	H_γ	18.7
	Westerlund et al (1963)	β vs. B-V	18.5
	Hutchings (1966)	H_γ (B stars)	18.1
		H_γ (A stars)	18.5
	Osmer (1973)	β vs. c_o	19.1

$$\langle m-M \rangle_V = 18.6 \pm 0.1$$

region diameters have the disadvantage that they are only proportional to H^{-1}. This suggests that radio or optical measurements of the brightness of HII regions might be a useful method for the determination of the extra-galactic distance scale.

3. Luminosity classification of galaxies. The brightest and the faintest known spiral galaxies differ in luminosity by more than a factor of 100. Such large differences in luminosity, and presumably also in mass, are reflected in the appearance of galaxies: intrinsically faint spirals, in which a single stellar association may provide a considerable contribution to the total luminosity, usually have a low surface brightness and short rather patchy spiral arms. Spirals with long well developed spiral arms are invariably objects of high intrinsic luminosity. This dependence of appearance on absolute magnitude may be used for luminosity classification of spiral galaxies (van den Bergh 1960 a,b).

Table III

Distance Moduli of M31, M33, NGC6822 and IC1613

Method	M31	M33	NGC6822	IC1613
Cepheids	24.6	24.8	24.8	24.5
Novae	24.5	-	-	-
W Virginis	24.7	-	-	-
Pop.II giants	24.4	-	-	-
$<m-M>_V$	24.6	24.6	24.8	24.5

From a comparison of the luminosity classifications of M31 and
M33 in the Local Group with the luminosity classification of more
distant galaxies a Hubble parameter $H=93^{+18}_{-15}$ km sec^{-1}Mpc^{-1} is obtained. The true uncertainty of this result may be somewhat larger
than the quoted mean error because M31 is a rather a-typical representative of luminosity class I-II. The reason for this might be
that the spiral structure in M31 is currently undergoing tidal
distortion by M32.

    4. The brightest globular clusters in galaxies. The brightest
and faintest known galactic globular clusters differ in luminosity
by a factor of more than 100. This shows that the mass spectrum
with which globular clusters are formed depends on the physical
conditions prevailing in the cluster forming regions of protogalaxies. It appears improbable that these conditions were similar
in the Galaxy (which is of the Hubble type Sb or Sc) and in distant
elliptical supergiant galaxies. It is therefore probably unrealistic

Table IV

Magnitudes and Distances of Key Local Group Members

	LMC	SMC	M31	M33	NGC6822	IC1613
$(m-M)_V$	18.6	18.9	24.6	24.6	24.8	24.5
V	0.1	2.3	3.5	5.7	8.5	9.6
M_V	-18.5	-16.6	-21.1	-18.9	-16.3	-14.9
A_V (foreground)	0.17	0.17	0.33	0.09	0.57	0.09
D(kpc)	49	56	710	800	700	760

to expect very different types of galaxies to form globular clusters with similar mass spectra. Additional evidence for such differences between globular clusters in different galaxies is provided by observations of the M31 cluster system (van den Bergh 1969) which show that: a) the most luminous globular cluster in M31 is about twice as luminous as is the brightest known globular in the Galaxy and b) the average metal abundance of the stars that make up the M31 globular clusters appears to be higher than it is in the case of galactic globulars. In view of these differences it seems unsafe to assume that the brightest globular clusters in different galaxies will exhibit the same absolute magnitudes. According to Hanes (1971) the brightest globular clusters in the supergiant elliptical galaxy M87 in the Virgo cluster have B$\approx$20.5. Comparison with ω Centauri, which is the brightest (M_B=-9.7) known galactic globular cluster yields $(m-M)_B \approx (m-M)_o$ =30.2 for the Virgo cluster. This value is, however, an underestimate because it does not allow for the fact that M87 contains approximately 10 times as many globular clusters as does the Galaxy. Reference to de Vaucouleurs (1970) suggests that the distance modulus of the Virgo cluster obtained in this way should be corrected by approximately 0.5 mag to take into account the size of sample effect. This yields $(m-M)_o$=+30.7 for the Virgo cluster. With this distance modulus and V_r=1175±64 km sec^{-1} (van den Bergh 1960c) it then follows that 5log(H/100)=-0.07 with an estimated uncertainty of $\sim$0.4. It should, of course, be emphasized that this uncertainty does <u>not</u> take into account possibly systematic differences between the mass spectrum of cluster formation in M87 and the Galaxy.

   5. <u>Mass-to-light ratios of galaxies</u>. Radio observations of neutral hydrogen gas may be used to obtain the total masses of galaxies. The masses obtained in this way are proportional to the assumed distances. The intrinsic luminosities of galaxies are, of course, proportional to the assumed distances squared. It follows that the mass-to-light ratios are inversely proportional to the assumed galaxy distances. Roberts (1969) has shown that the mean mass-to-light ratios of spirals and irregulars are almost independent of Hubble type and of galaxy luminosity. A comparison between the mass-to-light ratios of members of the Local Group (for which distances are well known) with the mass-to-light ratios of distant galaxies can therefore provide a determination of the Hubble constant. From Roberts' data and the distances quoted in Table IV a mean value <log(M/L)>=0.63±0.11 is obtained. Assuming that this same value also applies to the galaxies with D>4 Mpc, for which Roberts (1969) gives mass determinations, then yields H=105$^{+33}_{-26}$ km sec^{-1}Mpc^{-1}. The quoted mean error of this result is large because only 6 Local Group galaxies are available for the calibration of M/L. It is expected that it will soon be possible to strengthen this calibration considerably by using M/L values derived from the galaxies in the M81 and M101 groups.

   6. <u>Third brightest members of small clusters of galaxies</u>.
Perhaps half of all the galaxies in the sky are located in small

clusters such as the Local Group. From inspection of the Palomar
Sky Survey four small clusters were found which appeared to resemble
the Local Group. The brightest galaxies in these four small clusters
are NGC3227, NGC3368, NGC3627, and NGC4274. Comparison of the third
brightest members of these small clusters with M33, which is the
brightest member of the Local Group, yields $H=126^{+48}_{-35}$ km sec^{-1}Mpc^{-1}.
The rather large mean error that is quoted for this determination of
the Hubble constant is mainly due to the fact that the intrinsic
luminosity of the third brightest galaxy in small groups exhibits
rather a large dispersion. Furthermore, it should perhaps be emphasized that this determination of H involves a subjective choice
of clusters which are believed to be similar to the Local Group.

7. Comparison of galactic and extra-galactic supernovae. The
main advantage of using supernovae to calibrate the extra-galactic
distance scale is that very distant supernovae can be compared directly with supernovae in our own Galaxy without using the members
of the Local Group as intermediate stepping stones. The main draw-
backs of this method are: a) little is known about the intrinsic
dispersion in the luminosities of supernovae and b) only two galactic supernovae are known for which the absolute magnitudes at maximum light can be determined with reasonable accuracy. One of these
is the supernova that produced the Crab Nebula. According to
Minkowski (1966) the Crab Nebula was produced by a very unusual
supernova. It is therefore not quite clear which type of extra-
galactic supernova the Crab Nebula should be compared to. For
Tycho's supernova of 1572 van den Bergh (1970) obtained V_o(max)=
=-5.6±0.3. A discussion of the distance to Tycho's supernova
by Strom and Duin (1973) suggests that the distance to Tycho's
supernova is certainly greater than 3 kpc but less than 7 kpc.
Adopting $(m-M)_o$ =13.8±0.5 yields M_V(max)=-19.4±0.6. Combining this
value with the absolute magnitude calibration of type I supernovae
by Kowal (1968, 1969), who obtains M_V=-18.6+5log(H/100)=-0.8±1.0.

Unfortunately the supernova S And that occured in M31 in 1885
was also of an unusual type and it therefore cannot be used to estimate H.

8. The surface brightness of galaxies. Holmberg (1958) has shown
that there exists a loose correlation between the absolute luminosity
of galaxies and their surface brightness. A comparison of the surface
brightness of galaxies in the Virgo cluster with the surface brightness of M31 and M33 (both corrected for galactic interstellar redden-
ing) yields a Hubble constant $H=89^{+46}_{-30}$ km sec^{-1}Mpc^{-1}. The reasons for
the large quoted mean error of this determination are: a) individual
galaxies exhibit a large intrinsic dispersion about the mean relation
between luminosity and surface brightness and b) only two members
of the Local Group (M31 and M33 are available to calibrate the relation. This is so because the other members of the Local Group
are intrinsically fainter than are the galaxies of the Virgo cluster
for which the relation between surface brightness and luminosity is
known to be valid.

9. The diameters of galaxies. It is well known that there exists

a loose correlation between the intrinsic luminosities of galaxies and their diameters. Heidmann (1969) has shown that this relation may be used to estimate the Hubble constant. Comparison of Holmberg's (1958) data on the luminosities and diameters of galaxies in the Virgo cluster with the presumably known diameters of the members of the Local Group yields a Hubble constant $H=89^{+46}_{-30}$ km sec^{-1}Mpc^{-1}.

10. **The brightest stars in galaxies.** The brightest non-variable stars of Population I in members of the Local Group may be compared with the brightest stars in distant galaxies to obtain an estimate of the Hubble constant. A number of factors conspire to make this determination of the Hubble constant exceedingly uncertain: a) The luminosity of the brightest stars in galaxies depends on the total Population I content of these galaxies. The most luminous young stars would, therefore, be expected in supergiant galaxies of type Sc. b) At extra-galactic distances it is impossible to differentiate between the brightest individual stars and star clusters. (The Orion Trapizum would look like a single star at the distance of the Large Magellanic Cloud). c) The bright end of the stellar luminosity function is sparsely populated with stars so that the statistical separation of truly extragalactic stars of high luminosity from foreground field stars is uncertain. This difficulty is aggravated by the fact that the most luminous extra-galactic stars have colours that are quite similar to those of the brightest sub-dwarfs in the galactic halo. d) Outside the Local Group it is not possible to distinguish the brightest F and G type supergiants from bright globular clusters. From the very scanty observational material that is currently available on the brightest stars in distant galaxies it is estimated that $H=105^{+61}_{-39}$ km sec^{-1}Mpc^{-1}. A very considerable improvement in the accuracy of this determination of H is possible with currently available observational techniques.

11. **Dwarf galaxy colours.** It is well known (Baum 1959, McClure and van den Bergh 1968) that the colours of elliptical galaxies depend on their luminosities in such a way that the most luminous ellipticals have the reddest colours. This result may be used to estimate the Hubble constant by comparing the colours of dwarf systems in the Local Group with those of dwarf ellipticals in the Coma and Virgo clusters (Sandage 1972). After excluding M32, which is probably too faint for its colour (Faber 1973) due to tidal stripping by M31, the dwarfs NGC147, NGC185, NGC205, and the Fornax system yield $5\log(H/100)=1.12\pm0.65$ corresponding to $H=167^{+59}_{-43}$ km sec^{-1}Mpc^{-1}. This value is significantly greater than that obtained by any other method suggesting that the dwarfs in the Local Group are, for some unknown reason, anomalously red.

12. **Supernovae rate.** Van den Bergh (1972) has recently compared the supernova rate per unit luminosity in external galaxies with the galactic supernova rate that is determined from radio surveys of extended non-thermal sources in the Galaxy. With a galactic supernova frequency of one supernova per 50±25 years such a comparison yields $H=95\pm36$ km sec^{-1}Mpc^{-1}.

Most of the methods of determining the Hubble parameter that

Table V

Summary of Data on Hubble Constant

Method	$5\log(H/100)$	$H(\text{km sec}^{-1}\text{Mpc}^{-1})$
Supernova Expansion	-1.11 ± 0.55	60^{+15}_{-15}
Diameters of HII regions	-0.71 ± 0.73	72^{+30}_{-20}
Luminosity classifications of galaxies	-0.15 ± 0.38	93^{+18}_{-15}
Brightest globular clusters in galaxies	-0.07 ± 0.40	97^{+19}_{-16}
Mass-to-light ratios of galaxies	$+0.10\pm0.60$	105^{+33}_{-26}
Third brightest member of small clusters	$+0.50\pm0.70$	126^{+48}_{-35}
Tycho's supernova	-0.80 ± 1.00	69^{+41}_{-25}
Surface brightness of galaxies	-0.26 ± 0.90	89^{+46}_{-30}
Diameters of galaxies	-0.25 ± 0.90	89^{+46}_{-30}
Brightest stars	$+0.10\pm1.00$	105^{+61}_{-39}
Dwarf galaxy colours	$+1.12\pm0.65$	167^{+59}_{-43}
Supernova rate	-0.11 ± 0.87	95^{+36}_{-36}
Weighted mean	-0.16 ± 0.17	93 ± 7

have been discussed above, involve estimates of the distance modulus which is proportional to 5 logH rather than to H itself. The
most meaningful average value of the Hubble constant is, therefore,
obtained by forming the weighted mean value of 5 logH. From the
data in Table V it is seen that <5 logH>=1.84±0.17, which corresponds
to H=93±7 km sec^{-1}Mpc^{-1}. An unwanted mean of the data listed in
Table V yields <5 logH>=1.87 corresponding to H=94 km sec^{-1}Mpc^{-1}.
Past experience in determinations of the numerical value of the
Hubble parameter suggests that little trust should be placed in the
small formal mean error obtained for the value of H.

4. REGIONAL VARIATIONS IN THE HUBBLE PARAMETER

The estimates of the Hubble parameter that have been discussed in the previous section were almost exclusively based on observations of galaxies with redshifts smaller than approximately 2000 km sec^{-1}. The value H=93±7 km sec^{-1}Mpc^{-1} that is obtained from the data in Table V therefore refers to the volume of space that contains the huge Virgo cluster and other groupings of galaxies that together constitute what is sometimes referred to as the Local Supercluster. Since this region of space contains an above average number of galaxies, it is quite possible that the value of H that is deter- mined locally may not be representative of the Universe as a whole.

Recently Sandage (1968) has shown that the absolute magnitudes of the brightest members of rich clusters of galaxies exhibit a dis- persion of only ∿0.3 magnitude. A comparison of NGC4472, which is the brightest member of the Virgo cluster, with Sandage's data on the brightest galaxies in distant clusters then yield H(0)/H(∞)= =1.04±0.15, in which H(0) is the local value of the Hubble constant and H(∞) is the Hubble constant beyond the Local Supercluster. This result has been criticized by Abell (1970), who questions whether it is permissible to compare the brightest elliptical in the Virgo cluster (which is only a giant cluster) with the brightest members of huge supergiant clusters at large distances. Some support for the validity of this procedure is provided by Kowal's (1968, 1969) comparison of the supernovae in the Virgo cluster with more distant supernovae.

Furthermore a comparison of the galaxies in the Coma and Virgo clusters (Sandage 1972) yields Δ(m-M)=3.66±0.21 (m.e.). This is entirely consistent with the value Δ(m-M)=3.84 that is obtained from the mean radial velocities of the Virgo (van den Bergh 1960c) and Coma (Rood et al. 1972) clusters and the assumption that H(0)/H(∞)=1.00.

Finally Corwin (1971) has compared the distance module of the Virgo cluster and of the Hercules cluster. This comparison is par- ticularly significant because the Hercules and Virgo clusters have a similar galaxian population content. Furthermore, the Hercules cluster, which has almost ten times as large a redshift as does the Virgo cluster, is located well beyond the Local Supercluster. From luminosity classifications of a few of the brightest galaxies in the Hercules cluster Corwin obtains (after correcting for the difference in galactic absorption) a difference in the distance moduli of these two clusters of Δ(m-M)$_0$=4.0 mag. From a comparison of the apparent magnitudes and the diameters of the five brightest spiral galaxies and the five brightest ellipticals in the Virgo and Hercules clusters Corwin finds a difference of 4.6 magnitudes in the distance moduli of the Virgo and Hercules clusters. Combining these two results yields Δ(m-M)$_0$=4.3±0.5 where the quoted mean error should be re- garded as no more than a rough indication of the weight that should be assigned to this result. Corwin's value Δ(m-M)$_0$=4.3±0.5 yields H(0)/H(∞)=0.84±0.20. The conclusion from Sandage's comparison be- tween the brightest E galaxy in the Virgo cluster and the brightest galaxies in distant clusters and from Corwin's comparison of the Virgo and Hercules clusters appears to suggest that the difference

between the Hubble constant inside and outside the Local Supercluster is small.

REFERENCES

Abell, G.O., 1970, Lecture given at the University of Toronto.
Babcock, H.W., 1971, Carnegie Institution Year Book (Carnegie
Institute of Washington, Washington D.C.) p.418.
Baum, W.A., 1959, Publ. Astron. Soc. Pacific 71, 106.
Branch, D., and Patchett, B., 1973, Monthly Notices Roy. Astron.
Soc. 161, 71.
Buscombe, W., and Kennedy, P.M., 1962, J. Royal Astr. Soc. Canada
56, 113.
Corwin, H.G., 1971, Publ. Astron. Soc. Pacific 83, 320.
Divan, L., 1973, in Problems of Calibration of Absolute Magnitudes
and Temperatures of Stars, IAU Symposium No. 54, B. Hauck and B.E.
Westerlund (eds.), (Reidel Publishing Co. - Dordrecht, p. 78.)
Faber, S.M., 1973, Astrophys. J. 179, 423.
Feast, M.W., Thackeray, A.D., and Wesselink, A.J., 1960, Monthly
Notices Roy. Astron. Soc. 121, 337.
Gascoigne, S.C.B., 1974, Monthly Notices Roy. Astron. Soc. 166, 25.
Graham, J.A., 1973, IAU Colloquium No. 21, J.D. Fernie (ed.),
(Reidel Pub. Co. - Dordrecht, p.120).
Hanes, D.A., 1971, M. Sc. Thesis, University of Toronto (unpublished).
Heidmann, J., 1969, C.R. Acad. Sci. 268b, 1782.
Holmberg, E., 1958, Lund Obs. Medd. Ser. 2 No. 136.
Hubble, E.P., and Humason, M.L., 1931, Astrophys. J. 74, 43.
Hutchings, J.B., 1966, Monthly Notices Roy. Astron. Soc. 132, 433.
Kirschner, R.P. and Kwan, J., 1974, Astrophys. J. (in press).
Kowal, C.T., 1968, Astron. J. 73, 1021.
Kowal, C.T., 1969, Publ. Astron. Soc. Pacific 81, 608.
Mathewson, D.S., and Clarke, J.N., 1972, Astrophys. J. Letters 178,
L105.
Mathewson, D.S., and Clarke, J.N., 1973, Astrophys. J. 180, 725.
McClure, R.D., and van den Bergh, S., 1968, Astron. J. 73, 1008.
Minkowski, R., 1966, Astron. J. 71, 371.
Osmer, P.S., 1973, Astrophys. J. 181, 327.
Payne-Gaposchkin, E., and Gaposchkin, S., 1966, Smithsonian Contri-
bution to Astrophysics 9, (whole volume).
Roberts, M.S., 1969, Astron. J. 74, 859.
Rood, H.J., Page, T.L., Kintner, E.C., and King, I.R., 1972, Astrophys.
J. 175, 627.
Sandage, A.R., 1962, in Problems of Extra-galactic research, G.C.
McVittie (ed.) (The Macmillan Co. - New York) p. 359.
Sandage, A.R., 1968, Observatory 88, 91.
Sandage, A.R., 1968, Astrophys. J. 176, 21.
Sandage, A.R., and Tammann, G.A., 1968, Astrophys. J. 151, 531.
Sandage, A.R., and Tammann, G.A., 1971, Astrophys. J. 167, 293.
Sérsic, J.L., 1959, Observatory 79, 54.

Sérsic, J.L., 1960, Zs. fur Astrophysik 50, 168.
Strom, R.G., and Duin, R.M., 1973, Astron. & Astrophys. 25, 351.
Tifft, W.G., 1963, Monthly Notices Roy. Astron. Soc. 125, 199.
van den Bergh, S., 1960a, Astrophys. J. 131, 215.
van den Bergh, S., 1960b, Astrophys. J. 131, 558.
van den Bergh, S., 1960c, Monthly Notices Roy. Astron. Soc. 121, 387.
van den Bergh, S., 1969, Astrophys. J. Supplement 19, 145.
van den Bergh, S., 1970, Nature 225, 503.
van den Bergh, S., 1972, Astron. & Astrophys. 20, 469.
van den Bergh, S., 197?, Ch. XV in Galaxies and the Universe, A. and M. Sandage (eds.), (Chicago, University of Chicago Press).
Vaucouleurs, A de, 1970, Astrophys. J. 159, 435.
Westerlund, B.E., Danziger, I.J., and Graham, J., 1963, Observatory 83, 74.

GALAXIES AND THEIR EVOLUTION

Wallace L. W. Sargent

Hale Observatories, California Institute of
Technology, Carnegie Institution of
Washington

ABSTRACT. The methods by which the masses and mass-to
light ratios of galaxies are determined from the
velocity dispersions of stars and from dynamical
studies of clusters of galaxies are outlined. It is
concluded that while small clusters give no evidence
for "missing mass" there is a discrepancy of about a
factor of 5 in the case of large clusters. At present
it is not clear whether the extra mass lies in extended
envelopes of the luminous galaxies or whether it lies
in a background sea of low luminosity objects not
bound to any particular galaxy. Next an account is
given of simple theoretical ideas on the evolution of
the chemical composition of galaxies. Particular
attention is drawn to the role played by the initial
mass function in determining the yield - the efficiency
with which evolving stars convert interstellar hydrogen
into heavy elements.

1. THE MASSES OF GALAXIES - GENERAL

Knowledge of the total masses, mass distributions and
mass-to-light (M/L) ratios of a representative sample
of galaxies is obviously essential for theories of the
evolution of these objects. There are in essence five
ways in which the masses of galaxies can be determined:
a) From the study of rotation curves.
b) By statistical studies of binary galaxies.
c) From measurements of the velocity dispersion of the
   stars.

G. Setti (ed.), Structure and Evolution of Galaxies, 261–274. All Rights Reserved.
Copyright © 1975 by D. Reidel Publishing Company, Dordrecht-Holland.

d) From studies of the dynamics of clusters of galaxies
   - for example, by the virial theorem.
e) From virial studies of the dynamics of satellites of
   large galaxies - for example, globular clusters.

Method a) is mainly used for spiral galaxies while
method b) can be used, in principle, for pairs of any
type. Both are discussed in detail in E. M. Burbidge's
lectures in this volume. Here we shall give a brief
account of methods c), d) and e). In the past the
application of the virial theorem to clusters of galaxies has led to mass-to-light ratios for these objects
that are considerably higher than these deduced by the
other methods for individual galaxies. More recently,
Ostriker and Peebles [1] have argued that spiral galaxies are much larger and have much higher total
masses than has previously been supposed.

1.1 Masses from velocity dispersions of stars

This method is used primarily for elliptical galaxies
and for the central bulges of spirals. These objects
have spectra with many absorption lines due to the
predominantly cool stars which produce the major proportion of their visible light. It is possible to
determine the velocity dispersion σ_r of the stars in
the radial direction from the widths of the absorption
lines in the integrated light of the galaxy. The mass
of the system is obtained through an application of
the virial theorem which, for a bound, non-rotating
system, states that

$$2<T> + <\Omega> = 0 \qquad\qquad 1.1$$

where the brackets denote time averaged values of the
total kinetic energy T and the gravitational potential
energy Ω. Now

$$\Omega(r) = -G \int_0^r \frac{M(r')\,dM(r')}{r'} \qquad\qquad 1.2$$

where $dM(r) = 4\pi r^2 \rho(r)dr$ is the mass between radii r
and r+dr. In order to determine $\Omega(r)$ it is assumed
that the projected radial mass distribution is the
same as the projected light distribution - i.e., that
the system has a constant mass-to-light ratio with
radius. DeVaucouleurs has shown that the light distribution in an elliptical galaxy obeys the empirical
law

$$\log \sigma(r) = -Ar^{\frac{1}{4}} + B \qquad\qquad 1.3$$

where $\sigma(r)$ is the surface brightness at distance r from
the center and where A and B are constants which vary
from system to system. Poveda [2] showed that the

resulting expression for the gravitational potential
is

$$\Omega = -0.33 \, \frac{GM^2}{R'} \qquad\qquad 1.4$$

where M is the total mass of the galaxy and R' is the
"effective radius" - the radius of a circle which contains half of the light of the galaxy.

The kinetic energy T of the stellar motions is given
by

$$T = 3/2 \, M\sigma_r^2 \qquad\qquad 1.5$$

so that the total mass M is obtained from the observed
quantities σ_r and R' through the equation

$$M = 0.91 \, \frac{\sigma_r R'}{G} \qquad\qquad 1.6$$

In practice, to determine σ_r account must be taken of
the fact that the spectral lines in the K-type giants
that dominate the light of elliptical galaxies are
frequently blends. Moreover, the instrumental resolution is often comparable with the widths of the spectral features being measured. Accordingly, it is best
to evaluate σ_r by comparing the spectrum of the galaxy
with that of a K-type giant star whose spectrum
matches that of the galaxy over some particular wavelength interval. Let $I_s(\lambda)$ be the observed star spectrum. Assume that the stars which contribute to the
integrated light of the galaxy have a Maxwellian distribution of velocities. Then it is possible to compute an artificially broadened spectrum to be compared
with the observed galaxy spectrum:

$$I_G(\lambda, \sigma_r) = \frac{c}{\lambda (2\pi\sigma_r)^{\frac{1}{2}}} \int_{-\infty}^{\infty} I_s(\lambda') \, \exp\left[\frac{-c^2(\lambda'-\lambda)^2}{2\sigma_r^2 \lambda^2}\right] d\lambda' \qquad 1.7$$

Richstone and Sargent [3] have used this technique to
find the optimum value of σ_r for M32. More recently,
it has been suggested [4], [5] that an optimum way to
determine σ_r is to use the convolution theorem and the
Fourier Transform of equation 1.7. This procedure is
very suitable for analyzing data obtained by linear,
photon counting devices which make it very easy to use
fast Fourier Transform routines on a computer.

The values of σ_r so far obtained by various authors
range from 80 km/sec for M32 [6] to 500 km/sec for
M87 [7]. The mass-to-light ratios for elliptical
galaxies range from 6 for M32 up to 80 for M87. However, these values for M_{total} and M/L are strongly

dependent on the assumption that M/L remains constant
with radius and on the assumption that the value of σ_r
obtained from the motions of the K-giant stars is
representative of the stellar population of the galaxy
as a whole. Neither assumption is backed by any hard
empirical evidence.

2.3 Masses of compact groups of galaxies

It is convenient to discuss the application of the
virial theorem to three distinct types of clusters of
galaxies. First we consider the compact groups. These
are clusters containing a few galaxies in which the
characteristic separations between objects is comparable with their diameters - say less than 100 kpc.
Stephan's Quintet (Arp 319) and VV116 (Arp 321) are
typical examples of compact groups. There is no complete list of such objects but the Atlas's of Arp [8]
and Vorontsov-Velyaminov [9] each contain many examples.
Typical compact groups represent such large density
enhancements over the field that the probability of
comparably bright foreground or background galaxies
being superimposed on them by chance is small. However, it is well known that some compact groups contain
objects with wildly discrepant redshifts which cannot
be accounted for by normal virial motions [10].
We shall ignore the existence of these discrepant objects in the following discussion.

Given a group of galaxies we can measure their brightness, the mutual separations on the plane of the sky
of the member galaxies, and the dispersion in redshift
of the galaxies. In the method of analysis used by
Rood, Rothman and Turnrose [11], the virial mass M_{VT}
of a cluster or group is calculated from the relations:

$$M_{VT} = V^2 R / G \qquad\qquad 1.8$$

where

$$V^2 = \frac{1}{M_L} \sum_i m_i V_i^2 \qquad\qquad 1.9$$

and

$$R = M_L^2 \left[\sum_{pairs} \frac{m_i m_j}{r_{ij}} \right]^{-1} \qquad\qquad 1.10$$

In these relations M_L is the "luminous mass" of the
system obtained from the actual intrinsic luminosities
L_i of the component galaxies together with an assumed
mass-to-light ratio f_i for each system.

Thus

$$M_L = \sum_i m_i = \sum_i L_i f_i \qquad\qquad 1.11$$

V_i is the velocity of the i'th galaxy relative to the
center of mass of the system and r_{ij} is the separation
of any pair of galaxies i and j.   The projection
factors necessary to convert observed radial velocities
and separations measured on the plane of the sky into
3-dimensional values are given for spherical systems
by Limber and Mathews [12].

Rood [13] has analyzed data on 13 compact groups of
galaxies - mostly published by Burbidge and Sargent
[10] - and found a mean value of log M_{VT}/M_L = 0.35.
In this work M_L was calculated with an assumed mass-to-
light ratio of 7 for spirals and 50 for ellipticals and
SO's. The individual values range from log M_{VT}/M_L =
-0.14 for VV208 up to +1.13 for Arp 330. Now it is
important to stress that even when all the 3-dimensional
velocities and separations in a system are known exactly
the virial theorem is only true of time-averaged quantities. For systems composed of a small number of
particles quite large fluctuations about $\Omega/T = -2$ may
be obtained. In fact, Sargent and Turner [14] have
shown by numerical simulations that for clusters composed of six objects in which only projected separations and velocities are observed there is about a
fifty percent chance that the virial estimate of the
mass will deviate by more than a factor of three from
the true value.

Accordingly, the 13 compact groups analyzed by Rood
give no significant indication of excess mass in these
systems which does not lie within the normal confines
of the visible galaxies. In other words, for these
systems the virial theorem masses are consistent with
those estimated from the observed luminosities of the
component galaxies together with the assumed values of
f_i quoted earlier.

1.4 Masses from loose groups of galaxies

The Local Group, which contains M31, M33 and our own
Galaxy, and the somewhat more distant groups of galaxies around M81 and M101, are well-known examples of
loose groups of galaxies. Such systems contain a few
galaxies of comparable brightness, separated by characteristic distances many times their diameter. However,
they still represent density enhancements in excess of

a factor of ten over the field. Some years ago, de
Vaucouleurs [15] circulated a list of 54 groups in
the Local Supercluster. A virial analysis of the
available data on these groups was carried out by Rood
et al. [11]. They found that nearly all of the groups
had log $(M_{VT}/M_L)>0$; the median value was calculated to
be $\langle \log(M_{VT}/M_L)\rangle = 1.6$. The use of the virial theorem
supposes that the systems under consideration are gravitationally bound - i.e., that their total energy $E =
T + \Omega <0$. Turner and Sargent [16] have recently per-
formed a critical reanalysis of the deVaucouleur's
groups and have shown that more than half of the groups
have estimated crossing times for the member galaxies
in excess of the Hubble time, H^{-1}. On this basis these
groups are not bound systems. Turner and Sargent also
showed that the virial discrepancy M_{VT}/M_L is conspicu-
ously smaller for the groups with short crossing times
than for those with long crossing times. The problem
of the loose, nearby groups is under active current
investigation. Thus, Turner and Gott [17] have used a
computer to find groups of galaxies by an impersonal
procedure which demands knowledge only of the magnitudes and positions on the sky of a complete sample of
galaxies taken from the catalogues of Zwicky and his
collaborators. Moreover, Fisher and Tully [18] have
obtained redshifts of more than 1000 nearby galaxies
from observations of the 21 cm line: evidently these
redshifts lend to significantly smaller velocity dispersions and hence masses for some groups of galaxies
than do the older optical data [19].

1.5 Masses from large clusters of galaxies

Historically the application of the virial theorem to
large clusters of galaxies has led to values of M/L
much larger than those inferred from studies of individual galaxies. Recently, Oemler [20] has rediscussed
the problem using more sophisticated dynamical models
than previous workers. Specifically Oemler has tried
to allow for the fact that some clusters may still be
collapsing inwards and, therefore, not in virial
equilibrium.

Oemler notes that clusters of galaxies may be divided
into three main categories in which there is a correlation between the radial density distribution of the
galaxies and their distribution over morphological
type. At one end of the sequence lies the cD clusters
which are regular, centrally condensed systems and
which contain one, or occasionally, two dominant cD

galaxies. The Coma Cluster is a nearby example of a cD
cluster: in these systems the ratio Ellipticals: SO's;
Spirals is typically 3:4:2. At the other end of the
sequence lie the Spiral rich clusters, of which the
Hercules cluster is a typical example. These objects
are irregular in shape, have no pronounced central
concentration and have ratios E:SO:Sp ~ 1:2:3. Between
these poles lie intermediate objects which Oemler calls
spiral poor with ratios E:SO:Sp ~ 1:2:1. The clusters
Abell 194 and Abell 1314 are examples of this category.

Oemler determined the detailed density distributions
of several clusters in each of these categories and
compared them to evolutionary models of collapsing
clusters of galaxies computed by Aarseth. This permitted him to make two important improvements over the
usual assumption of virial stability. First, the ratio
Ω/T reaches the equilibrium value of 2 only after several crossing times and Oemler was able to estimate a
better value for each individual cluster. Secondly,
the models permit one to estimate radial dependent
values for the projection factors used to convert
radial velocities into 3-dimensional velocities.

Oemler's final results for 8 clusters are given in his
paper [20]. Here we may note that the mean value of
M/L for the 8 clusters is 170, with no significant
dependence on the type of cluster. For the Perseus
cluster, which is dominated by the radio galaxy NGC
1275, N. Bahcall [21] has shown that the value of M/L
is as high as 700. This is confirmed by unpublished
work by Sargent.

1.7 Summary of the present situation

A discrepancy between the mass to light ratios determined from dynamical studies of assemblies of galaxies
and those obtained by various methods for individual
galaxies seems to be significant only for the rich
clusters. Oemler observed that in the centrally condensed, symmetric clusters like Coma, the most luminous galaxies, presumably also the most massive, are
more condensed towards the center than the average
galaxy. He argued that these massive galaxies must
have M/L ~ 200 because the model computations showed
that there is insufficient time for this state of
affairs to be brought about if the individual galaxies
are less massive. However, tidal friction due to a
background sea of low mass, low luminosity stars, not
bound to any particular galaxy, would also segregate

the most massive galaxies first. Therefore, it does
not seem possible at the present time to say with certainty where the "missing mass" in the great clusters
lies.

2. THE CHEMICAL EVOLUTION OF GALAXIES

We are far from a fundamental theory of the way in
which the stellar populations and the composition of
the interstellar gas in galaxies evolve. However,
recently there has been some progress in reconciling
some of the main observed regularities in the Hubble
sequence of galaxy types. Let us first summarize the
gross observations. Along the Hubble sequence E,SO,
Sa, Sb, Sc, Irr I the proportion of interstellar gas
rises. The percentage of gas is essentially zero in
E and SO galaxies, rises to a few percent in the
spirals and is as much as 50 percent in the Irregulars.
Also along this sequence there are systematic changes
in the integrated colors of galaxies and in their integrated spectral types. For example, B-V changes from
+1.0 in the reddest ellipticals to +0.3 in the bluest
irregulars [22]. Moreover, ellipticals have spectral
type K while the bluest irregulars have spectral type
A. Morgan [23] has shown in detail that in the spirals
there is a correlation between the integrated spectral
type and the relative size of the nuclear bulge and
the spiral arms. Lastly there is a relation between
the form of a galaxy and its inferred M/L ratio.

In addition we have some knowledge of the compositions
of galaxies. In the disk of our own galaxy there has
been little change in the composition of the interstellar gas over most of its lifetime. [24]. Moreover,
when we examine the disks of other spiral galaxies
they all seem to have roughly the same heavy element
composition as our own Galaxy. However, Searle [25]
has shown that the relative strengths of emission
lines in H II regions indicate that there is a slight
but systematic radial dependence of the heavy element
abundance in spirals. This is in the sense that the
heavy element composition increases towards the center;
in most systems the ratio of interstellar neutral
hydrogen to stars increases outwards.

2.1 Theoretical quantities

If we suppose that the heavy elements are synthesized
in stars and ejected into the interstellar gas [26] then

we might suppose that three parameters would govern
the present interstellar heavy element composition of
a galaxy or, supposing there is no mixing, of a part
of a galaxy.

a) The age of the system
b) The initial mass function (IMF) which describes the
   distribution over mass of newly formed stars. The
   IMF might be time-dependent.
c) The overall rate of star formation as a function of
   time.

2.2 The colors of late-type galaxies

Recently Searle et al. [27] showed that the integrated
colors of late-type (Sb, Sc and Irr I) galaxies may be
understood by the following set of simple assumptions.
Suppose that the rate of star formation in a galaxy
depends on the stellar mass M and on time according to

$$\Phi(m,t) \sim M^{-\alpha} e^{-\beta t} \qquad\qquad 2.1$$

In the solar neighborhood such a power law in the mass
approximately satisfies the observed distribution over
mass of newly formed stars in galactic clusters with
$\alpha = 2.45$ [28]. Searle et al. calculated theoretical
(U,B,V) colors for galaxies for a variety of values of
the parameters α and β. The calculations employed
theoretical evolutionary tracks in the $(\log L, \log T_e)$
diagram for a wide range of stellar masses, together
with empirical conversions from $\log L$ and $\log T_e$ into
the observed quantities I_v, B-V and U-B. They concluded that the colors of Sb and Sc galaxies can be
understood on the basis that all the systems are 10^{10}
years old, and that all have the same value of $\alpha =$
2.45 found in the solar neighborhood. The range in
color is then interpreted as a range in β - the rate
at which the rate of star formation falls off with
time. On this hypothesis, the bluest irregulars have
experienced a roughly constant rate of star formation
with time while in the reddest spirals the present rate
of star formation is a few percent of the initial rate
10^{10} years ago. For our present purposes the most
significant result is the discovery that the colors of
late type galaxies are consistent with a constant value
of α - i.e., with a mass distribution of newly born
stars that is independent of the time.

2.3 Evolution of composition

A simple theory of the way in which the enrichment of

the interstellar gas is related to the rate of star
formation has been developed by Searle and Sargent [29].
In outline their analysis proceeds as follows. Suppose
that at any time t the mass of hydrogen in a galaxy
$M_H(t)$ is made up of two parts, that in stars and that
in the interstellar gas

$$M_H(t) = M_{H,*}(t) + M_{H,i}(t) \qquad\qquad 2.2$$

and suppose that initially $M_{H,*}(0) = 0$.

For the heavier elements, either singly or as a group,
we can write down a similar equation for the total
mass:

$$M_z(t) = M_{z,*}(t) + M_{z,i}(t) \qquad\qquad 2.3$$

It is convenient in this equation, however, to include
only the mass of heavy elements that circulate at some
time through the interstellar medium. We exclude those
heavy elements that are forever locked up in stellar
remnants.

Define the relative abundance of the heavy elements by
mass in the interstellar medium at time t to be

$$Z(t) = M_{z,i}(t)/M_{H,i}(t) \qquad\qquad 2.4$$

It is easy to show from the above equations that if
the rate of enrichment of the interstellar gas is slow
(the socalled instant recycling approximation) then

$$\frac{dM_{z,*}}{dt} = -Z(t)\,\frac{dM_{H,i}}{dt} \qquad\qquad 2.5$$

Talbot and Arnett [30] have shown that this is a
reasonable approximation in practical applications.
Now we introduce a quantity called the yield of heavy
elements - a measure of the net efficiency with which
a generation of newly formed stars turns interstellar
hydrogen into heavy elements:

$$y(t) = -\frac{dM_z(t)/dt}{dM_{H,i}(t)/dt} \qquad\qquad 2.6$$

The yield $y(t)$ depends only on the IMF and can, in
principle be calculated from stellar evolution theories.
Talbot and Arnett have estimated that with an IMF represented by $\alpha = 2.45$ then $y \sim 0.01$.

The above equations lead to an equation which relates
$Z(t)$ to the yield, namely

$$\frac{dZ(t)}{dt} = -y(t) \frac{d \log M_{H,i}}{dt}$$ 2.7

Since, by hypothesis, $Z(0) = 0$, we can solve 2.7 and find

$$Z(t) = <y> \log \left[\frac{M_{H,i}(0)}{M_{H,i}(t)}\right]$$ 2.8

where the time arranged yield is

$$<y> = \frac{\int_0^t y(t) \frac{d \log M_{H,i}(t)}{dt} dt}{\int_0^t \frac{d \log M_{H,i}(t)}{dt} dt}$$ 2.9

That is, we have

$$Z(t) = <y> \log \left[1 + \frac{\text{mass in stars (t)}}{\text{mass in gas (t)}}\right]$$ 2.10

Now we have quoted evidence to the effect that the exponent α in the IMF does not vary with time or from object to object among the late type spirals. Accordingly $<y>$ does not vary. Hence we have for small enrichments

$$Z(t) \propto \frac{\text{mass in stars}}{\text{mass in gas}}$$ 2.11

while for large enrichments we expect

$$Z(t) \propto \log \left(\frac{\text{mass in stars}}{\text{mass in gas}}\right)$$ 2.12

With a constant IMF and y, these results are independent of the particular history of the manner in which gas is converted into stars.

Two main conclusions emerge from the foregoing considerations: First, over the normal range of late galaxy types the ratio of the mass in gass to the total mass (μ) ranges from about 50 percent in irregulars to 5 percent. Accordingly, we expect from 2.12 that Z should vary by a factor of 5 and this is about what is observed. Secondly, in the solar neighborhood the fraction by mass of gas is about 10 percent and the present value of Z is about 0.02. Hence we can estimate $<y>$ from 2.10 to be ~ 0.01, which is the same as Tablot and Arnett's theoretical estimate of the <u>present</u> value of y. Thus we have a second line of evidence that leads to the result that the yield and hence the

IMF have been constant in the past.

2.4 Schmidt's objection

It was pointed out by Schmidt [31] that the distribution of metal abundance among the unevolved G-stars in
the solar neighborhood puts constraints on how the
rate of element enrichment has been related to the
rate of gas depletion. Thus, let N(<Z) be the number
density of unevolved G dwarfs with heavy element composition less than Z. Furthermore, let the present
value of Z in the interstellar medium be Z_1 and let
the present mass fraction of gas be μ_1. Then on the
assumption that the IMF and hence y are independent
of time we see that the ratio

$$\frac{N(<Z)}{N(<Z_1)} = \frac{1 - \mu}{1 - \mu_1}$$
                                                          2.13

It follows from 2.8 that

$$\frac{Z}{Z_1} = \frac{\log \mu}{\log \mu_1}$$
                                                          2.14

From these relations we can calculate the expected dis-

tribution of $\frac{N(<Z)}{N(<Z_1)}$ as a function of Z/Z_1. As Schmidt

showed, the observations show that there are too few
metal poor G-dwarfs to be accounted for by the theory.
There are several possible solutions to Schmidt's
difficulty:

a) Schmidt proposed that the yield y was very high
   when the disk of the galaxy first formed - i.e.,
   that the IMF was initially heavily enriched in
   massive stars. Generally, this is in conflict with
   earlier result that the present yield is equal to
   the past average, and Pagel [24] has shown that
   Schmidt's detailed proposal does not work.

b) Searle [32] and Talbot and Arnett [33] have proposed
   two versions of the basic idea that the interstellar
   gas is or was inhomogeneous and that star formation
   takes place preferentially in regions that are
   already enriched. Some physical justification can
   be advanced to support this type of hypothesis.

c) Ostriker and Thuan [34] have proposed a different
   theory based on the essential idea of inhomogeneity.
   In their version heavy elements ejected by stars in

a hypothetical massive galactic halo rain down on the disk, thereby enriching it from outside.

REFERENCES

1. J. Ostriker and P. J. E. Peebles, Astrophys. J., 186, 467, 1974.
2. E. E. Poveda, Astrophys. J., 134, 910, 1961.
3. D. O. Richstone and W. L. W. Sargent, Astrophys. J., 176, 91, 1972.
4. S. M. Simkin, Astron. & Astrophys., 31, 129, 1974.
5. G. Illingworth and K. C. Freeman, Astrophys. J., 187, L83, 1974.
6. W. L. W. Sargent, P. Schecter, K. Shortridge, and A. Boksenberg, 1974, unpublished.
7. I. R. King and R. L. Minkowski, 1973, unpublished.
8. H. C. Arp, Atlas of Peculiar Galaxies (California Institute of Technology: Pasadena) 1966.
9. B. A. Vorontsov-Velyaminov, Atlas and Catalogue of Interacting Galaxies (Moscow State University: Moscow) 1959.
10. E. M. Burbidge and W. L. W. Sargent, Nuclei of Galaxies, ed. D. J. K. O'Connell (Amsterdam: North Holland Pub. Co.) 1971.
11. H. J. Rood, V. C. A. Rothman and B. E. Turnrose, Astrophys. J., 162, 411, 1970.
12. D. N. Limber and W. J. Mathews, Astrophys. J., 132, 286, 1960.
13. H. J. Rood, Astrophys. J., 188, 451, 1974.
14. W. L. W. Sargent and E. L. Turner, 1974, unpublished.
15. G. de Vaucouleurs in Stars and Stellar Systems, Vol. XI, ed. A. and M. Sandage (Chicago: University of Chicago Press) 1975.
16. E. L. Turner and W. L. W. Sargent, Astrophys. J., 1974, in press.
17. E. L. Turner and R. J. Gott, Astrophys. J., in press, 1974.
18. J. R. Fisher and R. B. Tully, 1974, unpublished.
19. G. A. Tammann and J. Materne, preprint, 1974.
20. G. Oemler, Thesis, California Institute of Technology, 1973.
21. N. A. Bahcall, Astrophys. J., 186, 1179, 1973.
22. G. de Vaucouleurs, Ap. J. Suppl., 5, 233, 1961.
23. W. W. Morgan, Pub. Astron. Soc. Pacific, 70, 364, 1958.
24. B. E. J. Pagel, preprint, Chemical History of the Galaxy, 1974.
25. L. Searle, Astrophys. J., 168, 327, 1971.
26. F. Hoyle, Monthly Notices Roy. Astron. Soc., 106, 343, 1946.

27. L. Searle, W. L. W. Sargent and W. G. Bagnuolo,
    Astrophys. J., 179, 427, 1973.
28. E. E. Salpeter, Astrophys. J., 121, 161, 1955.
29. L. Searle and W. L. W. Sargent, Astrophys. J., 173,
    25, 1972.
30. R. J. Talbot and W. D. Arnett, Astrophys. J., 186,
    51, 1973.
31. M. Schmidt, Astrophys. J., 137, 758, 1963.
32. L. Searle, IAU Colloquium No. 17 "L'Age des Etoiles"
    1972.
33. R. J. Talbot and W. D. Arnett, Astrophys. J., 186,
    69, 1973.
34. J. P. Ostriker and T. X. Thuan, Astrophys. J.,
    1975 in press.

GALACTIC NUCLEI AND RADIO SOURCES*

G. Burbidge

Department of Physics
University of California, San Diego
La Jolla, California 92037 U. S. A.

In these lectures I shall discuss two topics. They are the general properties of galactic nuclei including their relation to QSOs and BL Lac objects, and the problem of the extended radio sources.

The subject of galactic nuclei is by now a very large one indeed, and I shall only outline some of the major areas of research. A very extensive review of the subject is given in Burbidge (1970).

Possible components which may be present in nuclei are:

>   Stars of normal types, gas, dust, supermassive
>   stars, and non-thermal energy sources which can
>   arise from a variety of processes.

In the majority of normal galaxies the central nucleus simply contains a high-density core of normal stars. Sometimes this nucleus is very small, bright, and distinct, so that the observer describes it as "stellar." However, this does not mean that the star density is so very high. For galaxies at distances of ~ 10 Mpc or more we are not able to resolve nuclei smaller

---

*An outline of lectures given at the Erice Summer School, July 1974.

G. Setti (ed.), Structure and Evolution of Galaxies, 275–283. All Rights Reserved.
Copyright © 1975 by D. Reidel Publishing Company, Dordrecht-Holland.

than about 50 pc in diameter, and within such systems the
mean star density is not greater than $\sim 10^3$ stars pc^{-3}. Very
much higher densities than this are required before the stars
begin to interact together to give rise to violent events.

When we observe the radiation emitted from galactic
nuclei, we are likely to be observing one or the other of two
different general types of effect. On the one hand, we may
simply be seeing the processes associated with the normal
slow evolution of stars. On the other hand, we may be seeing
effects due to violent activity often involving non-thermal
processes. The latter effects may have involved the late
stages of evolution of a very dense star cluster (stellar collisions,
multiple supernovae, supermassive stars and gravitational
collapse), or they may involve even more exotic processes.
Let us list the various types of phenomena that are seen and
try to decide whether they are due to normal stellar evolution
or to violent non-thermal activity.

OBSERVED PHENOMENON	NORMAL STARS	VIOLENT NON-THERMAL PROCESSES
Sharply peaked optical light distribution	Yes	Sometimes (if spectrum is non-thermal)
Strong emission lines	Yes	Yes
Very broad emission lines	No	Yes
Large infrared flux	Yes	?
Non-circular motions	Sometimes (M31)	Yes
Flux variability	No	Yes
Nuclear Radio Source	Possibly, if spectrum is thermal (M101)	Yes

OBSERVED PHENOMENON	NORMAL STARS	VIOLENT NON-THERMAL PROCESSES
Extended radio sources	No	Yes
Ejected gas (optical or 21 cm data)	No	Yes
Nuclear X-ray source	Conceivable (if bremsstrahlung)	Yes
Extended X-ray source	No	Yes (if due to Compton effect)

In the lectures examples of galaxies giving rise to each of these types of activity will be discussed. However, before I do this, let me discuss the possible connections between violent activity in galactic nuclei and the more extreme types of observed phenomena, namely BL Lac objects, N-systems and QSOs. Here we run directly into the problems of interpretation. Everyone agrees that these objects are related in some way to galaxies, but how?

They are each highly compact but in none of them can we certainly identify a population of stars. In general, we see a starlike object with fuzz around it and in general

$$z_{QSO} > z_N > z_{normal\ galaxy}$$

When we study the light or radio flux from these compact objects, we detect a non-thermal continuum, possibly a thermal continuum and strong emission lines. For QSOs and N-systems we can consider the following hypotheses.

(1) If we assume the z is cosmological in origin, then QSOs and N-systems are early or late phases in the evolution of galaxies. Both have been proposed. Such ideas must connect with the space distribution demonstrated by Schmidt and others, i.e., it must be argued that the rate of evolution is a strong function of $(1 + z)$ (cosmic time) in the sense that early in the history of the universe the space density of

such objects was small, it then rose steeply, and has
been decaying for the last 8 or 9 x 10^9 years.

(2) If we assume that the redshifts of QSOs and related objects
are of non-cosmological origin, then the association of the
objects with bright galaxies suggests that they are genet-
ically associated with them and have presumably been
ejected from galaxies. The absolute magnitudes of the
QSOs are then about -15, and in general they will not be
detected on the Palomar Schmidt plates beyond distances
~ 100 Mpc. If the QSOs are local, it is likely that they
are completely non-thermal objects and no stars are
present in them.

Attempts have been made to detect stars in the QSOs and
related objects. Positive evidence for a stellar population
underlying the non-thermal object would support the view that
we are looking at very violent phenomena in galactic nuclei.
Unfortunately, many well-known astronomers have not distinguished between a consistency argument and positive proof
that stars, and hence a galaxy, are present. For example,
Sandage has argued that the colors of N-systems are consistent
with their being a composite of an elliptical galaxy and a QSO.
It is then argued that by making the appropriate correction for
the non-thermal component the N-systems follow the Hubble
law. But nowhere is it proved that stars are indeed present,
or that other explanations of N-systems not involving the
presence of stars can be ruled out. Indeed the fact that the
N-systems with strong emission lines show a very strong
periodicity effect in their redshifts suggests a non-cosmological
origin.

Again Kristian has examined a sample of QSOs with some
faint fuzz around the starlike images and has argued that the
diameter of the fuzz is roughly inversely proportional to redshifts and that this indicates that we are looking at galaxies
at different distances. This again is a consistency argument.
By calling the fuzz a galaxy throughout his paper, but nowhere
proving it, the argument is highly prejudiced from the start.

Here I might interpolate a remark of V. L. Ginzburg, who
has often made it to me (and against some of my ideas), as
follows: "No evidence for, is evidence against." While this is
perhaps a little too strong, it is appropriate in this connection.

Probably the most baffling class of objects, which fit somewhere into the scheme involving explosive non-thermal objects, are the BL Lac objects. They vary extremely rapidly in optical, infrared and radio flux, but until very recently no spectrum lines had been reported. Thus, it had been suggested that they are lineless objects at the same distances as the QSOs (whatever those distances are), objects within our Galaxy, or objects with blueshifts.

Recently Oke and Gunn reported that they had detected absorption features due to stars in the fuzz around the stellar object BL Lac. They concluded that it was a "line-less" QSO embedded in an elliptical galaxy with a redshift z = 0.07. Earlier, using the colors and the distribution of light, others had estimated redshifts ranging from 0.02 to 0.07.

Before the conclusions of Oke and Gunn are accepted, their result, which is clearly marginal, needs to be confirmed by an independent group. *

A very different conclusion has been reached about another BL Lac object, PKS 0735+178, by Strittmatter and his colleagues. They can find no emission lines in the spectrum, but they have identified two sharp absorption lines as the Mg II double $\lambda\lambda 2796, 2803$, at z = 0.424. The sharpness of the lines and the magnitude of the redshift are similar to the situation found in the absorption spectra of QSOs with large emission redshifts.

Thus, studies of these objects have given no certain indication that a population of stars is present. Consequently, neither they nor the QSOs have been shown to be galaxies with very active nuclei.

We now turn to examples of various kinds of nuclear activity in galaxies. In this lecture I discussed the properties of the classical Seyfert nuclei, the radio, optical, infrared

---

*By now the Lick group, Baldwin, M. Burbidge, Robinson, and Wampler, have published a paper (Ap. J. Lett., January 15, 1975) in which they report that they <u>do not confirm</u> the results of Gunn and Oke.

and X-ray properties of NGC 1275 and the Perseus cluster,
the optical, radio and X-ray properties of M87 and its jet,
M82, and NGC 5128 (Centaurus A). Since a full discussion
with references is given by Burbidge (1970) for all of these
objects, it will not be repeated here.

I now consider briefly the origin of the large infrared
fluxes which are emitted by some galaxies. In the last few
years Low and Rieke and others have reported very large
infrared (~ 10 μ) luminosities for a number of galaxies.
Fluxes of order 10^{45} erg sec^{-1} are not uncommon. Some
Seyfert galaxies like NGC 1068 and NGC 4151 have been
observed frequently by Stein and Gillett. It is generally
agreed that the infrared flux may be of thermal origin. In
this case it is radiated from dust heated either by starlight
or non-thermal optical and ultraviolet sources. Alternatively
it may be of non-thermal origin and arise directly from the
synchrotron process. When it was originally reported that
comparatively rapid infrared flux variations were seen in
NGC 1068, it was argued that the flux must have a non-thermal
origin. However, at present there is no good evidence for
real flux variations.

We have the most detailed information about the infrared
flux from NGC 1068, and an analysis has recently been completed by Jones and Stein. Jameson et al. have shown that
the peak flux occurs near 18 μ and then decreases. Becklin,
Mathews and Neugebauer have shown that at least 70% of the
flux comes from a source which is resolved at about 1 arc
second (~ 50 pc). Knacke and Capps have found 2% and 3%
polarization at 3.5 μ and 10 μ, respectively.

Jones and Stein have shown, using the shape of the spectrum, that the observed size of the source is inconsistent
with the idea that the radiation is due to the incoherent synchrotron process. They conclude that more than 85% of the
radiation at 10 μ is radiation from dust grains in the nucleus.
These could be silicates with a 10 μ optical depth near unity,
but this is not uniquely established. The grains are heated
by sources which are responsible for ionizing the gas giving
rise to the emission lines. These sources could be thermal
or non-thermal.

We now turn to a survey of the theoretical problems associated with extragalactic radio sources. It is well established that all of these sources are radiating by the incoherent synchrotron mechanism. There are really three different types of source. These are the weak radio emitters similar to our own Galaxy, the highly compact high surface brightness sources seen in the nuclei of many different types of galaxy, and the very extended sources -- the typical radio galaxies. Dr. Ekers has given an account of the observed properties of these sources.

I shall not discuss the normal galaxies but concentrate first on the compact sources and then on the extended sources.

The compact sources in many QSOs and the nuclei of galaxies are often exceedingly small with sizes less than or equal to 10^{-3} - 10^{-4} arc seconds. Also, many of them are rapidly variable with timescales of months or years.

If we use the light travel time argument, the size of the source $R \lesssim c\tau$, and size limits set this way are independent of the distance of the source. If we derive R in this way, and also measure the angular size, then knowing a distance for the source and assuming that the source is expanding, we can measure the distance travelled and determine the velocity of expansion.

For the simplest models, we often find that very distant compact sources must be expanding relativistically.

For the simple expanding cloud models (Shklovsky, van der Laan, Kellermann and Pauliny-Toth) we can make rigorous calculations if we have angular size measurements, and spectra which turn over due to synchrotron self absorption. In nearly all cases in which such calculations have been made, we find that the particle energies are very much greater than the magnetic energies. Angular sizes which are calculated with the restriction that we do not get more X-rays from Compton scattering than are observed by UHURU, are in reasonably good agreement with the measured angular sizes.

For sources in objects with large redshifts, e.g. 3C 454.3, large relativistic effects are required unless it is supposed that the objects are much closer than the distances derived on

the assumption that the redshifts are of cosmological origin.
Highly relativistic effects mean that very large energies
$\sim 10^{58}$ erg per outburst are required.

Alternative models to account for rapid variability in compact sources have been proposed. The most popular of these
is the so-called "Christmas tree" model. In this scheme two
or more smaller components, spatially separated but static,
blink on and off to give the illusion of a simple expanding
source. However, there are severe difficulties with this
model. The number of components must remain small, and
their angular sizes must be exceedingly small. The total
solid angle is very small (< 5% of the whole source) unless
relativistic effects are invoked. However, in this case Compton
scattering would become important, thus giving rise to X-ray
fluxes greater than those which are observed, and of course the
energy requirements would be very large.

We now turn finally to the extended radio sources. In
considering the various possible models which have been proposed to explain these sources we must take into account the
new observations. In my view there are at least two developments of great significance. The first is that high-resolution
studies have shown that there are compact high surface-
brightness components in the nuclei of many sources, and even
in the extended components small scale structures are found.
The second is the realization that some sources are exceedingly
large, with dimensions ~ 1 Mpc (Centaurus A, Coma Cluster,
Virgo Cluster), and in the cases of 3C 236 and DA 240 dimensions of 2 and 6 Mpc.

We still have no real understanding of why sources are
double. This clearly has something to do with rotation but
there has never been any convincing quantitative solution to
the problem. It is now generally accepted that secondary
sources of energy continuously generate relativistic electrons
in the outer regions of the sources. Thus the simple ramjet
models are inadequate to explain the observations. Thus,
models have been proposed in which the secondary sources are
low-frequency electromagnetic waves or relativistic beams of
particles emanating from the central object. There are difficulties in each model, and all of them require that the sources
be confined by a hypothetical intergalactic plasma. In 1967 I
suggested that the most likely model was one in which coherent

objects (cannon balls) are ejected from the central object
and become secondary sources of particles. I still consider
that this model is very plausible.

The overall energetics of the sources are uncertain,
though it is nearly 20 years since the first equipartition calculations were made. I consider it unlikely that equipartition
between particles and magnetic fields really exists, and I
suspect that the particles dominate.

What are the requirements on the energy sources in the
nuclei of galaxies and QSOs which give rise to the radio
sources?

They must give rise to small nuclear sources which are
often variable. They may eject fairly massive coherent
objects. They must be able to generate relativistic particles
over scales of megaparsecs. The various theories for energy
production in nuclei have been reviewed many times. The
account that was given in these lectures was taken from my
article in Annual Review of Astronomy and Astrophysics,
Volume 8, 1970.

FORMATION OF GALAXIES, RADIO SOURCES AND RELATED PROBLEMS

M.J. Rees

Institute of Astronomy, Madingley Road,
Cambridge, England

SUMMARY. These three lectures deal with the following topics:
(i) The nature of radio galaxies and related objects. (ii) Inferences from statistical studies of radio galaxies and quasars
concerning the epoch of galaxy formation, and the redshift-de-
pendence of the properties of galaxies. (iii) Galaxy formation in
the context of "big bang" cosmologies. Although these topics
are not entirely unrelated, any of the lectures can be read
independently of the others.

1. EXTRAGALACTIC RADIO SOURCES

1.1 Observational situation

I shall primarily consider the strong extragalactic sources which
have radio luminosities P_{rad} in the range of 10^{42} - 10^{45} erg
sec^{-1} (10^3 - 10^6 times greater than the radio power of a normal
galaxy like our own). Although these are rare compared with
ordinary galaxies, they in fact comprise the bulk of the observed
sources at high galactic latitude (because sources can be de-
tected out to distances $\propto P_{rad}^{\frac{1}{2}}$, and so radio surveys are weighted
in favor of powerful sources). Several hundred of these sources
have now been optically identified. In most cases the associated
optical object is either a bright elliptical galaxy or else is
described as a "quasar". (As is discussed later, however, quasars
are probably the "optically hyperactive" nuclei of certain elliptical galaxies). The properties of these radio sources span a
wide range in all respects, especially in power output and linear
dimensions. However, there seems as yet to be no clear cut
correlation between any radio property and the nature of the

G. Setti (ed.), Structure and Evolution of Galaxies, 285–323. All Rights Reserved.
Copyright © 1975 by D. Reidel Publishing Company, Dordrecht-Holland.

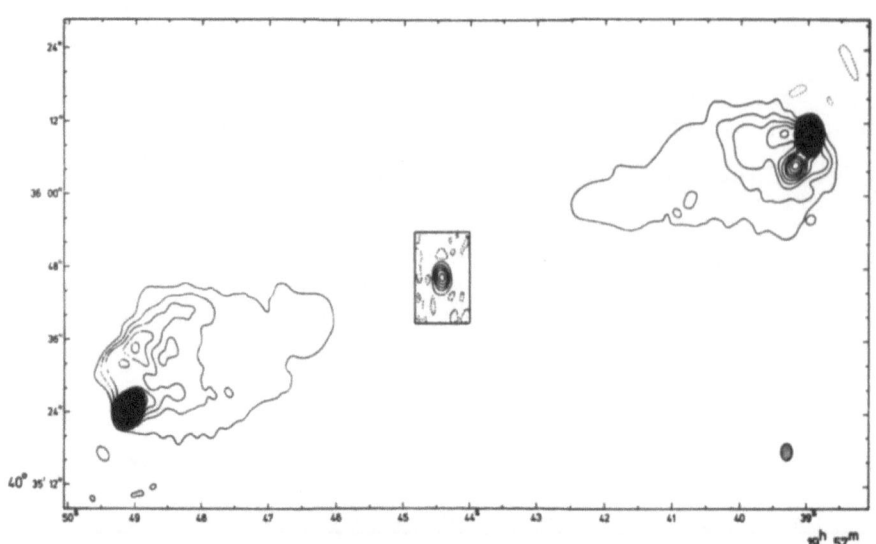

Fig. 1. Cygnus A at 5 GHz with an angular resolution of ∿2"
(from Hargrave and Ryle (1974)). The "hot spots" (blacked out in
diagram) are unresolved at 5 GHz.

optical counterpart.
   The linear sizes range from ⩽1 pc to ∿6 Mpc. The compact
sources are usually identified with the optical nuclei, and are
frequently variable on time scales comparable with the light travel
time across the source (i.e. years). In the powerful extended
sources the radio emission is generally concentrated in two regions,
one lying on each side of the optical object.
   The aperture synthesis technique-whereby measurements made
with an array of dishes can be combined to provide maps with
effectively the same resolution as a single dish whose diameter
equals the length of the baseline (see, for example, Ryle 1972)-
has, within the last five years, provided detailed structural
information about extended sources, and several surprising results
have emerged. The new Cambridge 5 km telescope has ∿2" resolution
at 5 GHz; and the Westerbork array in Holland, though it does not
have such good resolution, has higher sensitivity and can therefore
probe the structure of very extended sources of low surface brightness. More than a hundred sources have now been mapped using this
technique.
   Most features of double radio sources are exemplified by
Cygnus A (3C 405), shown in Fig. 1, which as well as having the
highest flux density of any extragalactic radio source, is also
one of the most intrinsically powerful. Its redshift is in fact
$z=0.056$, implying a distance ∿320 Mpc (assuming a Hubble constant
$H_0=50$ km sec^{-1} Mpc^{-1}). Similar sources would be detectable

even at the distances corresponding to $z \gtrsim 2$. Cygnus A was identified
by Baade and Minkowski (1954) and the potential importance of radio
astronomy for cosmology immediately became apparent, since it was
clear that many radio sources would be detectable whose optical
counterparts were too far away to be seen with even the largest optical telescopes.

The morphology of Cygnus A provides crucial clues as to the
nature of strong radio sources. The two features of highest surface brightness are fairly compact with linear sizes ≤ 2 kpc, and
are symmetrically disposed at distances ~ 80 kpc on either side of
the optical galaxy (which is itself also a weak radio source).
Surrounding the high brightness regions are two extended low
brightness regions that reach inwards towards the galaxy. The
radio emission from the whole of the source is substantially
linearly polarised.

It has been found observationally that this essentially simple double structure is characteristic only of the stronger sources;
weaker sources display more complex and varied, but no less interesting, morphologies (see the lectures by Ekers for some further discussion of these).

Recent observations with the Westerbork radio telescope
(Willis et al, 1974) have revealed the structure of 3C 236 and
this turns out to be the largest known double source (so far).
It is ~ 6 Mpc in linear extent and presents in many respects the
most stringent constraints on theory.

Frequently the optical objects identified with extended radio
components display continuing non-thermal activity in a small central region. Also the extended low brightness regions surrounding
the principle radio lobes often constitute a "bridge" connecting
them to the interior of the galaxy. It has therefore generally
been inferred that the double sources in some way have their origin
in active regions with dimensions ≤ 10 pc situated at the centers
of large galaxies, and have somehow propagated outward to their
present locations- no-one has yet been so perversely teleogical as
to seriously propose that the components are moving inward! Whilst
it must be emphasised that, for instance in the case of Cygnus A,
there is no direct evidence that any kind of ejection has actually
occurred, there is now a substantial amount of indirect supporting
evidence, from both optical and radio studies; and all of the
models we consider incorporate this premise.

1.2 Basic physical arguments

What then is understood about extended extragalactic radio sources?
One of the few points on which a general consensus exists concerns
the nature of the radiation mechanism. The spectra of the integrated radiation from double sources are distinctly non-thermal
and can usually be fitted by power laws; i.e. $S(\nu) \propto \nu^{-\alpha}$, where
α, the spectral index, is typically +0.7. Also, the emission
is generally substantially polarized (degrees of polarization in

excess of 60% have in fact been reported). This is just what
would be expected from "synchrotron" radiation emitted by ultrarelativistic electrons spiralling in a partially ordered magnetic
field. A simple calculation shows that if the electrons are
distributed in energy, E, as $N(E) dE \propto E^{-p} dE$, then $p = 2\alpha + 1$.
This explanation becomes even more credible when it is realized
that p is typically 2.4, a similar exponent to that measured for
locally observed cosmic rays. The degree of polarization for the
emitted radiation in a completely ordered magnetic field is ~70%,
with the electric vector normal to the projected field direction.
Alternative emission processes involving the scattering of low
frequency wave modes have been proposed (e.g. Rees 1971a), and
cannot at this stage be definitely excluded; but they do involve
a degree of contrivance not present in the synchrotron hypothesis,
especially in sources where high polarization is observed. Here-
after I shall assume the synchrotron mechanism.

The two radio lobes presumably contain relativistic electrons
and ordered magnetic field. The magnetic field strength B cannot
be directly measured. However the energies of the individual
electrons required to radiate at a given frequency are proportional
to $B^{-\frac{1}{2}}$, and the number of electrons needed to produce the observed
radio power $\propto B^{-1}$. The total energy contained in particles ($\propto B^{-3/2}$)
and magnetic field ($\propto B^2$) is minimised when there is approximate
equipartition between these two forms. After making an estimate of
the total emitting volume, we can therefore calculate the minimum
internal energy content for various sources. For Cygnus A this is
~10^{58} ergs and for 3C 236, ~10^{60} ergs (i.e. the rest mass energy
of 5×10^5 solar masses). The corresponding field strengths and peak
electron energies are respectively 3×10^{-4} G and 100 GeV for a high
surface brightness source like Cygnus A. The equipartition hypothesis may well yield a serious underestimate of the present internal energy: there are several plausible hidden contributions
(e.g. relativistic protons); there is as yet no convincing argument for equipartition to arise naturally; and there must be some
degree of inefficiency in the acceleration mechanism. The reali-
sation that extended sources entailed these colossal energy requirements was specially important when the estimates were first
made eighteen years ago by Burbidge (1956), because it was the
first evidence for extragalactic "violent events" on a large
scale—an idea to which we have become accustomed since the advent of
quasars and related phenomena.

In fact dynamical arguments tell us that the energies involved
in the formation of radio sources are almost inevitably at least
an order of magnitude larger than the present internal energies.
In the case of Cygnus A (Fig. 1), for instance, if the radio lobes
contained nothing but relativistic particles and magnetic field,
they would expand with speed ~$3^{-\frac{1}{2}}c$, the internal sound speed.
So even if the components moved away from the galaxy relativistically,
it is hard to see how they could have retained such small sizes in
relation to their separations. This ratio is about 1:30 in Cygnus A

and is even smaller in some other objects. Something must therefore be impeding the free expansion of the components, so that:

$$\frac{\text{expansion velocity of components}}{\text{velocity away from galaxy}} \lesssim \frac{d}{D} = \frac{\text{component size}}{\text{component separation}}$$

Three main possibilities have been considered (see Longair
et al (1973) or Rees (1972a) for general reviews). The first is
that the components contain a large mass of non-relativistic
plasma which provides extra inertia without much pressure. The
magnetic field would "glue" this plasma to the relativistic particles (yielding a non-relativistic sound speed for the composite
fluid) so that the expansion velocity could be much lower and therefore compatible with observations. (This hypothesis would require
$\sim 10^9$ solar masses of material in the case of Cygnus A). The
kinetic energy of this non-relativistic matter, however, would
exceed the internal energy by $\sim(D/d)^2$ -a factor ~ 1000 for Cygnus A.
A second possibility, somewhat more economical in terms of
energy, involves appealing to an external medium of density ρ_{ext}
enveloping the galaxy and the radio components. If the components
are moving supersonically with velocity V through this medium,
then they will be subjected to a "ram pressure" $\rho_{ext} V^2$ (De Young
and Axford 1967), independent of the external temperature. (There
would of course be a bow shock ahead of the component, behind
which the intergalactic gas would be heated to a temperature such
that its sound speed $\sim V$). However, work must be done pushing away
the external medium, and this work will exceed the present internal
energy by a factor $\sim D/d$, which is smaller than that associated
with inertial confinement. Consistent models for the most power-
ful sources can be devised with $V \approx 0.1c$ and $\rho_{ext} \approx 10^{-28}-10^{-27}$ gm
cm^{-3} (equivalent to particle densities $10^{-4}-10^{-3}$cm^{-3}). Evidence
that a density as high as this (which is $\gtrsim 100$ times too large to
be universal) can exist within a cluster of galaxies has recently
been provided by x-ray observations (see Field 1972 or Silk 1973
for a review); and it is quite plausible that comparable densities
surround any massive galaxy, perhaps as a result of a steady
accretion that has continued since it formed.
A third possibility is that the radio component moves
subsonically with respect to the external medium, its internal
pressure being balanced by the thermal pressure of the surrounding
gas (Gull and Northover 1973). Gas densities $\sim 10^{-28}$gm cm^{-3} and
temperatures $\sim 10^8$K are adequate to confine the low luminosity
sources (and the extended low brightness regions of the powerful
sources); and there is accumulating evidence that this is what is
actually occurring. However, there are several arguments which
indicate that the most compact high brightness structure cannot
be confined thermally, and thus must have moved away from the
parent galaxy supersonically.
The observed symmetry of the sources provides, curiously, both
upper and lower limits to the speed V. The upper limit can be

set because kinematic effects arising from the finiteness of the velocity of light would cause the two components-even if they were moving at the same speed-to appear at different projected distances from the parent galaxy, and at different stages in their evolution. (This effect would be expected unless the source axis were at right angles to our line of sight). The degree of symmetry generally observed in samples of strong optically identified double sources (Mackay 1971b) rules out highly relativistic bulk velocities: any quantitative upper limit to the velocity is, however, rather model-dependent. Unless the galaxy were completely at rest with respect to the surrounding medium, it should be somewhat displaced from the line joining the two components even if they were ejected in precisely opposite directions. In such clusters giant galaxies have random velocities $V_{random} \sim 500$ km s^{-1}, and the strongest sources viewed at radio frequencies show no evidence of such a systematic velocity. (The line joining the most intense regions of the components of Cygnus A in fact passes within 1.5 Kpc of the center of the optical galaxy). So it is probably reasonable to conclude that for these sources $V >> V_{random}$. Many of the weaker sources such as the "radio trail" 3C 129, however, show pronounced distortion, suggesting that in these cases the motion of the galaxy through the surrounding hot gas in fact dominates the dynamics of the radio source. Thus the value of V/V_{random} may determine the source morphology. In the case of 3C 236, an absolute limit $V \gtrsim 200$ km s^{-1} arises simply from the requirement that the source attains its present dimensions in the age of the universe.

A question somewhat disjoint from these dynamical considerations is the origin of the energy, or the nature of the "prime mover", for strong radio sources (and also for active galactic nuclei and quasars). Here the proposals neatly subdivide into two classes. On the one hand, the energy may be supplied over a time scale short compared with the total source lifetime; whereas, on the other hand, models have been constructed in which the energy is supplied continuously by the active nucleus, and must therefore propagate away from the galaxy to the components at a speed in excess of V.

As argued in, for instance, Rees (1971a), one might be inclined to favor the latter picture for the following reason. The bolometric luminosities of active nuclei and quasars are typically $\lesssim 10^{46}$ erg sec^{-1}. The radio luminosities of extended sources are never larger than this. Moreover, if a power supply $\sim 10^{46}$ erg sec^{-1} continued for the whole estimated source lifetime, it could cumulatively generate the energy content of extended sources. But if the energy was all released in a brief initial explosive phase (and specific theories of this type (e.g. Ryle and Longair 1967) have proposed a time scale $\lesssim 10^3$ years), then powers $\geq 10^{50}$ erg sec^{-1} have to be invoked, far exceeding anything normally contemplated even by the high energy astrophysicist! Why does one not see evidence for explosions of this violence, since even one such object on our light cone would blatantly outshine "ordinary" quasars and radio sources?

The most recent high-resolution radio maps are, however, beginning to reveal features which lend firmer support to the idea that the energy supply continues for much of the source's active lifetime. The discovery of "hot spots" at the outer edges of the components of Cygnus A (Fig. 1)-within which the energy density and equipartition magnetic field are higher than was previously suspected-aggravates the confinement problem for this source and raises severe problems for some alternative theories. If <u>thermal</u> gas within these regions were to provide the kinetic energy required in the ram pressure confinement model, the particle density would have to be 10^{-2} -10^{-1} cm^{-3}. According to Hargrave and Ryle (1974) the high polarization of the synchrotron radiation from these "hot spots" is incompatible with a thermal gas density as high as this. (Inertial confinement models would demand even more gas within the components). Also, the estimated synchrotron lifetime of the relevant relativistic electrons is less than the light travel time from the central galaxy (and therefore presumably less than the age of the source), implying continuous replenishment or re-acceleration of relativistic particles in the components themselves. Somewhat similar considerations apply in 3C 236, though the magnetic field is in this case much lower than in Cygnus A and so the dominant drain on the energy of the electrons probably results from inverse Compton scattering of the microwave background.

None of these arguments is impossible to circumvent (if one is prepared, for instance, to make suitable assumptions about the magnetic field geometry), so the conclusion that the supply of energy is continuous is not yet watertight. But one can readily envisage a whole zoo of hypothetical objects, any of which could energize to a strong radio source throughout its active life (probably 10^6-10^8 years). One type of model involves clusters of ordinary stars, pulsar-forming supernovae, stellar mass black holes, or evolved stars collapsing under rotational support; other ideas involve a few massive objects (with mass $\gtrsim 10^6$ solar masses) either collapsed to a black hole or in the form of a differentially rotating disc; and it would be quite easy to lengthen this list[†].

[†]Some insight might well result from drawing analogies with the Crab Nebula. Even though the pulsar's radiative output is small compared with that of the whole nebula, there is abundant evidence that the pulsar supplies all the relativistic particles, and probably also magnetic flux to the Nebula; but the actual mechanism involved is still a matter of controversy (see, for example, Rees and Gunn 1974). Thus one might suspect that an active galactic nucleus crudely resembled a million or so superposed Crab Nebulae, or else contained a single supermassive spinning object whose rotational energy is transformed into relativistic plasma by a mechanism similar to that which occurs in the Crab Pulsar.

I suspect that it will prove harder to settle the nature of the
energy source (except that it is probably ultimately gravitational)
than to understand the extended components.

By far the most mysterious aspect of the double sources is their
characteristic binary structure and the implied bifurcation of the
energy supply. It is too early to make a decisive choice in favor
of any particular model-in fact the history of radio astronomy
should warn us against this. Nevertheless it is worthwhile to
demonstrate that there are viable models capable of explaining
current observations within the ambit of conventional physical
ideas. (Such "existence theorems" are also valuable for the-pre-
sumably related-phenomena of quasars, because many people have
argued that these are so mysterious that "new physics" must be
invoked.

1.3 Specific models for double radio sources

Relativistic beams. One class of theory (Rees 1971a, Scheuer
1974, Blandford and Rees 1974) involves the idea that strong radio
sources are supplied continuously by two beams of "relativistic"
fluid travelling away from the nucleus in anti-parallel directions.
The beams are assumed to carry sufficient power to feed the radio
source components-the radio "hot spots" in, for instance, Cygnus A,
being identified with the regions where these beams impinge on the
intergalactic gas, and their bulk kinetic energy is randomised.
It is supposed that relativistic plasma is being continuously generated in an active galactic nucleus. If this central active
region is surrounded by a cloud of sufficiently dense lower
temperature plasma, perhaps originating in conjunction with the
active source, then the relativistic plasma will be unable to escape
isotropically. The cooler cloud need not in fact have much angular
momentum per unit mass relative to the galaxy to insure that it be
substantially rotationally flattened. The pressure, which falls
monotonically away from the center, will then decrease most rapidly
along the rotation axis. The cloud itself must be sufficiently
cool so that its internal sound speed is below the escape velocity
from the potential well in which it is situated, but this still
permits temperatures as high as $\sim 10^8$ K.

One can now examine the possible equilibrium flow patterns in
which the outflow rate of the relativistic plasma balances the production rate. (It is justifiable to treat the relativistic plasma
as a fluid provided that it possesses even a very weak magnetic
field. The Larmor radius for a proton of Lorentz factor γ would
be only $\sim 10^{-12} \gamma B^{-1}$pc, whereas the smallest length-scales associated with the flow are 1-10pc). If the flow is isentropic, it
will obey Bernoulli's equation. In normal laboratory applications
of Bernoulli's equation, one customarily considers gas flow through
a pipe of given cross-section. The pressure and velocity can then
be calculated. In the present application, however, it is the
pressure whose variation along the pipe is given (since it is fixed

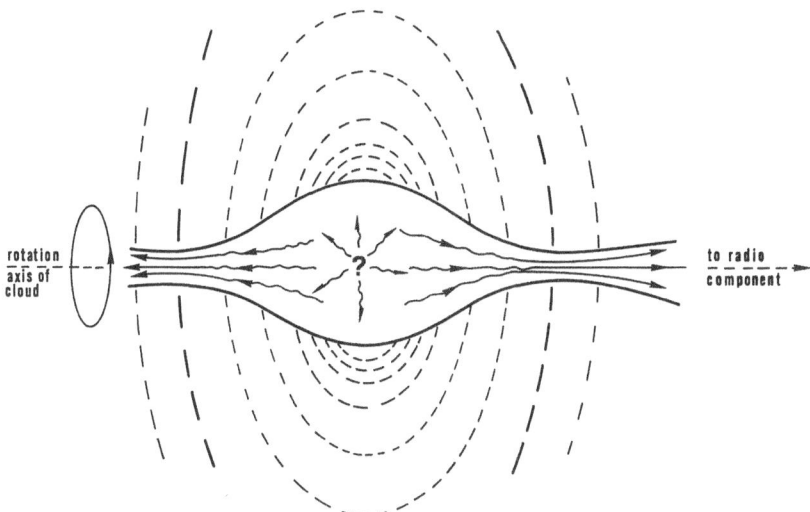

Fig. 2. The nucleus of a radio galaxy, as envisaged in the "beam
model" (Blandford and Rees 1974). "Hot" (probably relativistic)
plasma is assumed to be steadily generated in a central object
(indicated by "?") ≲10pc in size. The central object is surrounded
by a gravitationally confined rotating gas cloud, whose isobars
(oblate spheroids) are drawn as dashed lines. The thermal pressure
of this gas cloud constrains the outflowing relativistic plasma
within two oppositely directed tubes. An equilibrium flow is possible
only if the tube adjusts its shape so that nozzles form where the
external pressure had dropped to $\sim\frac{1}{2}$ its central value. (The
corresponding isobar is heavily drawn in the diagram). At the
nozzle the bulk velocity of the relativistic plasma is sonic (i.e.
$\sim c/3^{\frac{1}{2}}$). Beyond the nozzles, the tubes widen again as the external
pressure drops, but the flow is now <u>supersonic</u>. The outflowing
plasma thus gets collimated into a relativistic beam, its initial
internal energy being transformed into ordered kinetic energy.
The beam evacuates a channel for itself by forcing the ambient
extragalactic gas out of its way. Double radio source components
are, in this model, associated with places where these beams are
currently encountering the external medium and their energy is
being re-randomised (see Fig. 3).

     The "scale height" in the central thermal gas cloud is estimated
to be $\sim$100pc (i.e. very small compared to the overall size of the
optical galaxy). Its density may be up to 10^3 particles cm^{-3}, and
its temperature up to $\sim10^8$ K.

by the properties of the cool gas cloud); and the Bernoulli equation
then tells us how the cross section must adjust in order that the
pressure in the "relativistic" fluid can balance the given external
pressure, and a constant flow rate be maintained. The center of the

galaxy is a stagnation point, and so the ultrarelativistic fluid
will be able to flow subsonically down a "tube" of decreasing cross
section until the pressure has approximately halved. Thereafter
the flow will be supersonic and the tube will widen as the pressure
falls. The shape of the tube thus adjusts itself to establish
a "de Laval nozzle" (see Landau and Lifshitz, 1959), as illustrated
in Fig. 2. When the compact source is activated, copious quantities
of hot plasma will be liberated, capable of inflating a cavity
within the enveloping cloud. This configuration, with a light fluid
supporting a heavier fluid in a gravitational field, is well known
to be unstable; and as long as the hot fluid is being continuously
generated, it is plausible that the cavity open up to form two
oppositely directed tubes parallel to the rotation axis.

    This then constitutes a mechanism for transforming a small
volume of randomised energy into a larger volume of ordered motion
(isentropically and therefore in agreement with Liouville's theorem)
so that momentum can flow away from the galaxy, pushing against the
extragalactic medium whose presence must be invoked to provide
confinement at the head of the source.

    In the earlier model of Rees (1971a) the outflow was postulated
to be primarily in the form of low frequency electromagnetic waves,
whose "self-focusing" properties were invoked to achieve the colli-
mation into beams. The present mechanism however, is not specially
sensitive to the nature of the relativistic material in the beams.
(Indeed it need not necessarily be relativistic, the only essential
requirement being that it is substantially hotter than the surrounding
gas, its internal sound speed being high enough that its flow is
unaffected by gravity). Possible ingredients of the "relativistic"
fluid, apart from the plasma itself, are magnetic fields convected
along with the beam, and a variety of low frequency wave modes
unable to escape through the walls of the tubes. After the flow
has become significantly supersonic most of the energy is in bulk
kinetic rather than internal energy anyway. The tube widens as the
external pressure p drops, but only as p^{-n} with $3/8 \gtrsim n \gtrsim 1/4$. Thus
the beam remains well collimated even though p may be several orders
of magnitude lower in the extragalactic medium than in the central
gas cloud.

    The "hot spots" in the components of sources such as Cygnus A
are, in this model, identified with the regions where the beams
impinge on the intergalactic medium. The expected physical situation
in one of these regions is illustrated in Fig. 3. The beam passes
through a shock, where its energy is randomised again, this time
with a significant increase in specific entropy under conditions
ideal for particle acceleration and field amplification. The power-
law spectrum of the relativistic electrons would be established in
this region, rather than in the galactic nucleus. The energy supplied
by the relativistic beam over the source lifetime exceeds the current
internal energy by a factor $\sim (c/v) \, (D/d)$, where V is now the speed
of advance of the head of the source; and for this reason it has
been proposed that the outer parts of the channel be cocooned with

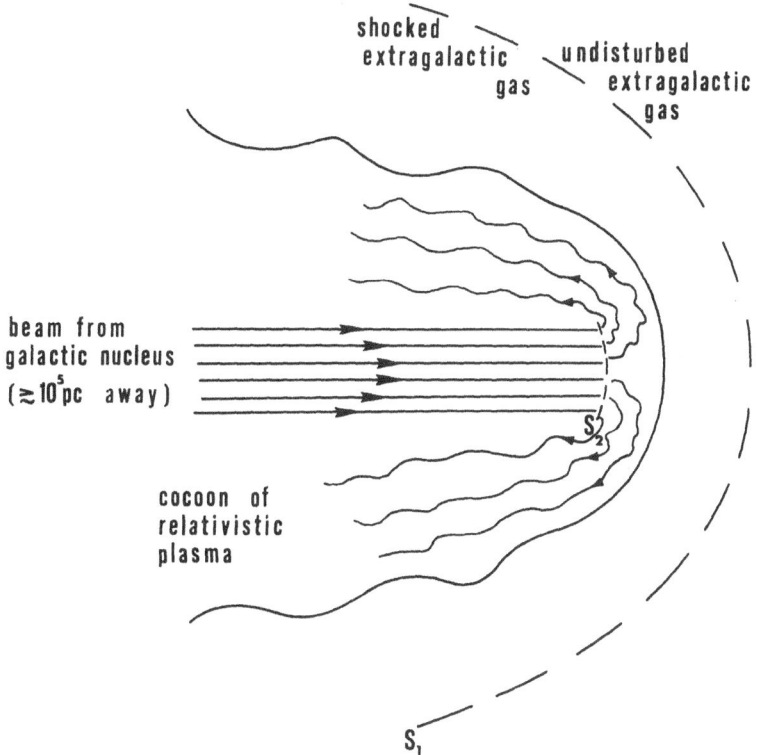

Fig. 3. Schematic diagram of the region where the beam (estimated to be ∿1 kpc in transverse dimensions) impinges on the extragalactic gas, whose density may be in the range 10^{-6}-10^{-3}cm^{-3} (perhaps depending on whether or not the galaxy lies in a cluster). A stand-off shock S_1 advances into the external gas (density ρ_{ext}) at a rate determined by the balance between the ram pressure $\rho_{ext}V^2$ and the momentum density in the beam. There is another shock S_2 where the ordered relativistic beam energy is randomised. The emission from the "hot spots" in the radio components of, for example, Cygnus A is attributed to the relativistic (and probably turbulent) plasma in the region just downstream from S_2. The relativistic plasma eventually forms a sheath or cocoon around the beam, which can be identified with the extended low surface brightness parts of the source components.

the "waste" energy (Scheuer 1974). This is one possible interpretation of the map of Cygnus A (Fig. 1) where it is found that more radio emission originates from the extensive low brightness features than the compact high brightness region where the beam is inferred to encounter the surrounding medium.

As V is supersonic, its value is given by balancing the ram

pressure of the external medium ($\sim \rho_{ext} V^2$) with the momentum flux in the beams ($\sim L/d^2 c$, where L is the luminosity of the central source). The diffuse features may however be confined by essentially thermal pressure ($2 \rho_{ext} kT_{ext}/m_p$). One questionable feature of this model is the stability of the beams. Shearing one fluid past another at high speeds tends to drive Kelvin-Helmholtz instabilities in the bounding surface, but whether such instabilities could develop sufficiently to completely disrupt the channel is still very much an open question. The inner parts of the tube, where the outflow is subsonic, may also be subject to Rayleigh-Taylor instability, but the growth-rate for this is low. Typical double sources are, of course, highly irregular and somewhat asymmetrical in appearance, and one would presumably need to attribute this to various kinds of instability.

In this scheme, complex intermediate-luminosity sources like the "radio trails" (Miley 1973) may represent situations where the relativistic wind fails to get collimated, and the velocity at which ram pressure balances the wind pressure is actually less than the velocity of the galaxy through the extragalactic gas.

The "slingshot" model. A very different theory involves the idea that the radio components themselves contain massive self-gravitating bodies which were flung out of the galactic nucleus (Burbidge 1967). Saslaw et al. (1974) propose that radio sources result from the dynamical interaction of three massive objects that each form without much orbital angular momentum within a galaxy, and fall towards the bottom of the potential well at the center. The first two objects form a stable binary pair; but when the third object comes within dynamical range, it is found that in a sizeable fraction of cases a binary and a single object are ejected from the nucleus along antiparallel directions, with or without exchange occurring. (A similar outcome would be expected if a single supermassive disc in the nucleus fragmented into three parts). These objects then "seed" the radio sources either by liberating their intrinsic rotational energy via magnetic torques or possibly fragmenting into more condensed objects (e.g. low magnetic field rotating neutron stars). If the massive objects had already collapsed to form black holes, they would have to be surrounded by a gaseous disc in order that accretion could provide a continuous power output. The shape and structure of the low surface brightness extensions of the radio components would still imply the presence of an extragalactic gas, but the required density would not be so high as in the "beam" model, because ram pressure confinement of the "hot spots" would not be necessary.

As the orbital angular momentum of the original three objects would presumably reflect that of the elliptical galaxy, this model would predict a correlation between the major axes of an elliptical galaxy and the radio source axis. The ejection speeds would depend on the random velocities within the original three-body system. For consistency, these would have to be much higher than the escape velocity (~ 1000 km sec^{-1}) from the centers of the elliptical galaxies.

A consequence of this theory is that one massive object is ejected in one direction, but two in the other. One might therefore expect to find some asymmetry between the two components.

Particle acceleration mechanisms. One aspect of the radio source phenomenon which still demands an explanation is the process whereby the relativistic electrons establish a power law spectrum. Because in Cygnus A (and probably other sources) the radiative lifetime of a high energy electron is comparable with the light travel time across the radio component, it is generally agreed that the electrons must be accelerated in situ. Fermi-type particle acceleration by a variety of wave modes (especially Alfven waves) possibly coupled with lifetime effects such as radiation losses and spatial diffusion, is most probably responsible, but as yet there is no fully worked out scheme. The processes operating in supernova remnants are probably similar. Much theoretical effort has been expended by Tsytovich and his associates on the so-called "plasma turbulent reactor" (PTR). Although this may indeed operate in pulsars or in galactic nuclei, the requirements for an efficient PTR certainly cannot be met in the relatively diffuse environment of an extended radio source (Norman and ter Haar, 1974). The ubiquitous occurrence of power law electron spectra with index $\sim$2.5 in magnetized cosmic plasma poses a very important problem, whose solution would give considerable insight into the nature of radio components.

1.4 General inferences and prospects

What are the prospects for an improved understanding of the radio source phenomenon in the near future? Several key observations are becoming possible as techniques in all wavebands steadily improve. As well as improved maps of intensity and polarization in the stronger and better resolved sources (in particular with a southern hemisphere interferometer) the following are especially important.

(i) Further evidence for extragalactic gas. So far the best evidence has been furnished by x-ray observations of clusters of galaxies, including that associated with Cygnus A (Longair and Willmore 1974); but these observations are capable of alternative interpretation and so the evidence is not as yet conclusive. Gas in clusters can also be detected from distortions in the microwave background, and inferred from studies of radio "trails" like 3C 129. If densities compatible with ram pressure confinement are confirmed, improved estimates of radio component velocities and hence ages will follow. Upper limits incompatible with ram pressure confinement would be specially interesting. By 1980 X-ray telescopes will provide maps of clusters with the same angular resolution as the best existing radio maps ($\sim$2"). Comparison of x-ray and radio structures could reveal many details of the gas dynamics in and around the components, especially the location of shock fronts.

(ii) Further evidence for thermal gas within radio components. Here the best evidence so far comes from radio polarization obser-

vations, but bremsstrahlung x-rays and optical emission might
be detectable from sufficiently dense gas.

(iii) Non-thermal infra red and optical emission from the
radio components. If such emission were detected it would indicate
the presence of very energetic and short lifetime electrons and thus
make stringent demands on the particle acceleration mechanism.

(iv) Detection of compact structure within external radio
components. So far there is no evidence of any radio structure on
length scale <<1 kpc except within galactic nuclei and quasars.
The detection of much smaller features via Very Long Baseline Interferometry (which can achieve angular resolutions better than
a milli-arc-second) would suggest that one or more massive objects,
located within the components themselves, power the radio source.

(v) Evidence on the orientation of source axes relative to
the associated optical galaxies. In "beam" models, the relativistic
plasma escapes preferentially along the rotation axis of the gas
cloud in the central part of the galaxy. This axis would presumably tend to be aligned with the minor axis of the whole elliptical
galaxy. The slingshot model, however, predicts that the radio
source axis should lie in the equatorial plane of the elliptical
galaxy. There is no clear evidence on this question at present
(Mackay 1971a, Bridle and Brandie 1973).

(vi) Observations of galactic nuclei and quasars. Current
theories claim to account for the properties of the extended components, and the origin of the characteristic double structure.
The nature of the central energy source, however, remains obscure.
Some fairly direct evidence on physical conditions in active galactic
nuclei comes from Very Long Baseline Interferometric studies of
compact radio sources. These sources are variable, and it should
in principle be possible to test (by comparing the location of
successive outbursts) whether a single supermassive object is involved, or whether successive radio outbursts are triggered by
the explosions of independent stellar-mass objects. Of special
interest are those sources which possess compact central components
as well as extended double structure. It would be interesting
to see if the central source also has a double structure. (There is
already evidence for this in 3C 236, and perhaps the compact double
is associated with the "nozzles" postulated in the "beam" models).

More generally, one would like to understand the relationship
between double radio sources and other forms of non-thermal emission
from galactic nuclei, and I shall return to this issue in the
second lecture.

2. EVIDENCE FROM STATISTICAL STUDIES OF RADIO SOURCES AND RELATED
   OBJECTS

2.1 The data

The basic inferences from radio source counts and related data are

very well known and I shall not repeat them here-for a review,
see, for example, Ryle (1968) or Longair and Rees (1973). The
form of the number-flux density relation, which display an "excess"
at intermediate flux densities and a fall-off at the faint end,
seems to imply that there were more powerful radio sources at early
cosmic epochs. The observations can be fitted by evolving $\rho(P)$
(the luminosity function per comoving volume) in innumerable ways.
Most authors who have discussed this question have restricted themselves to simple functional forms for the evolution, but a great
deal of freedom still remains. All the analyses, however, have
these gross features in common:

(i) The evolution is probably a cosmological effect, occurring
primarily at redshifts $z^* \gtrsim 1$. Many sources (e.g. Cygnus A) have
radio luminosities $\gtrsim 10^{44}$erg sec^{-1}, and would still appear in radio
surveys even if their redshifts were $\gtrsim 1$. If the excess occurs at
$z^* \gtrsim 1$, the eventual fall-off at low S can be attributed to the
redshift-dependent effects in the flux density-redshift relation
which become severe when z is much greater than unity. On the other
hand, if the "excess" occurred at $z^* \ll 1$, we would have to postulate
an intrinsic cut-off in the source population beyond z^*, or, in
other words, assume that the excess sources lie in a "shell" around
us, to explain the flattening at the faint end of the log N-log S
curve, because the cosmological effects are unimportant at redshifts
$\ll 1$.

(ii) The evolution involves predominantly the more powerful
sources. Even if the "excess" sources occupied a narrow redshift
range, the peak in the $N(>S)/N_o(>S)$ curve ($N(>S)$ being the number
of sources per steradian with flux densities exceeding S, and N_o
being the value of N expected in an "Euclidean" universe where
$N \propto S^{-3/2}$) would be broadened by any spread in P. Indeed a luminosity
function which had the same form at z^* as it has locally would neither
reproduce the sharp maximum in the curve nor be compatible with the
observed limits on the integrated background. The evolution cannot
therefore consist of a simple scaling of the local $\rho(P)$: instead,
only the more powerful sources can be permitted to evolve strongly.

(iii) The evolution must be very strong. The comoving space
density of powerful sources must be $\simeq 1000$ times greater at $z \sim 2$
than at the present epoch. Although the discrepancy in numbers
between observed and "predicted" counts is only a factor ~ 5 the
effects of evolution of an individual luminosity class is diluted
by the broad spread in $\rho(P)$.

Similar analyses have been performed on samples of quasars
(see Setti and Woltjer 1973 for a review). It turns out that knowledge of the redshifts helps very little. The broad scatter in the
intrinsic properties of quasars means that one must still-as for
radio sources-fall back on crude statistical tests, and one then
finds that quasars display strong evolution with cosmic epoch in
a similar fashion to radio sources. Even though radio sources and
quasars can be observed at much larger redshifts than can normal
galaxies, none of this work has really yet led to any progress in

"classical cosmology"; whichever cosmological model one choses, steep
evolution in $\rho(P)$ is required, and our complete ignorance of the
relevant astrophysics prevents us from subtracting out the "evolutionary correction" in the way this is done (or, at least, attempted
for brightest cluster galaxies.

2.2 Physical nature of the evolution

The most dramatic inference from these analyses is that the density
of powerful sources—both radio galaxies and quasars—must increase
by a factor $10^2 - 10^3$ per comoving volume (i.e. over and above the
$(1+z)^3$ dilution factor) between the present epoch and the epoch
corresponding to $z \approx (2-3)$. Drastic though this effect undoubtedly is,
we recall that the most remote known sources emitted the radiation
we now detect when the universe was ~ 20 per cent of its present age.
Maybe it should not altogether surprise us to find sharp differences
between such ancient epochs and the present time. By observing
discrete sources out to redshifts $z \approx 3$, we can in effect probe ~ 80
per cent of cosmic history. A hypothetical astronomer, observing
only $\sim 2 \times 10^9$ years after the initial singularity, would see a vastly
more active and dramatic cosmic scene. Whereas our nearest bright
quasar is 3C 273 (~ 800 Mpc distant), he would be likely to find a
similar object only ~ 15 Mpc away (~ 50 times closer), and as bright as
a fourth magnitude star. He would observe correspondingly greater
activity in the radio sky, the integrated flux at meter wavelengths
being perhaps 100 times higher than it is today.

This is the message of the radio source counts, reiterated by
the distribution of quasars, and it is obviously crucially important
to our understanding of the astrophysical evolution of the contents
of the universe. It is, however, disappointing that the discovery
of objects with very large redshifts has not led to any progress in
"classical" or "geometrical" cosmology. It is still studies of normal
bright galaxies (despite the smaller redshifts involved) which are
likely to yield the most significant limits on the deceleration parameter.

Since we can infer nothing about the geometry of the universe from
these observations, it is more interesting to invert the problem and
consider instead how much can be learnt about the epoch-dependence
of the source parameters. All the evidence indicates that radio
quasars and radio quiet quasars both display the same evolution as
had been inferred for strong radio sources as a whole from the radio
source counts. This suggests that the evolution is predominantly
due to a drastic secular decrease in the propensity of galactic
nuclei to undergo "violent events". The similarity of the optical
and radio evolution is, however, rather surprising. The optical
measurements refer to a compact object, with dimensions $\lesssim 10$ pc;
whereas the radio properties refer to structures on scales up to
$\gtrsim 100$ Kpc, and the relevant physical processes are quite distinct.
In particular, most models of extended radio sources invoke an
intergalactic gas to confine the source components. The higher

density of this gas at earlier epochs might be expected to cause some
variation of radio properties with z quite independent of the evolution of the optical properties. Ideally, one would like to have
enough observations to be able to subdivide the data and determine
the evolutionary behavior for different classes of objects separately.
At present such procedures result in rather small numbers of objects
in each category, and also introduce extra selection effects; so this
program must await the identification of many more quasars and radio
galaxies (and accurate magnitudes of the latter out to the largest
possible redshift). At present everything is consistent with a
similar evolution law for all powerful radio sources and quasars, either
an exponential decrease of $\rho(P)$ with cosmic time, or else a power law
$(1+z)^x$ with $x \approx 6$ and a truncation at $z \approx 2.5$.

More indirectly, quasars are of cosmological relevance because
the presence (or absence) of absorption features in their spectra
can tell us something about intervening intergalactic gas or galaxies
along the line of sight.

2.3 Implications for radio sources

The theories of extended radio sources discussed in lecture 1 all have
the consequence that a particular source starts off small, expands
outward from the parent object, and eventually fades. This suggests
that perhaps the wide scatter in observed source properties can be
interpreted as different stages in the evolution of a single more
standardised phenomenon. To this end, many authors have constructed
diagrams in which size, power, surface brightness, etc. are plotted
against each other. The results are discouraging, in that no very
clear-cut correlations emerge. This presumably means that there are
several different parameters–e.g. duration and power of activity in
the nucleus, gas density within and outside the galaxy, etc.–which
control the morphology and evolution of an individual source.
However, even if such diagrams did reveal some obvious correlations,
instead of looking more like "scatter diagrams", one should be wary
of over-interpreting the results, for the following reasons:
      (i) Some areas of such diagrams are bound to be blank because of
selection effects. For example, the higher-power sources in any
survey are bound to be all at substantial distances; and therefore
they are less likely to be resolved.
      (ii) The space density of the very powerful sources is so low
that some of those in the sample will have significant cosmological
redshifts. Therefore the correlation diagram should perhaps be con-
volved with some cosmological effect whereby the mean properties of
sources depend on cosmic epoch.
      (iii) To interpret any band in such a diagram as an evolutionary
track could be as misleading as interpreting a globular cluster H-R
diagram as representing the evolutionary track of an individual star!
      One common feature of most theories is that the active lifetime
of a strong source is $\lesssim 10^8$ years (though this obviously does not preclude repeated outbursts in the same galaxy, nor the possibility that

the sources may continue for a longer period at lower radio luminosity)
The fact that this is much less than the Hubble time means that the
sources observed at large redshifts are not individually younger-in
the sense of having experienced shorter active lifetimes-than nearby
sources.   Thus the apparent epoch-dependence of radio source properties
inferred from the radio source counts and from the redshift-distribution
of radio quasars, must be attributed to an epoch-dependence in the
source birth rate and/or in the form of the evolutionary track traced
out by a typical source.

Another direct inference from the short lifetime of individual
sources is that <u>dead sources must greatly outnumber the living</u>.   This
means that the integrated effect of all generations of radio sources
is important for the energy budget of intergalactic space.   This conclusion is even stronger when the higher inferred comoving density
of strong sources at early epochs is taken into account.   In particular
radio sources may produce a universal flux of cosmic rays, and a flux
of relativistic electrons, which, via inverse Compton scattering of
microwave background photons, contribute to the x-ray background.   In
all ram pressure confinement models the energy expended in pushing
against (and heating) the external medium exceeds, by a factor at
least $\sim(D/d)$, the integrated radio luminosity.   This energy-which is
provided either by the kinetic energy of thermal material in the
blobs or by the continuous thrust of the "beams" emanating from the
galactic nucleus-may therefore be important for heating the intergalactic medium, and could even make it so hot that it is no longer
confined within clusters . (Bollea and Cavaliere 1974).

2.4 "Violent activity" in general, and its relation to normal galactic evolution

If one accepts that quasars are indeed hyperactive galactic nuclei,
then the foregoing comments suggest that the key problem is to understand the general nature of this activity, and the reasons why its
frequency of occurrence is so strongly epoch-dependent.   The accumulation of a large mass of gas in the potential well at the center of
an elliptical galaxy is probably a prerequisite for any kind of
"violent activity".   It is completely unclear, however, what parameters
determine the form which the violent activity then takes.   (See Burbidge
1970, for a general review of galactic nuclei).   But even without
enquiring into the physical mechanisms which generate the non-thermal
energy, one can perhaps interpret the overall epoch-dependence along the
following lines.

The relevant gas could either be material which failed to condense into stars, or else could have been lost (via stellar winds,
planetary nebulae, or supernova explosions) from existing stars throughout the volume of the galaxy.   For this gas to form a concentrated
cloud in the galaxy's central potential well, the following requirements must be met:

(i) The mean angular momentum of the gas must be low, or else
there must be some efficient viscous transport.

(ii) The gas must not be expelled as a steady "galactic wind" (Mathews and Baker 1971).

(iii) The gas must not be swept out by the ram pressure of the apparent "intergalactic wind" which would occur if the galaxy was moving sufficiently fast through an extragalactic medium.

Plainly these conditions are more likely to be fulfilled by massive elliptical galaxies with low angular momentum and deep gravitational potential wells. But—more importantly—they are more easily fulfilled when the amount of uncondensed gas, and/or its rate of production from stars, is high. The detailed work of Larson (1974a,b) shows that this quantity decreases with time. By considering the resulting quantitative z-dependence of conditions (ii) and (iii), and convolving with the observed mass distribution and velocity distribution of ellipticals, Gisler (1975) shows that the expected frequency of "violent events" would indeed be very steeply dependent on cosmic epoch.

3. THE EVOLUTION OF PRIMORDIAL FLUCTUATIONS: GALAXY FORMATION

3.1 General "philosophy"

Observations of large-redshift objects obviously tells us something about the early evolution of galaxies. But one can also approach this subject from (as it were) the opposite direction, by adopting a particular cosmological model and considering how (and when) galaxies might have been formed. In this lecture I shall outline some results that have been obtained by following this deductive kind of approach.

Even if the universe contains inhomogeneities, the Friedmann models can still give a valid overall description provided that the typical fluctuations $\Delta\rho$ of the mean density satisfy

$$G(\Delta\rho)\lambda^2 << c^2 \qquad (3.1)$$

for all length scales λ. This ensures that the fractional departures from the Robertson–Walker metric are everywhere $<<1$, and that the peculiar velocities induced by the inhomogeneous mass distribution are $<<c$. Both for galaxies and clusters of galaxies, (3.1) is fulfilled by a margin of at least $\sim 10^4$; and the isotropy of the background radiation strongly suggests that it holds on all larger scales up to the "Hubble radius" $\sim c/H_o$.

The mean density of matter within galaxies is, very roughly, in the range of 10^{-1}-10^3 particles cm^{-3}. Obviously galaxies—and, a fortiori, clusters of galaxies—could not have existed in their present form when the <u>mean</u> density of the universe {now $\sim 10^{-5}\Omega(H_o/100$ km $sec^{-1}Mpc^{-1})^2$, where the density parameter Ω is defined as $4\pi G\rho/3H_o^2$} was higher than this. Protogalaxies may then have been merely regions of slightly enhanced density, whose expansion was retarded, and

eventually halted, by their excess gravitational field. Although
gravitational forces promote instability, pressure gradients (especially radiation pressure in the "plasma era" of the standard
"hot big bang" model) and viscous dissipation tend to counteract
this growth. I shall now describe the interaction between these
processes, and the consequent fate of various types of perturbation
in the "primordial fireball". The main aim of recent work on this
problem has been to try to understand the occurrence of structure
with the characteristic observed scales without artificially choosing
the initial conditions with this end in view; and perhaps (more
ambitiously) to understand the morphology and angular momentum distribution of galaxies and clusters. No very spectacular results have
yet been achieved, but the fact that these investigations can even
be contemplated illustrates how the discovery of the microwave background has enlarged the range of phenomena amenable to quantitative
discussion.

The view that present structural features evolved from small
perturbations in an initially almost homogeneous universe is, however,
no more than a working hypothesis: there is no firm evidence that, on
the scale of galaxies and clusters, the universe was ever smoother
than it is now. Ambartsumian, for instance, has long been voicing
the opinion that matter "emerges" locally from a high density singular
state, but this idea cannot be developed quantitatively in our state
of knowledge. A further incentive for pursuing studies of the fate
of small initial perturbations is that only in this approach is it
consistent to assume a homogeneous background cosmology: if galaxies
had always had the same contrast density $\Delta\rho/\rho$, then (3.1) would be
violated when ρ was large, so the mathematically tractable Friedmann
models would be irrelevant to the dense early stages.

3.2 Gravitational instability of "perfect fluid" Friedmann models

The first full relativistic treatment of perturbed Friedmann models
was given by Lifshitz (1946). There have been many subsequent discussions of this topic (see, for instance, Rees (1971b) or Field
(1975) for a summary). The results for the growth rate of small density perturbations are basically as follows (where for brevity we
limit attention to the cases $\Omega \lesssim 1$):

"Dust" universe. In a "dust" universe ($p << \rho c^2/3$) the scale
factor R, equal to $(1+z)^{-1}$ if we take $R \equiv 1$ at the present epoch, varies
with t roughly as $R \propto t^{2/3}$ (for $R \lesssim \Omega$) and $R \propto t$ (for $R \gtrsim \Omega$). Small-amplitude
density perturbations grow as

$$\frac{\Delta\rho}{\rho} \propto R \propto t^{2/3} \qquad (R \lesssim \Omega) \tag{3.2}$$

$$\frac{\Delta\rho}{\rho} \simeq \text{constant} \qquad (R \gtrsim \Omega) \tag{3.3}$$

This shows that an expanding dust universe is unstable to the growth
of those perturbations which can be regarded as fluctuations in the

local total energy, or local curvature. Any density irregularity-not merely a sine wave or a uniform sphere—would exhibit the same general type of growth. (One can also imagine initial irregularities in which the energy, or curvature, is unperturbed, but the expansion starts at different times. This type of fluctuation also emerges from a more elaborate mathematical treatment of the problem but is of little interest because it diminishes in amplitude in proportion to $R^{-3/2}$ during the expansion).

A useful concept in this subject is a hypothetical observer's "particle horizon", which consists of those points which have infinite redshift and thus, in some sense, delineates the region causally connected to a given observer. In all Friedmann (decelerating) models the horizon grows to encompass more and more matter as t increases.

Provided $\Delta\rho/\rho$ is interpreted as the fractional deviation from the background density which different comoving observers would measure at the same proper time t, (3.2) still holds even when the scale of the fluctuation is large compared to the particle horizon. Of course, this is true for any scale at sufficiently early times, because in dust models with $R \propto t^{2/3}$ the mass M_H within the particle horizon ($\sim \rho(ct)^3$) varies as

$$M_H \propto R^{3/2} \propto t \qquad (3.4)$$

Relation (3.2) obviously ceases to apply when the fluctuations attain order unity. However the subsequent behavior can still be easily visualized. In an Einstein-de Sitter ($\Omega=1$) background, any region of enhanced density will stop expanding (at a time when it has $9\pi^2/16$ times the background density) and will then start to collapse. Qualitatively similar considerations hold in any model, provided that $\Delta\rho/\rho$ attains order unity when $R < \Omega$.

Any fluctuations which start to recontract while still smaller than M_H would not really constitute a part of our universe. If $\Omega \ll 1$, a perturbation which still had $\Delta\rho/\rho \ll 1$ when $R \approx \Omega$ would retain almost constant amplitude thereafter. This is because self-gravitation of the perturbed region would then become unimportant, and the fluctuation would continue, like the unperturbed universe, in undecelerated expansion. These simple results for gravitational instability of spherical irregularities can be straightforwardly extended to more general perturbations, but the results are not qualitatively different.

If p is non-zero, the Friedmann dust solution still describes the overall dynamics if $p \ll \rho c^2$, but pressure gradients can then stabilize fluctuations on small scales. They will not grow, but instead oscillate like sound waves. Gravitation being a volume effect, will always dominate pressure on sufficiently large scales, and the minimum length scale over which gravitational forces overwhelm pressure is the so-called "Jeans length", first considered by Sir James Jeans in 1902:

$$\lambda_J = c_s \left(\frac{\pi}{G\rho}\right)^{\frac{1}{2}} \qquad (3.5)$$

where c_s is the sound speed. The corresponding "Jeans mass" is

$$M_J \sim \frac{c_s^3}{G^{3/2}\rho^{1/2}} \qquad (3.6)$$

In a cosmological model where curvature effects can be neglected (i.e. $R<\Omega$) this mass is such that the differential expansion velocity across it is $\sim c_s$. The effects of pressure are negligible for all fluctuations with $M>>M_J$, so in such cases the above discussion of dust models is applicable.

"Ideal" radiation universe. One can also derive simple approximations for the behavior of perturbations in a "radiation universe" containing a perfect fluid with $p=\rho c^2/3$. When $R<<\Omega$ the expansion of such a universe is given by $R \propto t^{\frac{1}{2}}$. If the radiation mean free path is small enough to confer fluid-like properties on the material- as in the case (on all relevant scales) during the "plasma era" of the canonical big bang-then the effective sound speed is $c_s=3^{-\frac{1}{2}}c$. Pressure therefore stabilizes all masses substantially smaller than the particle horizon (in other words $M_J \simeq M_H$). However this universe is unstable to perturbations with scales $>>M_H$ and one finds

$$\frac{\Delta\rho}{\rho} \propto R^2 \propto t \qquad (3.7)$$

Because M_H increases with t, a perturbation of given scale will start off by growing according to (3.7) but almost as soon as it comes within the horizon pressure gradients become important and it starts oscillating.

The limitations of this approach. The original studies of gravitational instability were motivated by the hope that galaxies and clusters might have condensed from the random $N^{-\frac{1}{2}}$ fluctuations which would naturally be expected in a universe composed of discrete atoms. For a galactic mass of $\sim 10^{11} M_\odot$, however, the statistical fluctuations would seem to be only $\sim 10^{-34}$, and these would not have condensed out by the present epoch unless the growth was initiated at a stage when the particle horizon encompassed only a few atoms. It is not clear that much physical significance attaches to this result. (A related idea has however been revived by Carlitz et al. (1973) in the context of an elementary particle theory which predicts that the primordial material was mainly in the form of "superbaryons" of $\sim 10^{15}$ gms.) The problem is that the overall expansion, which has a time scale equal to the "e-folding" time of the Jeans instability, transforms the growth rate of the density contrast from an exponential to a (much slower) power law. This means that one must either suppose that the fluctuations came into being at an exceedingly early epoch, or else assume larger initial amplitudes. For example, the existence of galaxies would be accounted for in a "dust" universe if there were perturbations with $\Delta\rho/\rho \approx 10^{-7}$ at the time when such a mass first came within the horizon. Though much larger than the "statistical" amplitude, these are still "small" perturbations in the sense that they would cause the geometry of the universe to deviate only negligibly from the strictly homogeneous case.

However, this type of hypothesis is unsatisfactory because it really explains nothing. The initial amplitude must be chosen in a purely ad hoc manner: if it were too small, galaxies would not yet have condensed; but if it were too large they would have stopped expanding at an early epoch, and would have higher mean densities than are observed. In addition, when p=0, perturbations of all scales grow at the same rate, given by (3.2). This means that there is no preferred scale of instability, but that the postulated initial spectrum must be weighted in favor of whatever masses are required. As T. Gold has put it: "Things are as they are because they were as they were". It is true that non zero pressure could stabilize scales which were $\leq M_J$ during certain stages of the expansion, with the result that the growth of large scales would be favored. However, this effect does not seem capable of impressing any prominent features on an initially smooth mass spectrum of fluctuations.

3.3 Perturbations in the canonical "hot big bang"

The detection of the thermal microwave background, and its interpretation as "primeval" radiation, stimulated renewed interest in this approach to the problem of galaxy formation. There are several reasons why this should have been so:

(i) This radiation constituted the first really compelling evidence for a hot dense, opaque "fireball phase" in the evolution of the universe.

(ii) The isotropy of the background, established to ≤ 0.1 per cent—unprecedented precision for a cosmological measurement— provided the first firm evidence that simple Friedmann models may provide an adequate dynamical description of the early universe.

(iii) The hot big bang model allows us to predict a definite equation of state, except perhaps during the "hadron era".

(iv) The fate of perturbations in the "fireball" is <u>not</u>, as in perfect fluid models, governed solely by competition between pressure gradients and gravitation: the coupling between matter and radiation is not perfect, and this leads to dissipative processes. The resulting damping acts preferentially on certain scales (and certain types) of perturbation. It would be gratifying if the irregularities most likely to survive and amplify bore some relation to the characteristic scales of observed structures in the universe. For a review of the physics of the "big bang" cosmology, see the review by Harrison (1973c) or the book by Weinberg (1972).

<u>Radiative damping of small amplitude oscillations</u>. In the "post-recombination era" ($t \gtrsim t_{rec}$) when $T \lesssim 3000$ K($R \gtrsim 10^{-3}$) the universe is essentially transparent[†], and motions of the matter are almost unaffected by the black body radiation. During the "plasma era", on

---

[†]This would no longer be true if the perturbations gave rise to a heat input which was sufficient to maintain a significant level of ionization after t_{rec} (see section 3.4).

the other hand, each photon is scattered many times by free electrons.
Thus the electrons must share the mean motion of the photon gas; the
nucleons, being electrostatically coupled to the electrons, are therefore dragged along as well. When the mean photon energy ($\sim 3kT$) is
$<< m_e c^2$, the Thomson cross-section $\sigma_T = 6.6 \times 10^{-25} cm^2$ is appropriate.
The mass $M_{\tau=1}$ of matter within a sphere of optical depth unity would
be $\lambda_{\tau=1}^3 \rho_m$, where $\lambda_{\tau=1} = (n_e \sigma_T)^{-1}$ is the photon mean free path between
Thomson scatterings and ρ_m the baryon density. At t_{rec}, when
$n_e \approx 10^4 \Omega$ cm^{-3}, this mass is

$$M_{\tau=1} \simeq 10^7 \Omega^{-2} M_\odot, \tag{3.8}$$

and it diminishes rapidly ($\propto T^{-6}$) as we extrapolate further back into
the past. Thus, at least for perturbations on galactic scales
($\sim 10^{11} M_\odot$), the matter and radiation behave as a single fluid for
$t \lesssim t_{rec}$. The sound speed in this composite fluid (evaluated taking
the inertia of both matter and radiation into account, but neglecting
the pressure of the matter) is

$$c_s = \frac{c}{3^{\frac{1}{2}}} (1 + \frac{3}{4} \frac{\rho_m}{\rho_\gamma})^{-\frac{1}{2}} . \tag{3.9}$$

In the "hot big bang" model, $\rho_m \propto R^{-3}$ and $\rho_\gamma \propto R^{-4}$. Before the
"epoch of equal densities" t_c (i.e. when $\rho_\gamma > \rho_m$), $c_s \approx c/3^{\frac{1}{2}}$ and $M_J \approx M_H$.
Between t_c and t_{rec}, M_J remains constant, at a value

$$M_1 \simeq 3 \times 10^{15} \Omega^{-2} M_\odot; \tag{3.10}$$

After t_{rec}, M_J drops drastically because the thermal pressure of the
particles is only $\sim 10^{-7} \Omega$ of the radiation pressure, so M_1 is the
largest mass which can ever be stabilized by pressure forces[†].

Non-uniformities in ρ_γ on scales $<< M_J$ would, in effect, behave
like sound waves. The effective sound speed for wavelengths λ is
actually $c_s(1 - \lambda^2/\lambda_J^2)^{\frac{1}{2}}$, the effect of gravity being to "soften"
the fluid, by an amount which becomes negligible as $\lambda/\lambda_J \to 0$. If
the matter and radiation were perfectly coupled (no viscosity or
damping) we could readily infer the behavior of the amplitude
$\Delta\rho_\gamma/\rho_\gamma$ by applying the usual theory of adiabatic invariants to
"standing waves"-the energy within one wavelength varies inversely
as the period $\mathcal{N}$ of the oscillation. For perturbations with a given
comoving wavelength $\lambda (<< \lambda_J)$, the oscillation energy varies during the
expansion as

(amplitude)^2 x (sound speed)^2 x $\left(\begin{array}{c}\text{mass-energy within a cubic} \\ \text{comoving wavelength}\end{array}\right)$,

---

[†]If $\Omega \lesssim 0.05$, $t_c \gtrsim t_{rec}$, (3.10) is then inapplicable, and the largest
scale which can be stabilized by pressure then contains a particle
mass of $\sim 3 \times 10^{19} \Omega M_\odot$.

i.e. it is proportional to

$$\left(\frac{\Delta\rho_\gamma}{\rho_\gamma}\right)^2 c_s^2 \{(\rho_m + \rho_\gamma)R^3\} \tag{3.11}$$

The period of oscillation ((wavelength)/(sound speed)) varies roughly as

$$\mathcal{N} \propto R \qquad (t \lesssim t_c)$$
$$\mathcal{N} \propto R^{3/2} \qquad (t_c \lesssim t \lesssim t_{rec}). \tag{3.12}$$

So, approximately,

$$\frac{\Delta\rho}{\rho} \propto \begin{cases} \text{constant} & (t \lesssim t_c) \\ R^{-1/4} \propto t^{1/6} & (t \gtrsim t_c) \end{cases} \tag{3.13}$$

The physical interpretation of the behavior when $t \lesssim t_c$ is that the inertial mass within a comoving value varies as R^{-1}, so the reduction in energy ($\propto R^{-1}$) required by the adiabatic invariant can be absorbed by the decrease in mass, without necessitating any decrease in the amplitude.

The photon mean-free-paths, however, are not completely negligible, and a more accurate treatment allowing for photon diffusion and viscosity reveals that the oscillations would be damped (Silk 1968, Weinberg 1971). For length scales $<<\lambda_J$ one can treat small amplitude oscillations as acoustic waves, neglecting general relativistic effects. The damping can be attributed to two effects:

(i) The photons tend to leak out of the compressed regions, and there is a consequent phase lag between the matter oscillations and the radiation fluid (i.e. pressure oscillations, leading to attenuation of the waves).

(ii) The radiation field becomes slightly anisotropic in sheared regions, and this leads to viscous dissipation.

Taking both these effects into account, a linear analysis by Weinberg (1971) shows that oscillations of wavelength λ are damped at a rate

$$\tau_{damp.}^{-1} = \frac{2\pi^2 c}{3} \frac{\lambda_{\tau=1}}{\lambda^2} \frac{\rho_\gamma}{(\rho_m + \frac{4}{3}\rho_\gamma)} \left\{ \frac{16}{15} + \frac{\rho_m^2}{\rho_\gamma(\rho_m + \frac{4}{3}\rho_\gamma)} \right\} \tag{3.14}$$

The two terms in the braces represent the contributions from effects (i) and (ii) respectively. When $\rho_m << \rho_\gamma$ ($t << t_c$) the viscosity term dominates, but when $\rho_m >> \rho_\gamma$ ($t >> t_c$) the "phase lag" effect is more important. In all cases, however, τ(damp.) is comparable with the time a photon takes to "random walk" a distance λ. The damping is more important on smaller scales, and at any epoch t there will be a minimum undamped length scale λ_D for which τ(damp.)$\propto$t. In fact $\lambda_D \simeq (\lambda_{\tau=1}\lambda_H)^{\frac{1}{2}}$. By the end of the "plasma era", oscillatory motion involving all masses up to a certain mass M_2 would have been severely attenuated.

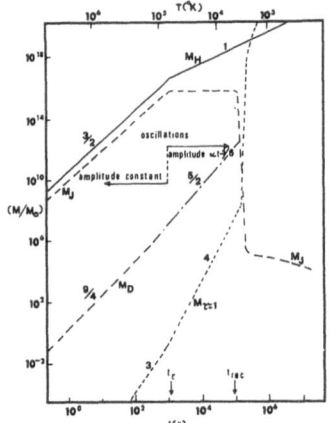

Fig. 4. Some characteristic masses for a cosmology with $\Omega=1$ and
present temperature 3 K. M_H denotes the mass of particles within
the horizon (note that M_H attains approximately a galactic mass
after one year). $M_{\tau=1}$ is the mass within a sphere with unit optical
depth to Thomson scattering. Adiabatic perturbations on scales
$<M_D$ are severely damped. During the expansion, the various masses
increase as the powers of t which are given alongside the curves.
Analogous (but somewhat more complicated) diagrams can be constructed
for cases when primordial vorticity and non-linear effects are important.

We find

$$M_2 \simeq \begin{cases} 6\times10^{12}\,\Omega^{-5/4}M_\odot & (\Omega\gtrsim0.03) \\ 1.5\times10^{13}\,\Omega^{-1}M_\odot & (\Omega\lesssim0.03) \end{cases}. \qquad (3.15)$$

Fig. 4 illustrates how various characteristic masses behave
during the "plasma era", in a universe with $\Omega=1$. The mass M_H of
the <u>matter</u> within the particle horizon is plotted (we observe that
a whole galactic mass first comes within the horizon when $t\simeq1$ year).
Also plotted are M_J, $M_{\tau=1}$ and the minimum undamped mass $M_D\simeq\rho_m\lambda_D^3$.
A region in which the photon density is perturbed therefore behaves
according to (3.7) until the baryon mass M associated with it falls
below M_H. If $M<M_1$ (relation (3.10)), the perturbation comes within the
horizon while the universe is still radiation dominated (and $M_J\simeq M_H$),
so it then starts to oscillate according to (3.13). The oscillations
will be attenuated if $M\lesssim M_D$. So for all mass scales with $M>M_2$ (relation (3.15)), $\Delta\rho_\gamma/\rho_\gamma$ will be reduced exponentially by the time the
plasma recombines. (Fluctuations with mass $M>M_1$ would never, of
course, oscillate).

The relevant parameter for galaxy formation is $\Delta\rho_m/\rho_m$, and not
$\Delta\rho_\gamma/\rho_\gamma$; and it is only when the initial perturbations preserved the
photon/baryon ratio (i.e. $\Delta\rho_\gamma/\rho_\gamma=(4/3)\,\Delta\rho_m/\rho_m$) that the damping of

the radiation distribution reduces $\Delta\rho_m$ to zero also. These <u>adiabatic</u> <u>perturbations</u> are very much a special case: in the more general situation where $\Delta\rho_\gamma/\rho_\gamma$ and $\Delta\rho_m/\rho_m$ are independent (and the entropy per baryon itself is perturbed), a residual <u>isothermal perturbation</u>

$$\frac{\Delta'\rho_m}{\rho_m} = (\frac{\Delta\rho_m}{\rho_m} - \frac{3}{4}\frac{\Delta\rho_\gamma}{\rho_\gamma}) \qquad (3.16)$$

will remain even after the radiation distribution has smoothed out. Isothermal perturbations on scales $<<M_J$ are in effect "frozen in" throughout the "plasma era" because radiative drag constrains the matter to expand at the same rate as the photon gas. A perturbation with $M<<M_1$ which was isothermal when it first came within the horizon would never oscillate, but would maintain the same amplitude until t_{rec}.

There is at present no special motive for postulating one type of initial perturbation rather than another. Precisely adiabatic perturbations would occur if, for some reason, the entropy per baryon were strictly constant. One process leading to almost purely isothermal fluctuations was first discussed by Harrison (1967). He points out that, in a universe possessing overall symmetry between matter and antimatter, small non-uniformities (1 part in $\sim10^7$) in the matter-antimatter ratio in the hadron era could yield isothermal fluctuations of order unity at later times. This is also a general feature of the "matter-antimatter symmetric" cosmology being considered by Omnès and his associates (see Omnès (1973) for a "status report" on this subject).

<u>Primordial vorticity</u>. One can also envisage initial perturbations involving <u>vorticity</u>. If the associated peculiar velocities are $\Delta V<<c_s$ and the scales $<<M_H$, then it may be valid to apply the theory of incompressible turbulence because the fractional induced density inhomogeneities would be $\sim(\Delta V/c_s)^2$. The characteristic rotational velocities V_{rot} on a given comoving scale vary as

$$V_{rot} \propto \begin{cases} \text{constant} & (t\lesssim t_c) \\ R^{-1} & (t\gtrsim t_c) \end{cases}. \qquad (3.17)$$

For motions on scales $<<M_H$, (3.17) follows immediately from conservation of angular momentum $[\propto R^5\omega(\rho_m+\rho_\gamma)\propto R^4V_{rot}(\rho_m+\rho_\gamma)]$ within a comoving volume. Photon viscosity would however damp out small eddies. In a linear situation where there is no energy input from larger eddies, the smallest eddy to survive viscous damping until t_{rec} has a mass (Ozernoi and Chernin 1968, Harrison 1973a,b)

$$M_2' \simeq 10^{10}\Omega^{-11/4}M_\odot \qquad (3.18)$$

The reason why, for $\Omega\approx1$, this is much less than M_2, is that when $t>t_c$ the momentum is carried mainly by the nucleons, whereas the viscosity involves only the photons. (The calculation of M_2 assumes that the eddy is roughly spherical: damping of <u>flattened</u> eddies would be important on even larger scales).

Other linear processes. Other processes which affect irregularities before the recombination epoch include:

(i) Neutrino viscosity, which is maximally effective when the
ν_e mean free path is $\sim ct$ (i.e. at $t \simeq 1$ sec, $T \simeq 10^{10}$ K) but has no direct
effect on galactic-scale irregularities because M_H is then $\lesssim 10^{-4} M_\odot$.
As Misner (1968) was the first to emphasize, however, neutrino viscosity
may play a key role in reducing any initial anisotropy which the
universe might have possessed.

(ii) Adiabatic oscillations can in principle be damped if the
bulk viscosity is non-zero. The bulk viscosity is most significant
when the electron-positron pairs are no longer ultrarelativistic
but have not yet all annihilated, i.e. when $T \simeq 5 \times 10^9$ K. Even then,
the effect is unlikely to be important because the electron collision
rates are much shorter than the oscillation frequencies of interesting
scales.

(iii) Field (1971) has discussed a non-gravitational instability,
first pointed out by Gamow, which arises because the electron temperature is always slightly below T_γ (owing to the expansion) and
each particle tends to shield its neighbors from the intense ambient
radiation. The resulting "mock gravity"-an inverse square law
attraction-does not seem strong enough to be of interest.

The development of small-amplitude irregularities on the scales
which seem directly relevant to galaxy formation is thus governed,
when $t \lesssim t_{rec}$, by gravitation, radiation pressure, and radiative viscosity. If the amplitudes are small enough for a linear analysis to
be applicable, we can therefore relate the amplitude at t_{rec}, as a
function of mass, to the "initial" spectrum. There is no a priori
reason for postulating any particular form for the initial spectrum.
Since there appears to be no effective mechanism for damping the
largest scales- and indeed all masses $>M_1$ (Eq. (3.10)) grow continuously
in amplitude-the apparent large-scale uniformity of the universe at
the present time forces us to assume that the corresponding initial
amplitudes were very small. The least artificial spectrum is one
in which, at a given instant, the "power" is concentrated in smaller
scales (or higher wave-numbers). Many authors have adopted a spectrum
in which the amplitude of adiabatic fluctuations, measured at an
epoch before any relevant scales have been stabilized by pressure
or attenuated by damping, is proportional to (wave-number)^2 or,
equivalently, $\propto M^{-2/3}$. This choice of spectrum which is further
discussed in section 3.5, has the property that all scales M have
equal amplitudes at the time when $M \simeq M_H$ (the smaller masses come
within the horizon first). For this spectrum, the fractional metric
fluctuation $G(\Delta\rho)\lambda^2 c^{-2}$ (c.f. relation (3.1)) has the same value for
all scales. The necessary amplitude must be fixed a posteriori, but
the possibility that some perturbations may be selectively damped
or amplified raises the hope that certain preferred scales of irregularity may be especially prominent when they emerge from the fireball.
Peebles and Yu (1970), have followed such fluctuations through to
the completion of recombination, but find no appreciable increase in
the minimum undamped mass above M_2. An important conceptual

uncertainty in this work concerns the way one formulates the behavior of fluctuations larger than M_H. Some such scales may, in effect, be "damped" by viscous effects even before entering the horizon, because locally the expansion would mimic that of an anisotropic but almost homogeneous model, and the calculations of Misner (1968) and others might be relevant.

Non-linear effects. Granted that some non-infinitesimal initial perturbations must be present to account for the present-day structure of the universe, there is no reason why their amplitude should not be so large that a linear analysis is inadequate. Recently several authors have made more ambitious investigations of the non-linear interactions between different scales of irregularity in the plasma era.

Adiabatic oscillations of large amplitude would develop shock waves (oscillations of wavelength λ develop harmonics of wavelength $\lambda/2$ which...). Peebles (1970) shows that this process would reduce the amplitude of adiabatic oscillations with mass M to

$$\frac{\Delta\rho}{\rho} \lesssim \frac{\left(1+\frac{3}{4}\frac{\rho_m}{\rho_\gamma}\right)^2}{\left(1+\frac{\rho_m}{\rho_\gamma}\right)\left(1+\frac{21}{16}\frac{\rho_m}{\rho_\gamma}\right)} \left(\frac{M}{M_H}\right)^{1/6} \qquad (3.19)$$

It would therefore affect fluctuations of masses $>M_2$ if their amplitudes before recombination were sufficiently large.

If the velocities V_{rot} associated with rotational perturbations on scale λ are such that $V_{rot}t \gtrsim \lambda$, transfer of energy between eddies of different scale will modify the initial spectrum. Ozernoi and his coworkers have shown that initial rotational perturbations could establish a Kolmogorov turbulence spectrum over a range of eddy scales corresponding to masses up to

$$5\times10^{15}\left(\frac{V_{rot}}{c}\right)^3 \Omega^{-2} M_\odot \qquad (3.20)$$

This spectrum (for which the rotational velocities $\propto M^{1/9}$) would continue down to a mass less than M_2' (Eq.(3.18)) because energy fed in from larger scales will compensate for viscous dissipation. Even though the character of the primordial irregularities is unspecified and "ad hoc", these non-linear processes tend, irrespective of the initial spectrum, to establish a definite form for the spectrum at t_{rec} even on the range of scales unaffected by exponential damping. It is this spectrum which, albeit perhaps indirectly, determines the mass distribution of galaxies and clusters and their angular momentum. Even before recombination, the sound waves inevitably associated with turbulence may be significant enough to complicate the picture (e.g. Jones 1973). Harrison (1973b) has discussed the possibility that primordial turbulence may generate a magnetic field, and amplify it to such an extent that its pressure becomes comparable to that of the matter, and therefore dynamically important after t_{rec}.

3.4 Processes after recombination

The work described above shows that all purely adiabatic fluctuations with masses up to those of the largest galaxies, or even somewhat larger, would have been attenuated by radiative viscosity at $t \lesssim t_{rec}$. However, isothermal fluctuations would survive with their initial amplitude, as would the isothermal component of any more general initial perturbation. Also, primordial turbulence-even though, for $V_{rot} \ll c_s$, it would have been essentially incompressible before t_{rec}-would quickly generate density irregularities as soon as the recombination had reduced the effective sound speed.

Any density fluctuations whose amplitude just after t_{rec} was already $\gtrsim 1$ would form condensations with density $\gtrsim 10^4 \Omega$ atoms cm^{-3}. Therefore any agglomerations of galactic mass which are destined to form ordinary galaxies (typical densities $10^{-1}-10^3$ atoms cm^{-3}) probably had $(\Delta\rho/\rho) \ll 1$ just after decoupling. This provides some reassurance that a linear treatment may be appropriate before t_{rec}; and may also, as discussed by Peebles (1971a), raise problems for theories of primeval turbulence (e.g. Ozernoi and Chibisov 1971, Oort 1970, Tomita et al. 1970, Harrison 1973a).

After t_{rec}-or at least by the time T_γ has fallen to $\lesssim 1000$ K-the primordial plasma would,in the canonical model, have recombined sufficiently that there would be no dynamical coupling between radiation and matter, as far as fluctuations with $M \lesssim M_H$ are concerned. Both the radiation drag acting on optically thin fluctuations and the pressure due to the high frequency tail of the radiation spectrum in the Lyman lines or Lyman continuum are easily shown to indeed be negligible (but see Bonometto et al.1974). Unless magnetic fields are dynamically significant, gravitational instability after t_{rec} is thus apparently inhibited by gas pressure alone, the corresponding Jeans mass being only

$$M_3 \approx 3 \times 10^5 \Omega^{-1/2} M_\odot. \qquad (3.21)$$

Masses in the range that interests us can therefore perhaps condense unimpeded by pressure. Their behavior would therefore be governed by (3.2). One uncertainty is, of course, that dissipative processes associated with the fluctuations may maintain a sufficient level of ionization to prevent the matter from completely decoupling from the radiation after t_{rec}.

Note also that a non-interacting background of, for example, gravitons or degenerate neutrinos which dominated the mass-energy density would reduce the overall expansion timescale $(\sim (G\rho_{total})^{-1/2})$. Density perturbations in the matter (growing on a timescale $\sim (G\rho_m)^{-1/2} > (G\rho_{total})^{-1/2})$ then maintain almost constant amplitude instead of growing according to (3.2). Meszaros (1974) has calculated the accurate growth rate under these conditions. His results are also of interest in rather different contexts-such as, for instance, a situation where a certain fraction of the primordial material has collapsed into black holes, whose motion is unaffected by radiation drag.

Since all perturbations $>>M_J$ grow at essentially the same rate, the scale of the first condensations to separate out would be that for which, just after t_{rec}, $\Delta\rho_m/\rho_m$ was greatest. If the initial fluctuations were strictly adiabatic and irrotational the smallest scales surviving at t_{rec} would have mass $\sim M_2$. If the amplitude decreased towards larger scales, the first condensations to form would be of mass $\sim M_2$, and it is tempting to identify them with large galaxies. If we knew the maximum radius attained by protogalaxies, we could infer the epoch at which they stopped expanding. Studies by Eggen, Lynden-Bell and Sandage (1962) of stars in our Galaxy with highly eccentric orbits-stars whose low heavy element content identifies them as among the oldest in the Galaxy-suggest that these stars formed during a free-fall collapse from a radius ~ 50 kpc. The mean density corresponding to the maximum radius is $\sim 10^{-2}$ particles cm^{-3}, and one can easily calculate that if $\Omega\approx 1$ the expansion of our Galaxy must have halted at a redshift $z\approx 5$. The collapse would then have been completed by $z\approx 2.5$. Relation (3.2) then tells us that, at t_{rec}, $\Delta\rho/\rho$ must have been ~ 0.5 per cent.

If the original perturbations were not precisely adiabatic and the initial spectrum favored smaller masses, we would expect condensations of $\sim 10^6 M_\odot$ to develop first. (Peebles (1969) has shown that, for various assumptions about the initial spectrum of the perturbations, a typical condensation would be within an order of magnitude of M_3). Peebles and Dicke (1968) suggest that some of these might form globular clusters. This hypothesis accounts for the allegedly 'standardised' properties of these objects, and predicts that there should be large numbers of them in intergalactic space. On this picture, galaxies would result from inelastic collisions between clouds of $\sim 10^6 M_\odot$ occurring before they had condensed into stars, or perhaps from tidal disruption of globular clusters which had already aggregated into objects of galactic mass.

Doroshkevich et al. (1967) have suggested a more complex chain of events. They speculate that the "primary" condensations of $\sim 10^6 M_\odot$ do not develop in the manner envisaged by Peebles and Dicke, but instead form unstable "superstars" which explode and heat up their surroundings. Even if only one part in $\sim 10^4$ of the material in the universe exploded in this fashion, enough heat would be generated to raise the temperature of all the remaining gas to $\geq 10^6$ K. If the redshift at which this happened was $z\lesssim 20$-as would be the case if $\Delta\rho/\rho$ were ≤ 2 per cent at t_{rec}-the gas would not have time to cool down again. The consequent increase in the Jeans mass acts like a thermostat and inhibits the formation of further superstars. The irregular heating would, however, create inhomogeneities on scales 10^9-$10^{12} M_\odot$ (corresponding to the "sphere of influence" of each superstar) even if such scales were completely absent from the initial spectrum of perturbations. These could then condense to form galaxies. More recently, the Moscow group

(see Doroshkevich et al. 1974) have suggested that the first condensations have masses comparable to clusters of galaxies. These
condensations would generally be asymmetrical, collapsing towards
a pancake-shaped configuration, and shock waves would develop along
sheets or "caustic surfaces". In this scenario, radiative cooling
would prevent the gas from rebounding elastically, and galaxies
would be presumed to form in the region of greatly enhanced density
behind these shocks. Binney (1974) has emphasised that such non-
planar shocks would generate vorticity. An adequate theory must
indeed explain why galaxies possess vorticity as well as angular
momentum. Authors who do not invoke primordial vorticity but
attribute the angular momentum of galaxies to tidal interaction
have somewhat glossed over this problem; on the other hand, there
are ways of circumventing Kelvin's circulation theorem without
invoking shocks.
      In the "primordial turbulence" model, the perturbations be-
fore recombination involve primarily the velocities and not the
density. However, at t_{rec} the effective speed of sound drops by
a factor $\sim 3 \times 10^3 \Omega^{-1/2}$, so the motions (previously subsonic) become
supersonic and generate density irregularities. Ozernoi and
Chibisov (1971) have discussed the subsequent supersonic turbulence
in general semi-quantitative terms. Their work suggests that density irregularities are, after recombination, stabilized by an
effective pressure (>>thermal gas pressure) contributed by turbulent
eddies on scales smaller than the given irregularity, and that
proto-galaxies would not isolate themselves from the expanding
background until $z \approx 100$. It is argued that all the angular momentum
of galaxies can, in this approach, be ascribed to the primordial
vorticity. For initial perturbations that are irrotational (whether
adiabatic or isothermal), the present spin of galaxies must be
attributed to tidal interactions between neighboring proto-galaxies
or to turbulence generated after the density perturbations become
$\gtrsim 1$. There has been some debate in the literature about whether
such processes can be effective enough (Oort 1970, Peebles 1971b),
and whether they can account for vorticity (Binney 1974, Doroshkevich
et al. 1974).
      At the moment, the only firm achievement of these studies of
perturbations in the "fireball" has been to show that if the initial
fluctuation spectrum is "smooth", but decreases in amplitude as the
scale increases, three characteristic scales are favored:
M_1: the minimum mass whose growth is never interrupted by pressure
      forces ($\sim 10^{15} M_\odot$).
M_2: the mass of the smallest adiabatic perturbation which escapes
      severe damping during its oscillatory phase before t_{rec}
      ($\sim 10^{12} M_\odot$).
M_3: the minimum mass able to condense freely after t_{rec} ($\sim 10^6 M_\odot$).
If the initial perturbations involve vorticity, some characteristic
masses still emerge (Harrison 1973a) but they are numerically somewhat different from the above.
      The nature and spectrum of the initial irregularities is

conjectural; but, if the amplitudes are large enough, non-linear processes occurring before t_{rec} may set up a calculable spectrum that is insensitive to the initial conditions. It should then be feasible to calculate the mass-density relation (and perhaps also the mass-angular momentum dependence).

At present, however, it is little more than speculation even to associate M_1, M_2, and M_3 with observed scales of agglomeration (clusters of galaxies ? galaxies? globular clusters?) in the universe. Moreover, even if the processes operative at $z \gtrsim 10^3$ ($t \lesssim t_{rec}$) seem fairly straightforward, the conditions between $z \approx 10^3$ (when the sudden reduction in c_s may give rise to new non-linearities and dissipative effects) and $z \approx 10$ (which may still correspond to an epoch before galaxies really "formed") are very obscure, despite the recent efforts of the Moscow and Padua groups, and others.

3.5 "Confrontation" with observations

What (if any) are the "characteristic masses" in the universe? The luminosity function of galaxies is very broad. Most of the light in extragalactic space seems to be contributed by "typical" galaxies of absolute magnitude ~ -21 (and estimated masses $10^{11}-10^{12} M_\odot$), but the brightest galaxies are ~ 2 magnitudes brighter than this, and-at the other end-the luminosity function probably extends right down to objects comparable to globular clusters.

Several recent pieces of evidence-some observational, and some based on theoretical considerations-suggest that galactic masses and mass-to-light ratios have been seriously underestimated (Ostriker et al. 1974 and references cited therein). Giant elliptical galaxies extend to much larger radii than was previously thought; and it is possible that spiral galaxies are surrounded by halos with properties similar to the outlying regions of giant ellipticals. Ostriker and Thuan (1974) and Gott (1974) suggest that the disc of a spiral galaxy forms as a mainly "secondary" structure from gas shed by halo stars. If this important idea were valid, it would imply that the apparent mass and angular momentum of spiral galaxies was only very indirectly related to that of the original protogalactic cloud. There are very few data on the rotation of elliptical galaxies. One general feature of elliptical galaxies is the characteristic and fairly standardised way in which the stellar density falls off in the outer regions ($\propto$(distance from center)$^{-2.8}$, approximately). This could be a consequence of "violent relaxation" (Gott 1973, 1974). But it could, on the other hand, be explained by a rather different theory in which elliptical galaxies condense around point masses (e.g. massive primordial black holes) which exist as part of the initial conditions (Hoyle and Narlikar 1967; Ryan 1972).

The characteristic properties of clusters of galaxies are even less well understood. Of the reality of large clusters like Coma there can be no doubt, but Peebles (1974) has recently argued that there is actually no evidence for any preferred scale of clustering-the covariance function for the distribution of galaxies varies on a simple power law over a wide range of length scales, from individual galactic dimensions right up to aggregations the size of "superclusters". This lends support to a conceptually simple picture where "point masses" of 10^6-$10^9 M_\odot$, randomly distributed at t_{rec}, gradually undergo clustering on progressively larger scales under the action of gravitational forces only (Press and Schechter 1974).

At the moment it would not be possible to subject a detailed theory of galaxy formation-even if it existed-to many observational tests! The main features of galaxies may be determined by what happens to them after formation, and by whether they are in a rich cluster or not. A further uncertainty is that the present masses of galaxies may be different from the masses of the protogalactic clouds from which they formed: very massive galaxies may have accreted a substantial amount of extra surrounding material since their formation; on the other hand, some galaxies may expel a large fraction of their initial gaseous content before it has time to be converted into stars. Moreover, most of the mass in the universe may not even be in the form of galaxies at all!

When did galaxies form? The most distant normal galaxies yet detected only have $z<0.5$. However, quasars are known with redshifts exceeding 3. If the quasar phenomenon is something which occurs in galaxies that have already formed, this obviously sets a lower limit to the redshift at which galaxies (or, at least, some galaxies) formed. One might think that estimates of the age of our Galaxy based on nuclear cosmochronology and stellar evolution arguments would be relevant. But time and redshift in pressure-free Friedmann models are related by

$$dt = \frac{-dz}{H_0 (\Omega z+1)^{\frac{1}{2}} (1+z)^2} \qquad (3.22)$$

The fractional difference between the look-back times to $z \simeq 2$ and (say) $z \simeq 10$ is much less than the uncertainty in H_0, so galactic age estimates are not very useful.

It has been argued (Partridge 1974 and references cited therein) that young galaxies might have been highly luminous, and should be optically detectable as extended objects even at redshifts $z \simeq 10$. Failure to detect "young galaxies" would then imply that they formed very early. Any quantitative conclusions, however, depend on the assumed cosmological model (which sensitively affects the angular size and apparent magnitude of objects at large redshifts) and on the astrophysical assumptions (e.g. if the light output from young galaxies were very centrally concentrated, they might appear stellar rather than extended, and could not then be distinguished from stars without some spectroscopic criterion such as the presence of highly red-shifted Lyman emission lines of H or He). The upper limit to the possible redshift of galaxy formation is perhaps $z \simeq 50$-at earlier epochs, the dynamical timescale for a typical galaxy would have exceeded the age of the universe.

The amplitude of the primordial fluctuations on various scales.
It is clear from (3.2) that, for galaxies to have formed by (say)
$z \approx 3$, the density perturbations on the appropriate scale must have
had amplitudes $\gtrsim 0.005$ at the epoch of recombination. Larger amplitudes would be required if $\Omega << 1$, or if galaxies formed much
earlier; also, the amplitudes would need to be larger if turbulent
pressure, or radiative drag, prevented the density contrast from
growing according to (3.1) since t_{rec}. On the scale of clusters
of galaxies, similar arguments tell us that $\Delta\rho/\rho$ must have been
$\gtrsim 0.001$ at recombination.
   What evidence do we have about the amplitudes of initial
fluctuations on other scales?
   (i) On scales larger than clusters, the best evidence comes
from the isotropy of the microwave background (see, e.g., Longair
and Rees (1973) for a review of these considerations).
   (ii) Adiabatic perturbations on scales <<galactic masses,
dissipating before recombination, would have augmented the microwave
background radiation. Any energy which was dissipated at a sufficiently early stage would have been completely thermalised; however
scales $\gtrsim 10^6 M_\odot$ dissipate at a later stage when the thermalisation
is no longer so efficient, and there would be a consequent distortion
of the background spectrum (Sunyaev and Zeldovich 1970).
   (iii) Although no very significant limits can be placed on
primordial fluctuations scales $1-10^6 M_\odot$, it is interesting that
there may be some constraints on scales $<< 1\ M_\odot$. These latter
fluctuations would come within the horizon during the so-called
"hadron era" of the big bang, when the equation of state is rather
uncertain. If the true equation of state for hadronic matter were
"soft" (i.e. $p << (1/3)\rho c^2$) then M_J would be $<< M_H$, and a perturbation
might eventually collapse to form a "mini" black hole even if its
amplitude was small when it first came within the particle horizon.
If more than one part in 10^8 of the primordial material had collapsed
in this way, the resulting black holes would contribute a mean density $\gtrsim 10^{-29}$ gm cm^{-3} to the present universe. It was originally
conjectured by Zeldovich and Novikov that any primordial black hole
would accrete rapaciously from its surroundings for the whole
duration of the radiation era, eventually attaining a mass $\sim 10^{15} M_\odot$;
but recent work by Carr and Hawking (1974) shows that this cata-
strophic growth cannot occur except for very special initial conditions. (These "mini-holes", being almost unaffected by radiation
drag, would undergo progressive gravitational clustering throughout
the lepton and radiation era. Such clusters of mini-holes might
then be able to initiate galaxy formation (Meszaros 1974) even if
galactic scales were completely absent from the initial fluctuation
spectrum).
What is the spectrum of the initial fluctuations? Unless
some characteristic scales are fed into the initial spectrum, this
spectrum must be in the form of a power law. If the initial perturbations are adiabatic curvature fluctuations, then the typical
value of $|\Delta\rho/\rho|$ over a region of mass M would be proportional to

M^{-n} (the constant of proportionality depending on cosmic time). Some authors have tried to "predict" the value of n on the basis of physical arguments pertaining to the very early universe. I shall not discuss these arguments here. The considerations mentioned in the previous subsection show that there are at least some observational constraints on n. The case n=2/3 is especially attractive because it has the property that the fluctuation amplitude on a scale M, measured at the time when that scale first comes within the particle horizon (i.e. $M=M_H$) is the same for all M. This is, moreover, the only choice of n for which one can consistently assume that the universe is approximately "Robertson-Walker" on all scales: if n<2/3, the curvature fluctuations on scales comparable with the present Hubble radius may be large (given rise to overall anisotropies on the observable universe); if n>2/3 the early universe (where M_H was very small) would have been chaotic and non-Friedmannian.

Peebles (1974) claims that his analysis of clustering rules out n=2/3; however it is possible that his results tell us less about the initial fluctuation spectrum than about the present density distribution within clusters of galaxies (which is presumably mainly determined by "violent relaxation" after the clusters have formed).

Sunyaev and Zeldovich (1970) have shown that the assumption of an n=2/3 initial spectrum is just about compatible with all the constraints, provided that curvature fluctuations on each scale M have an amplitude of $\sim 10^{-4}$ when $M \simeq M_H$. They have also argued that the most natural type of metric fluctuations would give rise primarily to density perturbations, and that the "primordial turbulence" approach—which demands that the velocity perturbations are dominant—is more artificial.

3.6 Some misgivings

The main conceptual problem in cosmology is not, perhaps, the presently observed inhomogeneity (galaxies, clusters, etc.) but, on the contrary, the overall large-scale homogeneity and isotropy. The difficulty is that in Friedmann models the mass M_H within a particle horizon shrinks to zero as one extrapolates back towards the initial instant (equation 3.4). The "canonical" models therefore necessitate the unpalatable postulate that all parts of the universe are accurately synchronized, and start expanding at the same time and with the same entropy and curvature, even though there was then no communication or causal connection between neighboring regions. In these Friedmann models, therefore, the observed overall uniformity is postulated, never explained; and is even unexplainable. Matter at one point, however willing to live and evolve in conformity with matter at another point in the universe, has no means whatever of knowing the "conditions of life" beyond its own horizon. The parts of the "last scattering surface" in different directions, from which we receive microwave photons agreeing in temperature to better than 0.1 per cent, were causally quite unconnected at the time this radiation was emitted. Thus the main "problem" is perhaps not (as is often alleged) the origin of the initial fluctuations, but why their amplitude is not so large that the universe is completely chaotic (Misner 1968; see also Rees 1972b).

But there is no direct evidence that the Robertson-Walker metric is even remotely applicable arbitrarily close to t=0, and the horizon may behave quite differently in more general models. Though it is not feasible to calculate the dynamics of a fully inhomogeneous universe, much attention has recently been devoted to anisotropic models, whose horizon structure is indeed found to be qualitatively different from in Friedmann models. One would then seek damping processes (e.g. neutrino or photon viscosity) which could homogenise the universe on large scales. These more general cosmological models may indeed be more realistic, but their adoption would make quantitative studies of the origin of galaxies very much more difficult.

REFERENCES

Baade, W. and Minkowski, R., 1954, Astrophys. J. 119, 206.
Binney, J., 1974, Mon. Not. R. Astr. Soc. 168, 73.
Blandford, R.D. and Rees, M.J., 1974, Mon. Not. R. Astr. Soc. (in press).
Bollea, C. and Cavaliere, A., 1974, Astr. Astrophys. (in press).
Bonometto, S.A., Danese, L. and Lucchin, F., 1974, Astr. Astrophys. (in press).
Bridle, A.H. and Brandie, G.W., 1973, Astrophys. Lett. 15, 21.
Burbidge, G.R., 1956, Astrophys. J. 124, 416.
Burbidge, G.R., 1967, Nature 216, 1287.
Burbidge, G.R., 1970, Ann. Rev. Astr. & Astrophys. 8, 369.
Carlitz, R., Frautschi, S. and Nahm, W., 1973, Astr. Astrophys. 26, 171.
Carr, B.C. and Hawking, S.W., 1974, Mon. Not. R. Astr. Soc. 168,399.
De Young, D.S. and Axford, W.I., 1967, Nature 216, 129.
Doroshkevich, A.G., Sunyaev, R.A. and Zeldovich, Y.B., 1974, "Confrontation of Cosmological Theory with Observational Data" ed. M.S. Longair (Reidel, Holland) p. 213.
Doroshkevich, A.G., Zeldovich, Y.B. and Novikov, I.D., 1967, Sov. Astron. 11, 233.
Eggen, O.J., Lynden-Bell, D. and Sandage, A.R., 1962, Astrophys.J. 136, 748.
Field, G.B., 1971, Astrophys. J. 165, 29.
Field, G.B., 1972, Ann. Rev. Astr. & Astrophys. 10, 227.
Field, G.B., 1975, Stars and Stellar Systems, vol. 9 ed. A. and M. Sandage (in press).
Gisler, G.R., 1975, in preparation.
Gott, J.R., 1973, Astrophys. J. 186, 481.
Gott, J.R., 1974, Astrophys. J. (in press).

Gull, S.F. and Northover, K.J.E., 1973, Nature 244, 80.
Hargrave, P.J. and Ryle, M., 1974, Mon. Not. R. Astr. Soc. 166, 305.
Harrison, E.R., 1967, Phys. Rev. Lett. 18, 1011.
Harrison, E.R., 1973a, Cargese Lectures in Physics VI, ed. E. Schatz-
man, (Gordon and Breach, London and New York) p.581.
Harrison, E.R., 1973b, Mon. Not. R. Astr. Soc. 165, 185.
Harrison, E.R., 1973c, Ann. Rev. Astr. & Astrophys. 11, 155.
Hoyle, F. and Narlikar, J.V., 1967, Proc. Roy. Soc. 290A, 177.
Jones, B.J.T., 1973, Astrophys. J. 181, 269.
Landau, L.D. and Lifshitz, E., 1959, "Fluid Mechanics" (Pergamon,
London) p.349.
Larson, R.B., 1974a, Mon. Not. R. Astr. Soc. 166, 585.
Larson, R.B., 1974a, Mon. Not. R. Astr. Soc. (in press).
Lifshitz, E., 1946, Zurn. Eksp. Teor. Fis. 10, 116.
Longair, M.S. and Rees, M.J., 1973, Cargese Lectures in Physics VI,
ed. E. Schatzman (Gordon and Breach, London and New York) p.269.
Longair, M.S., Ryle, M. and Scheuer, P.A.G., 1973, Mon. Not. R.
Astr. Soc. 164, 243.
Longair, M.S. and Willmore, A.P., 1974, Mon. Not. R. Astr. Soc.
168, 479.
Mackay, C.D., 1971a, Mon. Not. R. Astr. Soc. 151, 421.
Mackay, C.D., 1971b, Mon. Not. R. Astr. Soc. 154, 209.
Mathews, W.G. and Baker, J.C., 1971, Astrophys. J. 170, 241.
Meszaros, P., 1974, Astr. Astrophys. (in press).
Miley, G.K., 1973, Astr. Astrophys. 26, 413.
Misner, C.W., 1968, Astrophys. J. 151, 431.
Norman,.A. and ter Haar, D., 1974, Astr. Astrophys. (in press).
Omnes, R., 1973, Proc. 6th Texas Conference (N.Y. Acad. Sci.) p.339.
Oort, J.H., 1970, Astr. Astrophys. 7, 381.
Ostriker, J.P., Peebles, P.J.E. and Yahil, A., 1974, Astrophys. J.
(in press).
Ostriker, J.P. and Thuan, T.X., 1974, Astrophys. J. (in press).
Ozernoi, L.M. and Chernin, A.D., 1968.
Ozernoi, L.M. and Chibisov, G.V., 1971, Sov. Astron. 14, 615.
Partridge, R.B., 1974, Astrophys. J. 192, 241.
Peebles, P.J.E., 1969, Astrophys. J. 157, 1075.
Peebles, P.J.E., 1970, Phys. Rev. D.1, 397.
Peebles, P.J.E., 1971a, Astrophys. & Sp. Sci. 11, 443.
Peebles, P.J.E., 1971b, Astr. Astrophys. 11, 377.
Peebles, P.J.E., 1974, Astr. Astrophys. 32, 197.
Peebles, P.J.E. and Dicke, R.H., 1968, Astrophys. J. 154, 891.
Peebles, P.J.E. and Yu, J.T., 1970, Astrophys. J. 162, 815.
Press, W.H. and Schechter, P., 1974, Astrophys. J. 187, 425.
Rees, M.J., 1971a, Nature 230, 312 (errata, p.510).
Rees, M.J., 1971b, General Relativity and Cosmology, ed. R.K. Sachs
(Academic Press) p.315.
Rees, M.J., 1972a, Proc. 1st I.A.U. European Assembly Vol. 3,
(Springer-Verlag, Berlin-Heidelburg,-New York) p.190.
Rees, M.J., 1972b, Phys. Rev. Letters 28, 1669.
Rees, M.J. and Gunn, J.E., 1974, Mon. Not. R. Astr. Soc. 167, 1.

Ryan, M., 1972, Astrophys. J. (Letters) 177, 79.
Ryle, M., 1968, Ann. Rev. Astr. & Astrophys. 6, 249.
Ryle, M., 1972, Nature 239, 435.
Ryle, M. and Longair, M.S., 1967, Mon. Not. R. Astr. Soc. 136, 123.
Saslaw, W.C., Valtonen, M.J. and Aarseth, S.J., 1974, Astrophys. J. 190, 253.
Scheuer, P.A.G., 1974, Mon. Not. R. Astr. Soc. 166, 513.
Setti, G. and Woltjer, L., 1973, Proc. 6th Texas Conference (N.Y. Acad. Sci.) p.8.
Silk, J.I., 1968, Astrophys. J. 151, 469.
Silk, J.I., 1973, Ann. Rev. Astr. & Astrophys. 11, 269.
Sunyaev, R.A. and Zeldovich, Y.B., 1970, Astrophys. & Sp. Sci. 9, 368.
Tomita, K., Nariai, H., Matsuda, T. and Takeda, H., 1970, Prog. Theor. Phys. 43, 1511.
Weinberg, S., 1971, Astrophys. J. 168, 175.
Weinberg, S., 1972, "Gravitation and Cosmology" (Wiley, N.Y.).
Willis, A.G., Strom, R.G. and Wilson, A.S., 1974, Nature 250, 625.

SUBJECT INDEX